Frontiers in Aerospace Science

(Volume 2)

(High Order Large Eddy Simulation for Shock-Boundary Layer Interaction Control by a Micro-ramp Vortex Generator)

Authors

Chaoqun Liu, Qin Li, Yonghua Yan, Yong Yang, Guang Yang, Xiangrui Dong

Center for Numerical Simulation and Modeling, Department of Mathematics, University of Texas at Arlington, Arlington, Texas, TX, USA

Frontiers in Aerospace Science

Volume # 2

High Order Large Eddy Simulation for Shock-Boundary Layer Interaction Control by a Micro-ramp Vortex Generator

Authors: Chaoqun Liu, Qin Li, Yonghua Yan, Yong Yang, Guang Yang and Xiangrui Dong

ISSN (Online): 2468-4724

ISSN (Print): 2468-4716

ISBN (Online): 978-1-68108-597-5

ISBN (Print): 978-1-68108-598-2

Bentham Science Publishers Ltd.
Executive Suite Y - 2
PO Box 7917, Saif Zone
Sharjah, U.A.E.
Email: subscriptions@benthamscience.org

BENTHAM SCIENCE

CONTENTS

PREFACE

Shock boundary layer interaction (SBLI) control is a very important topic in aerospace science and engineering. SBLI could cost flow separation, pressure fluctuation, noise generation, engine stop, energy efficiency reduction, and even structure destruction. As a passive SBLI control tool, micro-ramp vortex generator (MVG) is a robust and efficient device and received wide attention by scientific researchers and engineers. This book mainly introduces high order large eddy simulation (LES) for SBLI control by MVG. Since Li and Liu published their milestone LES work in 2010, which shows that the vortex structure around MVG is not simply a pair of streamwise vortices, but consists of a train of spanwise vortex rings, the vortex structure around MVG becomes a hot topic for scientific research and more and more research papers have been published since then. However, we have no intention to include all LES work about MVG in this book, but are focused on the LES work which was conducted by the research team at Center for Numerical Simulation and Modeling (CNSM) of University of Texas at Arlington (UTA) under support of US Airforce Office of Scientific Research, Grant No. FA9550-08-1-0201.

In this book, an implicitly implemented large eddy simulation (ILES) by using the fifth order bandwidth-optimized WENO scheme is applied to make comprehensive studies on ramp flows with and without control at Mach 2.5 and Req=5760. The work is mainly contributed by Dr. Chaoqun Liu and Fellows in Center of Numerical Simulation and Modeling at University of Texas at Arlington. Flow control in the form of microramp vortex generators (MVG) is applied. The results show that MVG can distinctly reduce the separation zone at the ramp corner and lower the boundary layer shape factor under the condition of the computation. A series of new findings are obtained about the MVG-ramp flow including the structures of the surface pressure, the three-dimensional structures of the re-compression shock waves, the complete surface separation pattern and the new secondary vortex system, *etc*. The mechanism about the formation of the momentum deficit is deeply studied. A new mechanism on the shock-boundary layer interaction control by MVG is discovered as associated with a series of vortex rings, which are generated by the high shear layer at the boundary of the momentum deficit zone. Vortex rings strongly interact with the flow and play an important role in the separation zone reduction. In addition, the governing equation, boundary condition, high quality grid generation, high order shock capturing scheme, DNS inflow condition are all introduced in details. These will provide a powerful tool for researchers to use LES to study shock boundary layer interaction and supersonic flow control including shock induced separations and noise reduction. This book is organized as follows: In chapter1, an introduction of shock-boundary layer interaction (SBLI) is given. A paper review about the driven mechanism of pressure fluctuation caused by shock induced flow separation on a ramp corner is presented and a research history on micro vortex generator (MVG) is briefly reviewed. In particularly, a series of new findings made by the LES team at Center of Numerical Simulation and Modeling led by Dr. Chaoqun Liu under the support of US Air Force Office of Scientific Research (AFOSR) are briefly reported. These new findings are milestone work which have made breakthrough in understanding the physics of SBLI, shock-vortex ring interaction and control of shock induced flow separation and noise generation. In Chapter 2, the dimensional and non-dimensional 3-D time-dependent Navier-Stokes equations in curvilinear coordinates are given in details. The flux splitting scheme and non-reflecting boundary conditions are discussed and provided. In Chapter 3, a brief introduction of algebraic grid generation, transfinite interpolation and elliptic grid generation is given. In particular, a two-step elliptic grid generation method developed by Spekreuse (1995) is introduced, which can generate high quality and orthogonal grids near the boundary. The smooth and orthogonal grid for the LES case of MVG and Ramp is generated by the above method. In Chapter 4, a series of high order shock capturing schemes including WENO scheme, weighted compact scheme, modified upwinding compact scheme are introduced, which can get sharp shock capturing but keep high order accuracy in the smooth area. These schemes are particularly useful for LES of shock-turbulence interaction where both high resolution and sharp shock capturing are important. All these schemes are some kind successful in high order LES for SBLI. An efficient shock detector is introduced as a part of weighted compact and modified upwinding schemes. In addition, the accuracy, truncation errors, dissipation and dispersion of these schemes are analyzed. In Chapter 5, a high order direct numerical simulation (DNS) is applied to generate fully developed turbulent inflow. The fully developed inflow was taken from a case of DNS for flow transition. The inflow condition is carefully checked with the velocity profile and ratio of boundary layer thickness and made sure it is a fully developed turbulent flow in order to compare with the wind tunnel test. In Chapter 6, a series of new findings by UTA's high order LES is presented including spiral points near the leading edge, momentum deficit origin, a train of spanwise vortex rings, flow separation topology is given and the physics of the observation is discussed and concluded. In Chapter 7, interaction of shock and vortex rings which were generated by MVG is described and analyzed in details. The change of vorticity, vortex, shock strength and location is described in details. The physics of shock and vortex interaction and its influence on the flow structure is also discussed. In Chapter 8, a new mechanism of SBLI control by MVG is presented. Traditionally people believe the reduction of shock-induced separation is caused by streamwise vortex which mixes the boundary layer flow and make the velocity profile be more capable to resist the flow separation. The new mechanism shows the spanwise vortex rings are critical which are quickly moving and fast rotating to destroy or weaken the shock and significantly reduce the shock-induced flow separation. Potentially, this study could lead to some technology revolution in control of SBLI, flow separation and noise reduction. In Chapter 9, the correlation between density and pressure fluctuation and vortex motion is studied. The correlation between flow separation and vortex motion is studied as well. The correlation clearly shows the vortex ring generation, size, frequencies, and strength are closely related to the fluctuation of density and pressure. This would provide a powerful tool for SBLI control and noise reduction. In Chapter 10, the results are validated against the TU Delft experiment on the same MVG geometry and Mach number. Several techniques are used to analyze the coherent structures in the MVG wake. The vortex system in the MVG wake is visualized using vortex identification method and it is found that two primary counter rotating streamwise vortex pair are induced by the MVG, which would further lead to a train of vortex rings through K-H instability. The average distance between adjacent vortex rings is determined by a spatial auto-correlation of vorticity, which is estimated as 1.5 MVG height. Two dimensionality reduction algorithms, namely Proper Orthogonal Decomposition (POD), and Dynamic Mode Decomposition (DMD) are applied to a set of flow field sequences on the spanwise symmetry plane. POD modes identifies structures contain most of the turbulent kinetic energy and DMD modes capture single frequency structures most essential to the unsteady dynamics.

The book is mainly written by Dr. Chaoqun Liu, the Director of CNSM Center at UTA and co-authored by former CNSM research fellows including Dr. Qin Li, Dr. Yonghua Yan, Dr. Yong Yang, Dr. Guang Yang and Ms. Xiangrui Dong. The authors would like to thank Dr. Frank Lu, Dr. Yinling Dong, Dr. Maria Oliveira and Dr. Hua Shan for their assistances.

ACKNOWLEDGEMENTS

This work was supported by AFOSR grant FA9550-08-1-0201 supervised by Dr. John Schmisseur. The authors are grateful to Texas Advantage Computing Center (TACC) for providing computation hours.

CONFLICT OF INTEREST

The author declares no conflict of interest, financial or otherwise.

Chaoqun Liu
Center for Numerical Simulation and Modeling,
Department of Mathematics,
University of Texas at Arlington,
Arlington, Texas 76019, TX,
USA
E-mail: cliu@uta.edu

CHAPTER 1

Introduction and Paper Review

Abstract: In this chapter, an introduction of shock-boundary layer interaction (SBLI) is given. A paper review about the driven mechanism of pressure fluctuation caused by shock induced flow separation on a ramp corner is presented and a research history on micro vortex generator (MVG) is briefly reviewed. In particular, a series of new findings made by the LES team at the Center of Numerical Simulation and Modeling led by Dr. Chaoqun Liu under the support of US Air Force Office of Scientific Research (AFOSR) is briefly reported. These new findings are milestone work which have made a breakthrough in understanding the physics of SBLI, shock-vortex ring interaction, control of shock induced flow separation and noise generation.

Keywords: Flow separation control, Large eddy simulation, Micro vortex generator, Shock boundary layer interaction, Shock vortex interaction.

1.1. A SHORT REVIEW ON STUDY OF DRIVEN MECHANISM OF UNSTEADINESS OF SBLI IN A COMPRESSED RAMP

Shock wave and boundary layer interaction (SBLI) is a common phenomenon (Fig. **1.1**) which occurs in transonic airfoils, supersonic inlets, wing-fuselage junction in missiles, nozzles, and *etc*. There are many adverse effects caused by SBLI, for instance, a major source of pressure fluctuation causing structural fatigue and even damage, the inclination to induce flow separations, the reduction of total pressure, the degradation of velocity profile unfavorable in the engine combustion, large acoustic noises, so on and so forth. SBLI is a typical topic that has been extensively studied over the past decades [1]. There are many review papers on this topic [2 - 8]. Dolling [1] gave a list of issues which needed to be studied, *e.g.*, the source of low-frequency unsteadiness. In the following, a review will be given on this topic.

Low-frequency unsteadiness of SBLI is a common observation and still remains un-clarified in the society of fluid dynamics, *i.e.* the driven force to cause the low frequency unsteadiness is not fully understood. Because most realistic inflow is turbulent, the scenario of current discussion mainly concerns about the shock-turbulent flow interaction. There are several theories which try to explain the

Chaoqun Liu, Qin Li, Yonghua Yan, Yong Yang, Guang Yang, Xiangrui Dong

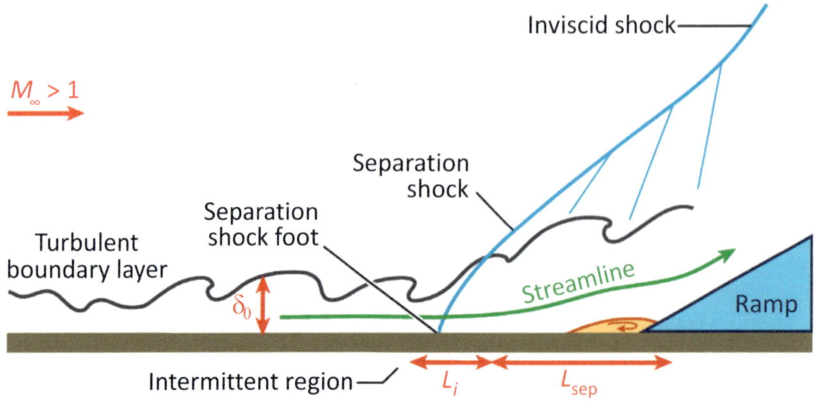

Fig. (1.1). SBLI in a compress corner (Clements *et al.*, 2014).

generation mechanism of low-frequency oscillations. The first one is that the low-frequency unsteadiness is caused by upstream incoming flow fluctuations. Plotkin [9] believed the shock low frequency motion should be caused by incoming turbulence fluctuation with high frequency. In the experiment of Andrepoulos and Muck [10], it was found that the non-dimensional frequency of shock motion, *i.e.* Strouhal number $St_\delta = f\delta_0/U_\infty$, was in the same order of magnitude of turbulence bursting frequencies. A lot of work has been done by Dolling and his group. Erengil and Dolling [11] discovered a high correlation between the shock-foot velocity and the pressure fluctuations in the upstream of boundary layer. Unalmis and Dolling [12] gave a thickening and thinning mechanism. Ganapathisubramani *et al.* [13] conducted a plan-view (streamwise-spanwise plane) PIV imaging in the boundary layer upstream of a compression ramp interaction with Mach number being 2. The long regions of low-velocity fluid in the log region which remained coherent were observed and were similar to turbulent inflow structures. Ganapathisubramani *et al.* [14] demonstrated that the coherent superstructures can have a length as large as 40δ. A separation line surrogate (based on a particular low-velocity contour) was defined and tried to find the correlation between these superstructure and the fluctuations in the SBLI. Humble *et al.* [15 - 18] have made extensive planar and 3-D tomographic PIV measurements in an impinging shock interaction at Mach2.1. Humble *et al.* [17] specifically investigated the influence on the upstream boundary layer. They used 3D technique and found that the hairpin-type vortex structures were related with low-momentum regions, which are similar to hairpin-packet model structures [19]. Like Ganapathisubramani *et al.* [13], they showed a high correlation between velocity fluctuations of upstream and the surrogate separation location. Apparently, investigations of these researchers seemed to show the low-frequency unsteadiness in SBLI is originated by turbulent fluctuations of incoming flows.

In the meanwhile, other investigators provided evidence that the low-frequency unsteadiness of SBLI was caused by downstream instability. Thomas *et al.* [20] studied the role of burst-sweep events in the upstream boundary layer for a compression ramp interaction at Mach1.5. A filtering approach was adopted to isolate burst-sweep events and tried to find the correlation of events with the shock motion. Their work demonstrated that there was no correlation between the two occurrences. In addition, they found that the fluctuations of reattachment region were highly correlated with the fluctuations in the intermittent region which implied a downstream source responsible for the unsteadiness. Touber & Sandham [21] conducted large-eddy simulations of SBLI with a configuration simulating the experiments by Dupont *et al.* [22] where an 8°shock generator was used to form the impinging shock. They analyzed the incoming boundary layer structures at y/δ_0 = 0.2 and found that the superstructures observed by Ganapathisubramani *et al.* [13] did not exist. Therefore, they concluded that the superstructures played a minor role in the SBLI unsteadiness. Dupont *et al.* [22] measured the correlation between the fluctuations of the wall pressure near the vicinity of the reflected shock and locations near the reattachment region, and they found that the value was greater than 0.8. This high correlation led to suggestion that the separation bubble and the separation shock oscillation could be correlated as a quasi-linear system. Dupont *et al.* [23] conducted a detailed study on the spatial and-temporal organization of the separated flow. They confirmed the earlier results of Dupont *et al.* [22] again about the coherence and phase between the separated shock and separated bubble. Piponniau *et al.* [24] introduced a simple model in terms of the unsteadiness of separation bubble, where the shear layer played an important role by its entrainment nature. They assumed that the shear layer can entrain the low-momentum fluid out of the separation bubble and caused the mass depletion. A similar mechanism based on entrainment considerations was also suggested by Wu & Martin [25] for compression ramp interactions. Piponniau *et al.* [24] made an assumption that the Strouhal number regarding the frequency depended on the shear layer entrainment rate and the mass (and hence size) of the separation bubble. Wu & Martin [24] studied the mechanism introduced by Ganapathisubramani *et al.* [13, 14], *i.e.*, the conclusion of a high correlation between the upstream boundary layer fluctuation and the motion of the separated flow. Wu & Martin pointed out that the magnitude of the correlation might be overestimated significantly by the separation surrogate, and their correlation coefficient between separation shock oscillation and the separation point was approximately 0.8. Wu & Martin's DNS also revealed the presence of upstream superstructures, which really influenced the separation line, but their effect was restricted to inducing the spanwise wrinkling of the separation line with a smaller scale. According to their observations, Wu & Martin believed that the motion of large scale structures was driven by the separated flow pulsations. Apparently, these two mechanisms oppose each other and both cannot be correct. Clemens and Narayanaswamy [8] concluded that these two mechanisms were always present in all shock-induced turbulent flows with separation, while the downstream mechanism took action for strong separated flows, and a combined mechanism dominated for weak separated flows. Apparently, this is a compromise, may or may not be true. There were still many other people who believed SBLI unsteadiness was a stochastic process but not deterministic [26].

In short, what is the real mechanism to drive the low frequency unsteadiness is still uncertain. Apparently, we must understand the physics of the driven force of SBLI low frequency unsteadiness if we want to control SBLI efficiently. Therefore, investigation of SBLI unsteadiness is important to both science and engineering applications.

1.2. A SHORT REVIEW ON MICRO-VORTEX GENERATORS (MVGS) FOR CONTROLLING SHOCK/BOUNDARY LAYER INTERACTIONS (SBLIS)

The control of the boundary layer, such as reducing or removing separation zones, drag reduction and improving of flow, is as old as the boundary layer concept. Considering the practicality and reality of aircraft applications, the incoming boundary layer to control is usually chosen as turbulent flow. Practical requirements of the flow control devices, such as the robustness, ease of installation, simplicity and light weight, *et al.*, tend to favor passive devices like vortex generators, vanes, fences and the less developed large-eddy breakup unit or riblets, although active devices also have their advantages [27 - 39]. A recent study introduced a technique of boundary layer flow control which was to arrange an array of micro vortex generators (MVGs). These MVGs had the height less than thickness of incoming boundary layer and were installed ahead of the region with adverse flow conditions. Different from the conventional vortex generator with a height comparable to the thickness of incoming boundary layer, an MVG has a height of around 20% to 40% of the boundary layer thickness. As discussed above, shock boundary layer interactions (SBLIs) in high-speed inlets can result in many adverse effects and has a negative influence on the performance of the engine, therefore is important to employ techniques to control including MVGs.

MVGs are a hopeful new device which can alleviate or overcome the negative influence of SBLIs, thus to improve the boundary layer. There are different kinds of MVGs, such as micro-vanes and micro-ramps, and the latter seem to be more preferable since they are more robust than micro vanes especially in high speed flows. Recently, some intensive experimental and computational studies have been done on MVGs. MVGs have a number of advantages compared to conventional vortex generators due to their small size and nonintrusive appearance, which bring about structural reliability, and low drag. Initial studies were conducted at low speeds [30 - 39] with the application to practical configurations [40 - 46].

Recently, the study of MVGs has moved to the supersonic flows in terms of adverse effects reduction of the separation induced by SBLIs. These studies on high speed problems were related to an impinging or reflecting shock generally [47 - 65]. As a side note, there has been a previous study of conventional sized vortex generators which demonstrated a reduction of unsteady pressure loads in SBLIs [66]. Some numerical simulations of MVGs have been made for comparative studies and to support applications. Ghosh *et al.* [55] performed the computations in detail under the experimental conditions proposed by Babinsky [51]. RANS computations and hybrid RANS/LES computations with immersed boundary techniques were also included in these numerical studies. Lee *et al.* [62] also did the simulations on MVGs by using monotone integrated large eddy simulations (MILES). In their computations, the MVG were located at a position that was the same as in a wind tunnel. The waves around the MVG, which including the main shock, expansion waves and re-compression shock similar to as Babinsky's *et al.* experiment [51], were reproduced in their studies. The momentum deficit was also captured.

From the results of experiments and computations, MVGs were introduced to reduce the scale of the separated flow. However, few satisfactory explanations on how the MVGs affect the separation have been achieved. It has been suggested that the boundary layer is energized by the MVGs by a system of streamwise counter-rotating vortices. Recently such a mechanism has been studied in low-speed experiments in detail. The presence of embedded longitudinal vortices shed from sub-boundary layer vanes was found by Velte *et al.* [67]. Logdberg *et al.* [68] made experiments by using hotwire anemometry and smoke visualization. In their visualizations with an exposure period of 10ms, a pair of turbulent vortical structures was manifested to convect the downstream. The presence of strong counter-rotating vortices generated from the MVGs was indicated by the hotwire measurements, while there was no evidence or information on unsteadiness provided. However, Angele and Grewe [69] gave the details of the unsteadiness of the flow downstream of MVG array. They suggested that the Reynolds stresses around the mean center of vortex would be increased by the unsteadiness, and the increase would be disappeared if the vortices changed to steady. Other measurement also indicated that the vortices were unsteady. Duriez *et al.* [70] concluded that a self-sustaining process between the streamwise velocity streaks in a turbulent boundary layer and the streamwise vortices generated by an MVG array was produced by the steady spatial forcing at different wavelengths.

The detailed flow physics of MVGs at high speeds has not been fully or properly investigated. Although there were suggestions regarding the flow entrainment by two pairs of streamwise, counter-rotating vortices from MVG, the effects of compressibility and baroclinic torque have not been understood.

In the study of the mechanism on the shock and boundary layer interaction and its control, the use of RANS and corresponding time-averaging interpretation seems to be problematic because of the inherent unsteadiness in such flows. In this regard, it is appropriate to view RANS as mainly an engineering tool. While experiments currently have difficulties to provide a three-dimensional, time-resolved solution, the high-order LES has become an important tool to study flow mechanisms, especially considering the advances in computer and code capability. In the meanwhile, LES results must be checked by experiment, namely, the so-called validation and verification process. Thus, the integration of high-order LES and experiment is extremely important and powerful for gaining insight into the physics of MVGs in SBLI control.

Recently, the University of Texas at Arlington has carried out investigations on SBLI and MVG control with the support of AFOSR, which has suggested a complex physical mechanism based on the unsteady shedding of ring vortices arising from the interaction of streamwise vortices trailing from the MVG [71 - 76]. This phenomenon appears to be a consequence of Kelvin-Helmholtz (KH) instability. The mechanism can be simply described as follows. The MVG wake is a cylindrical region of momentum deficit which generates a high shear layer. As a result, a cylindrical inflection arise in the shear layer and causes KH instability, which results in the breakdown of the high-shear (HS) layers into vortex rings [77]. This evidence is obtained by UTA LES team through high-resolution large eddy simulations in 2009 [73, 74]. This new LES finding was confirmed by experiment [76, 77]. In their computation and experiment, the inflow is fully developed turbulent, and the breakdown of KH instability was found to strongly depend on the momentum deficit, the inflow turbulence level and the Reynolds number. As a parenthetical note, Blinde *et al.* [52] suggested hairpin vortices instead of vortex rings.

To effectively apply MVGs for controlling SBLIs, it is necessary to well understand the physics. Three issues which should be clarified are listed as follows:

a. What are the structures of the wave system generated by a MVG? In previous studies, only information about the two-dimensional structures was available and confirmed by experiment.

b. What is the relation between the flow structure and the momentum deficit and where is the low-speed fluid from?
c. Are there any new mechanisms on the pronounced transportation and exchange of momentum by streamwise vortices?

The new numerical, experimental, and analytical study at UTA has yielded a new mechanism and follow-on work should examine the detailed vortex structure downstream of the MVG, conditions that influence Kelvin-Helmholtz stability and ring generation, kinematics and dynamics of vortex rings, vortex ring/shock interaction, unsteadiness and separation zone dynamics arising from vortex ring/shock interaction induced by a ramp and a cylinder.

1.3. SOME NEW FINDINGS BY LES AND EXPERIMENT

A number of new findings are obtained through recent LES study and confirmed by experimental in University of Texas at Arlington and supported by US AFOSR. These new findings can be described below.

1.3.1. Detailed Flow Structures around MVG

Flow structures around MVG (Fig. **1.2**) found by our LES and experiment show the same qualitative agreement. Major features include: (a) leading-edge separation, (b) 5 pairs of trailing vortices and (c) spiral foci at the trailing edge of the MVG, (d) K-H instability and vortex ring generations by momentum deficit caused by MVG.

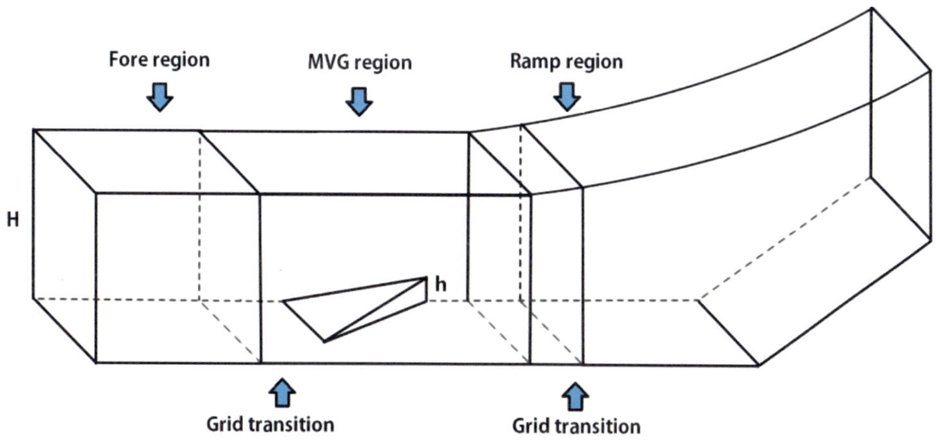

a. Computational domain for ramp with MVG

Fig. 1.2 contd.....

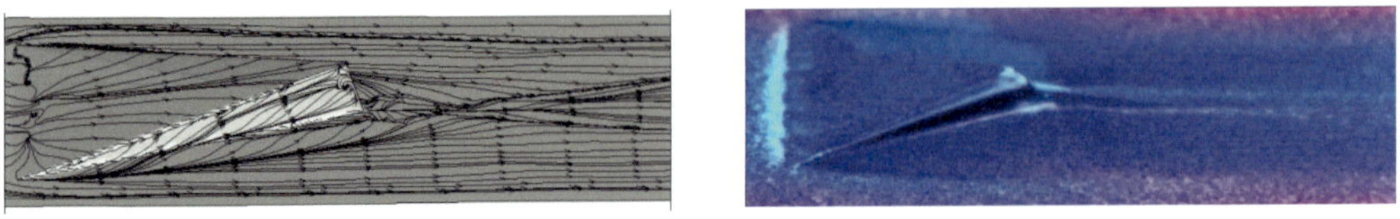

b. UTA LES (Li & Liu [73]). **c. UTA Experiment (Pierce et al. [78]).**

Fig. (1.2). Side view of surface streamlines.

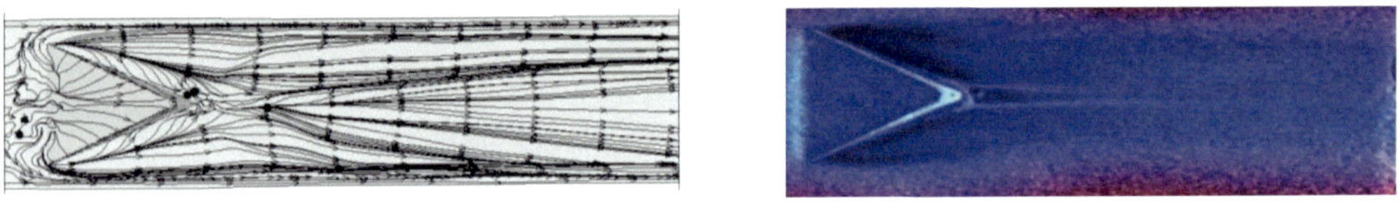

a. UTA LES (Li & Liu [73]). **b. UTA Experiment (Pierce et al. [78]).**

Fig. (1.3). Top view of surface streamlines.

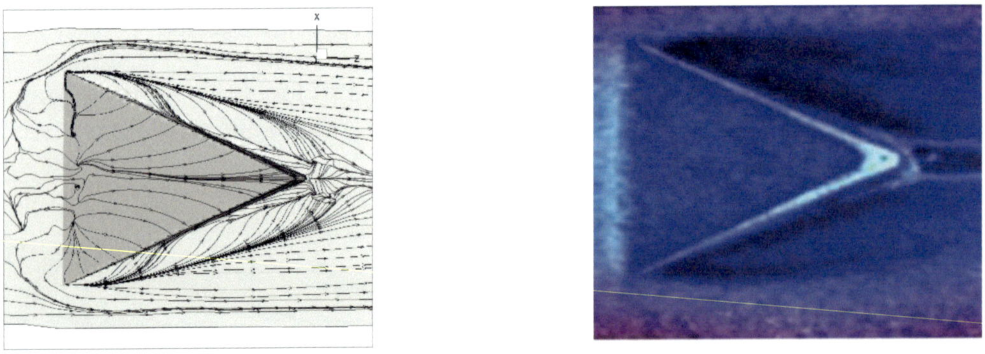

a. UTA LES (Li & Liu [73]). **b. UTA Experiment (Pierce et al. [78]).**

Fig. (1.4). Enlarged top view of surface streamlines.

Leading edge separation is found by both LES and experiment (Figs. **1.2**-**1.4**). Such three-dimensional separation gives rise to a horseshoe vortex system which can be revealed by surface streamlines, numerically or experimentally [75]. The primary difference between the numerical and experimental result is the geometric scale due to the lower Reynolds number of the numerical simulation.

1.3.2. Spiral Points around MVG

The complete separation topology has been obtained by UTA LES [73, 79] and supported by UTA experiment [75]. The structure can be explained in the following. The flow passes the MVG surface and generates a pair of strong primary trailing vortices, one on each side of the MVG. The strong primary vortices generate two pairs of secondary vortices, one on side surface of the MVG and one on the surface of the flat plate. The two secondary vortices on each side travel for a very short distance before being separated from the surface to form a complex three-dimensional trajectory under the influence of the strong primary vortex. Further, the strong primary vortices continue their travel and induce a new pair of secondary vortices.

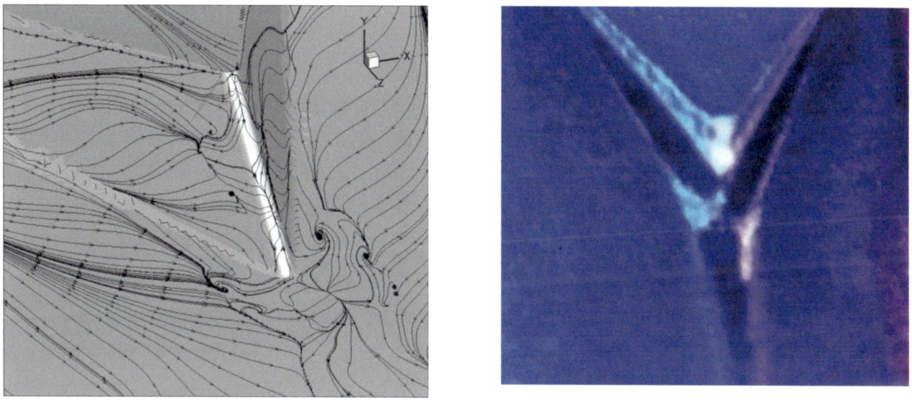

a. UTA LES. **b. UTA Experiment.**

Fig. (1.5). Side view of the flow structure on MVG surface and surrounding area.

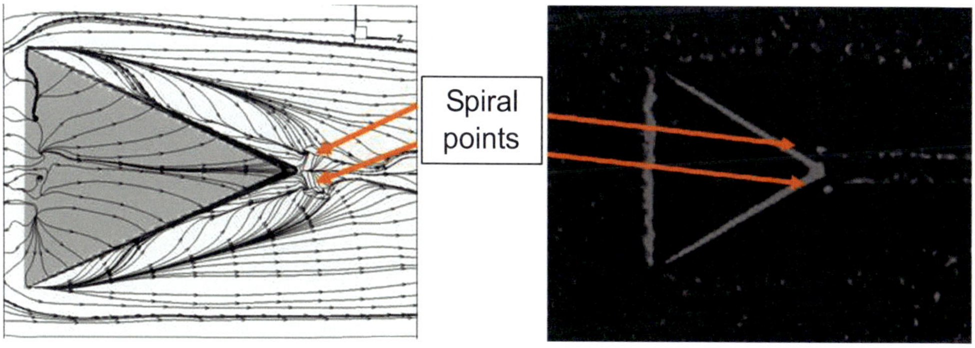

a. UTA LES. (A repetition of Figure 1.4(a)) **b. UTA Experiment (Pierce et al. [78]).**

Fig. (1.6). Top view of spiral points behind MVG.

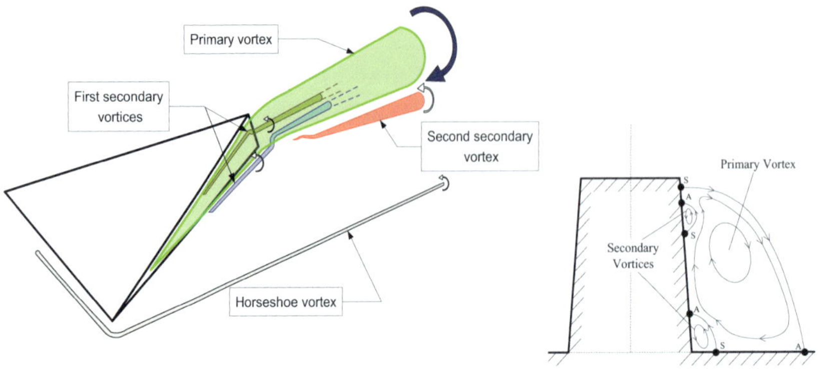

a. Schematic of new five pairs of vortices given by UTA LES and experiment.

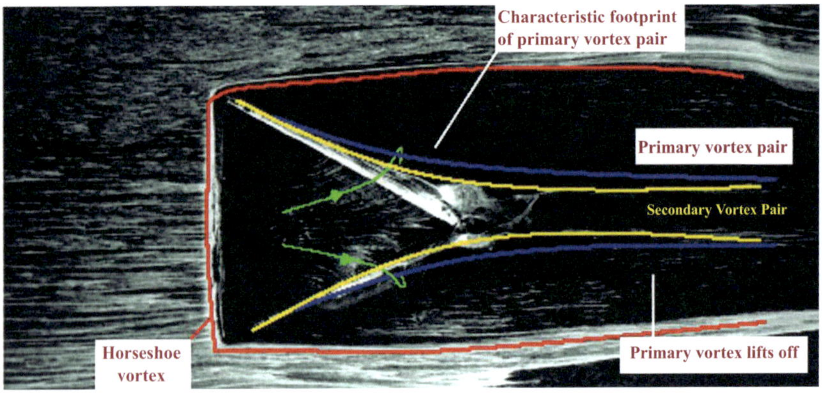

b. Schematic of three pairs of trailing vortices given by Babinsky at NASA SBLI Workshop in 2008.

Fig. (1.7). Schematics of two different vortex structures.

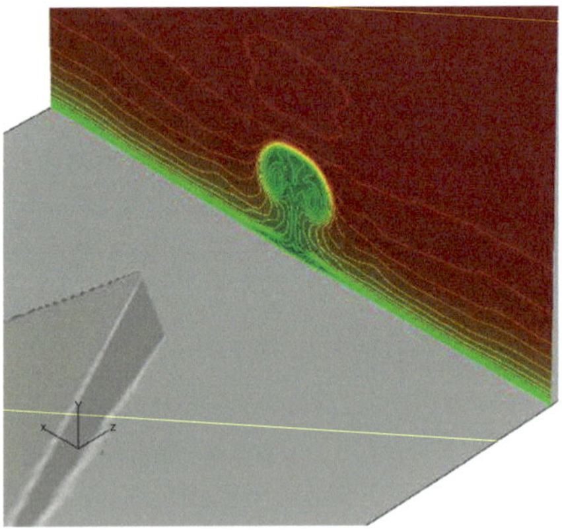

Fig. (1.8). Mushroom-shaped structure (Li and Liu [72]).

After analyzing our LES and experimental results, we present a new flow structure consisting of five pairs of trailing vortices (Fig. **1.7a**): a horseshoe vortex generated by leading edge separation, a pair of primary trailing vortices, two pairs of secondary vortices as also observed previously, and a new pair of secondary vortices after the spiral points. This structure is consistent with topological rules [80]. Current model differs from that of Babinsky in two aspects: one is that the first secondary vortex should be two pairs not one, *i.e.*, one is generated by the MVG side surface and the other by the plate; the second is that the original secondary trailing vortex must be lift up from the wall surface. This will lead to an end of surface separation through a spiral point in our surface visualization (Figs. **1.5-1.6**). More discussions will be continued in Section 1.7.

1.3.3. Mechanism of Momentum Deficit Formation

The momentum deficit behind the MVG was observed by Babinsky *et al.* [51] and re-confirmed by the computation of Ghosh *et al.* [55] and Lee *et al.* [61]. In the region near the MVG, the shape of deficit in cross-section appears like a circle, which was connected with a root coming from the boundary layer, appears as a mushroom shape (Fig. **1.8**). Underneath the head of the mushroom, there are two regions with high value of streamwise velocity. Babinsky held the view that the deficit showed the wake of the MVG. Lee *et al.* mentioned in their study that "a larger tube has two counter-rotating vortices inside which is created by two vortical tubes".

What is the relation between the flow structure and the momentum deficit? Where is the low-speed flow from? Few studies provide a clear explanation about the deficit up till now. Someone might suppose that the momentum deficit behind the MVG is generated by streamwise vortices through entraining the low-speed fluid from the bottom of the boundary layer out to form a mushroom structure (Fig. **1.8**). However, further investigation by UTA LES finds that the momentum deficit mainly comes from the low speed zone. The fluid of boundary layer which passes over the MVG is entrained by the streamwise vortices to generate the circular structure (Fig. **1.9**). This is clearly demonstrated by tracking the streamlines around the deficit region of the boundary layer. The low momentum fluid with reduced streamwise velocity forms the momentum deficit. This momentum loss is essentially caused by the boundary layer ahead of the MVG.

1.3.4. Recompression Shock Structure Around MVG

Fig. (**1.10**) shows the three-dimensional recompression shock envelopes around the MVG. The arc shape of the recompression shock grows downstream. The re-compression shock wave is not an integrated one initially, *i.e.*, the top or head of

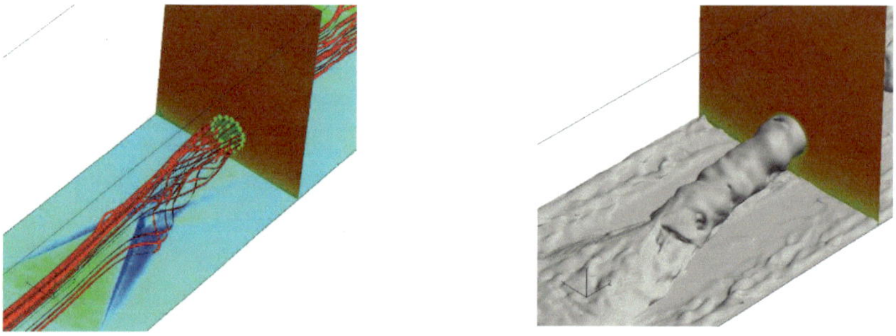

a. Spiral stream track of flow passing MVG. **b. Iso-surface or streamwise velocity contour.**

Fig. (1.9). Side view of the flow structure on MVG surface and surrounding area by UTA LES.

the shock is separated from the lower part or leg, and afterwards the head and leg are connected. The curved shock creates entropy and vorticity *via* Crocco's theorem. Further, the misalignment between the density and pressure gradients creates baroclinic torque which is tentatively suggested to add to the vorticity in the MVG wake.

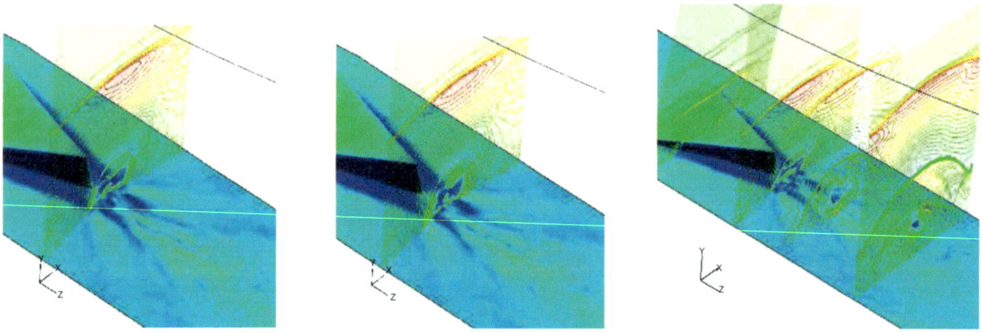

a. First station. **b. Second station.** **c. Third station.**

Fig. (1.10). Surface pressure and shape of re-compression shocks at three stations behind MVG by UTA LES (light blue color represents lower pressure).

1.3.5. Kelvin-Helmholtz Instability and Generation of the Vortex-Ring in MVG Wake

The inviscid Kelvin-Helmholtz (KH) instability is a well-known instability existed in shear layers [77]. In our case, we believe that the momentum deficit in the wake of the MVG will generate a cylindrical shear layer that breaks down into a train of vortex rings (also known as vortex loops). The train of vortex rings at low or high speed exhibit complex behavior of leapfrogging and merging, as evidenced from animations of both the LES and experimental images [81 - 84].

As verified by computation and experiment, the MVG will generate a deficit of the streamwise velocity. The deficit appears to be roughly circular in cross section and has strong streamwise velocity shear in its deficit boundary (Fig. **1.11**). An examination of the streamwise velocity distribution reveals inflection points within the shear layer at the boundary of the deficit (Figs. **1.12** and **1.3**). The inflection points cause the shear layer to become unstable to produce ring-like vortices (named by Kerswell [85]) (Fig. **1.14**). Hence, vortex rings are generated within the shear layer and have strong interactions with the flow. In order to visualize the vortex rings, the iso-surface is drawn using the so-called λ_2 introduced by Jeong and Hussain [86] in1995, which is the second eigenvalue of the 3×3 matrix comprised of velocity gradient $M_{ij} = \sum^3_{k=1} (\Omega_{ik}\,\Omega_{kj} + S_{ik}\,S_{kj})$, where $\Omega_{ij} = 1/2(\partial u_i/\partial x_j + \partial u_j/\partial x_i)$ and $S_{ij} = 1/2(\partial u_i/\partial x_j - \partial u_j/\partial x_i)$. A train of ring-like vortices generated behind MVG is shown in Fig. (**1.15**).

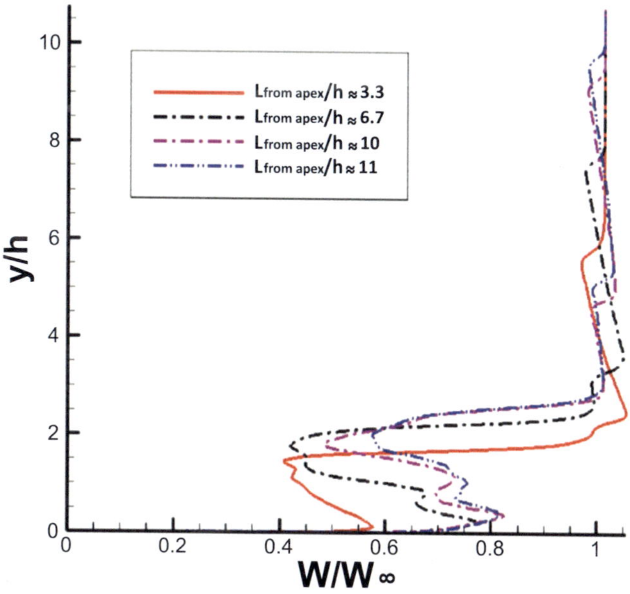

Fig. (1.11). Distribution of streamwise velocity at different locations.

Experimental evidence indicating the presence of large structures, ostensibly vortex rings, from laser lightsheet visualization [87, 88] with an exposure of 1 μs, is shown in Figs. (**1.16** and **1.17**). The lightsheet is aligned in the streamwise direction and normal to the flat plate, it is also incident at the centerline of the

MVG. The bright object near the center is the MVG and the bright surface indicates stray reflection of the laser sheet. Fig. (**1.16b**) is a global visualization where the seed particles are introduced in the plenum chamber of the wind tunnel. Fig. (**1.17b**) is a local visualization where the seed particles are introduced from a surface pressure tap ahead of the MVG.

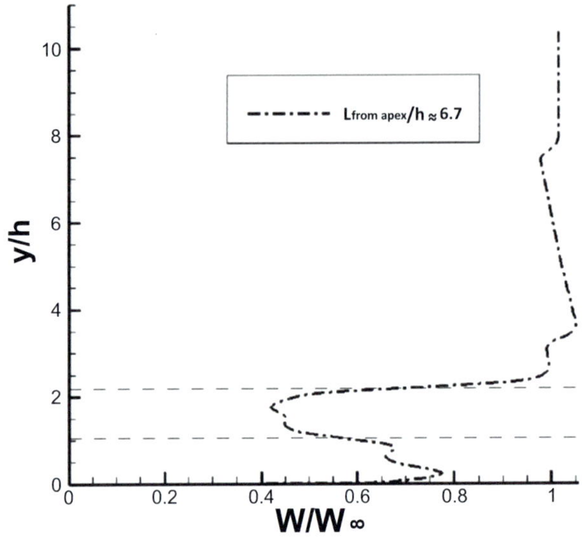

Fig. (1.12). Streamwise velocity distribution $L_{from\ apex}/h$=6.7.

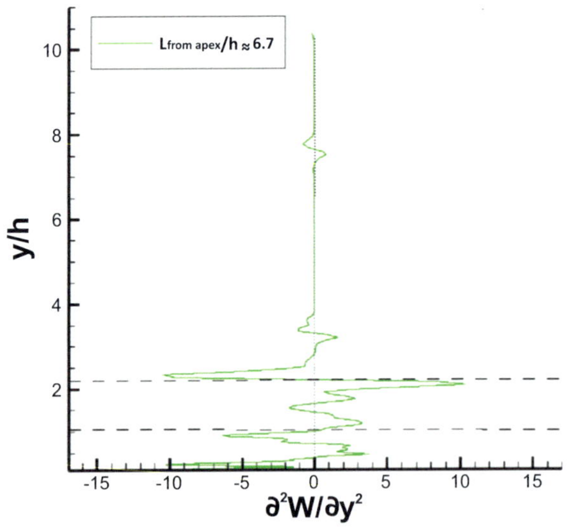

Fig. (1.13). Second derivative of the streamwise velocity to axis y at $L_{from\ apex}/h \approx 6.7$.

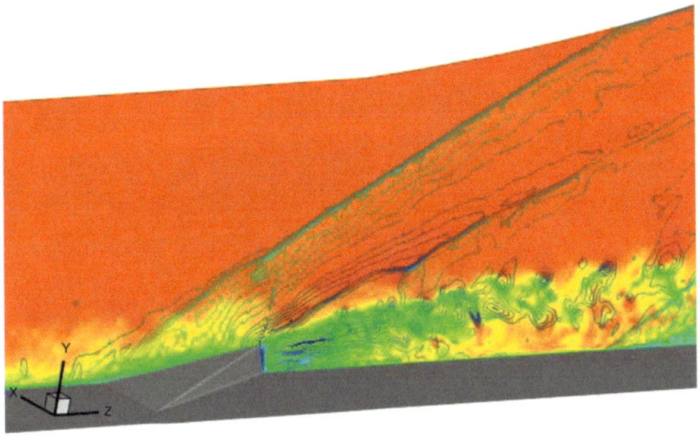

Fig. (1.14). Instantaneous view of pressure and streamwise velocity contour by UTA LES.

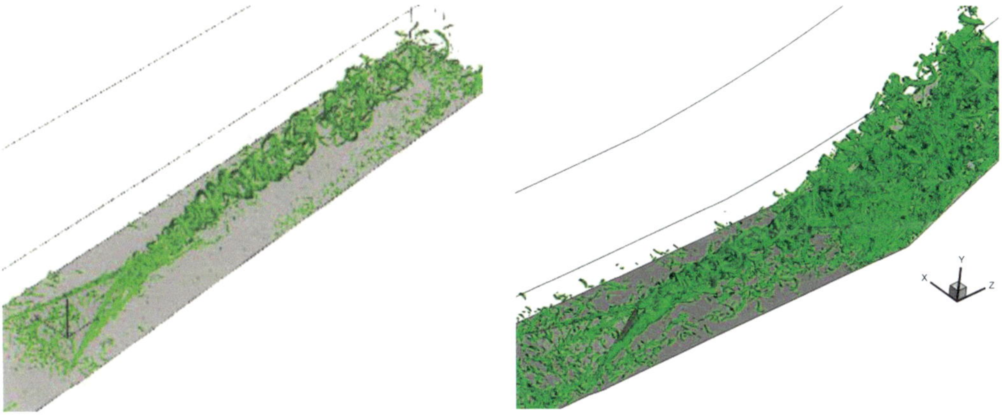

a. Flat plate with MVG **b. Ramp with MVG**

Fig. (1.15). The ring-like vortices behind MVG shown by the iso-surface of λ_2.

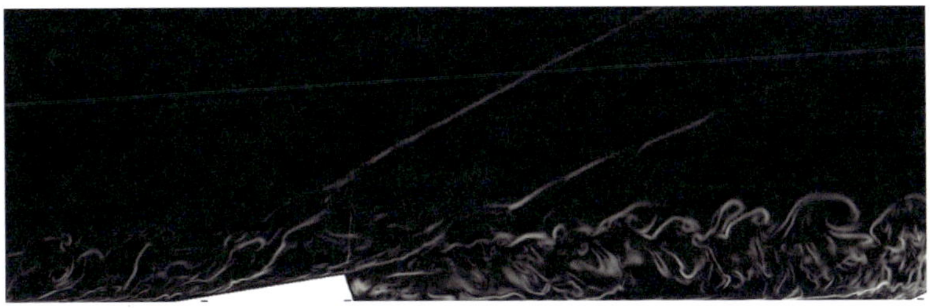

(a) UTA LES

Fig. 1.16 contd.....

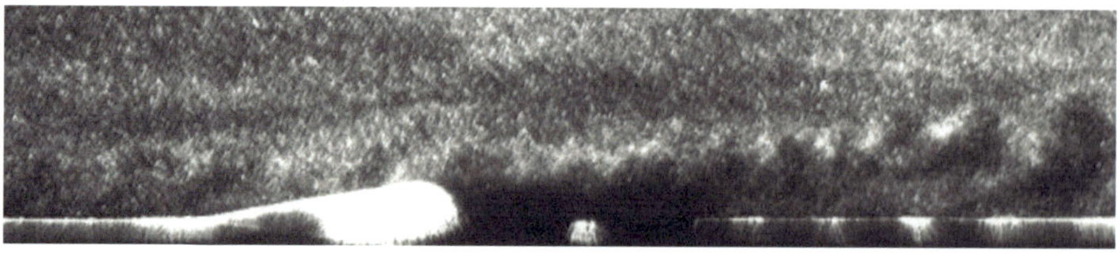

Fig. (1.16). Large structures downstream of MVG from global lightsheet visualization.

(a) UTA LES

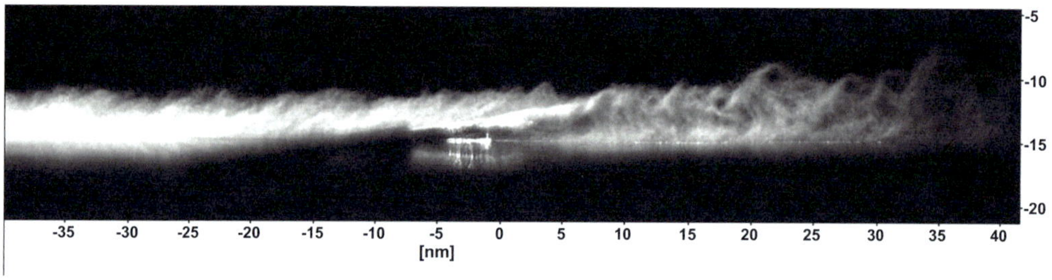

(b) UTA Experiment

Fig. (1.17). Large structures downstream of MVG from focal lightsheet visualization.

Both figures show a distinct difference between the large structures from the approaching undisturbed boundary layer and those downstream of the MVG even though the visualization is found along the centerline of an MVG. This evidence helps to support the LES results of K-H instability and large vortex rings formed by the momentum deficit. The bright feature to the right of the MVG in Fig. (**1.17**) is the recompression shock system. Visualization in other streamwise and crosswise planes are ongoing to provide further evidence of the large vortex rings. This is a new MVG flow mechanism not previously reported in the literature.

1.4. EFFECTS OF MVG TRAILING EDGE ANGLE

Different trailing edge angles of MVGs are designed to study the effect of the trailing edge angle, *i.e.* β = 70° and 45°. For both configurations, the basic structures such as the wave system, vortex rings, the momentum deficit and the separation pattern are similar. However, the vortices including the initial streamwise vortices and a chain of vortex rings generated by the MVG with smaller trailing edge angle at β = 45° are closer to the wall (Fig. **1.18**). Therefore, this kind of configuration is more favorable for boundary-layer separation control.

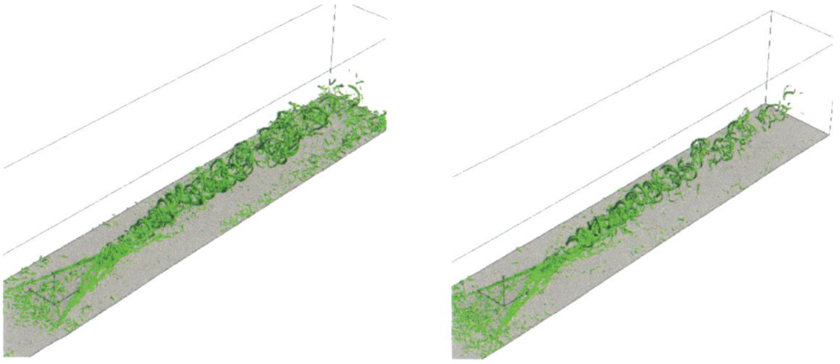

Fig. (1.18). Iso-λ_2 surface **for** β=70° (left) and β=45° (right) by UTA LES.

1.5. SEPARATION ZONE REDUCTION BY MVG

Fig. (**1.19**) shows the time-averaged experimental and instantaneous numerical schlieren images at the SBLI region. Due to the difficulty on the determination of the attachment location, the separation size is defined as the length from the beginning of the separation to the ramp corner. In figure, the separation zone appears to be a shape of "V". At two spanwise sides of the domain, the separation zone length is about 6.8~7*h* and the length on the center plane is around 2.6*h*. From Figs. (**1.20** and **1.21**), another mechanism to reduce separation zone can be identified, namely, the vortex ring/shock interaction. Since the vortex ring shows a low pressure area in the vortex core and a high pressure area behind ring center, the shock breaks up and the separation zone is reduced.

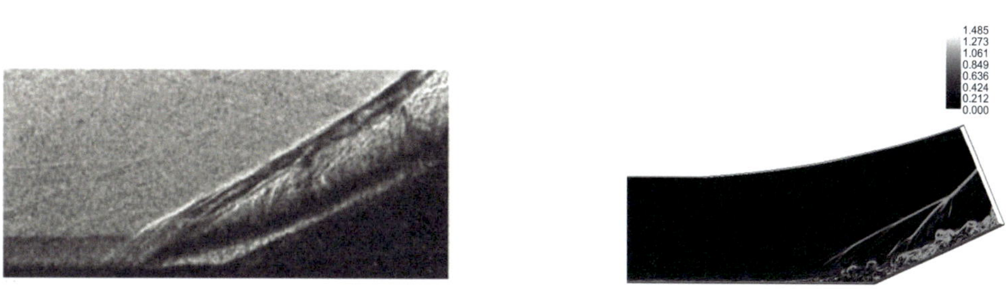

a. Schlieren by experiment **b. Numerical schlieren**

Fig. (1.19). Flow separation on a supersonic ramp with MVG.

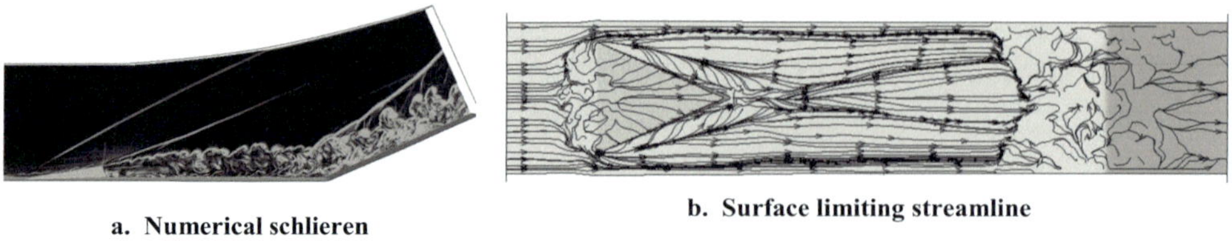

Fig. (1.20). Schlieren for flow separation on a supersonic ramp without MVG.

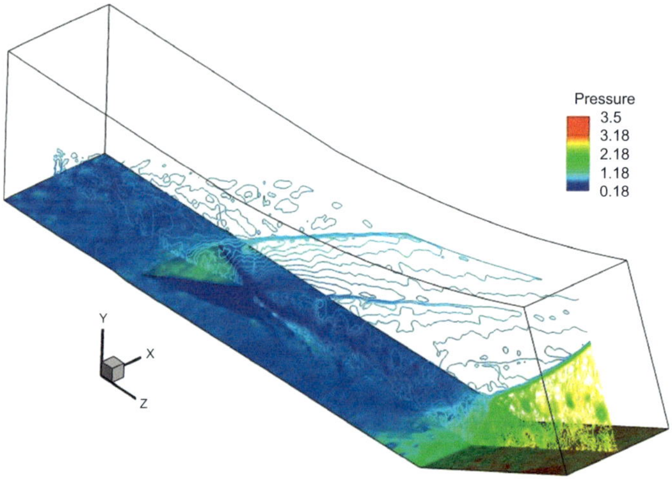

Fig. (1.21). Ring vortex/shock interaction behind MVG.

1.6. SUMMARY OF INTEGRATION OF NUMERICAL, EXPERIMENTAL AND THEORETICAL STUDIES

UTA LES [73, 79] demonstrated a tight integration of numerical, experimental and theoretical studies in understanding the physics of MVG on SBLI. They have revealed features that have not been previously observed. To summarize:

Consistency in surface topology between LES and experiment, especially the existence of spiral foci on the surfaces, the streamlines of the momentum deficit region were tracked by using CFD and it was found the momentum deficit is mainly caused by drag on the flow immediately by the boundary layer ahead of the MVG. LES showed K-H instabilities and breaking-down structures as ring vortices with convincing experimental evidence. A theoretical model has been proposed consisting of 5 trailing vortex pairs from an MVG, breaking-down in a complex manner to produce ring vortices which then interact with the downstream shock-induced separation region.

1.7. SUPERSONIC FLOW AROUND MVG

Some studies on MVG of supersonic flow focused on ramp and vane types, especially the former, considering to their robustness. Surface flow visualizations of MVGs in supersonic flow at Mach 1.4 and 2.5 are shown in Figs. (**1.22a** and **1.22b**). These figures show that a separation bubble is generated, due to a regional, two-dimensional shock/boundary layer interaction, and the evidence is an accumulation of oil which is shown as a bright zone in these figures. On the other hand, the LES results (Fig. **1.22d**) detect separation more clearly by the surface streamlines directed upstream [73, 79]. They proposed a model of the flow passing the MVG in which the separation leading edge (corner) produces a pair of weak trailing vortices according to the initial results of Babinsky *et al.* [51]. A primary vortex pair shed from the ramp top edges of the MVG. Several pairs of weak secondary vortices are also discovered. Fig. (**1.23**) shows the separation actually occurs below the incipient criterion. It is can be understood that the separation is induced by a local Mach number of the boundary layer, which is far lower than the freestream. As observed by Settles *et al.* [89], accepted two-dimensional incipient separation criteria depended on observations such as the incipient presence of a plateau ("kink") in the surface pressure distribution or incipient accumulation of oil in surface oil-flow visualization [90]. These authors held the view that the separation bubble is present all the time, even a miniscule one at the region of unseparated interactions. This suggestion is further supported by present observations. The flow separation at the leading edge of MVG is weak and generates a weak horseshoe vortex with a height of approximately the thickness of boundary layer [91]. Experimental evidence of the horseshoe vortex is given only from interpreting surface flow visualization, thus far while numerical evidence comes from LES with high order scheme albeit at a lower Reynolds number, also from surface topology Fig. (**1.22d**). Fig. (**1.22a**) shows the surface flow visualization that there is no obvious interference between the MVGs although they are close. The horseshoe vortex system, which is regarded as a secondary vortex pair bifurcating from the leading edge of the MVG, is extremely weak as it arises from the extremely weak separation of leading edge; so there is little significantly effect on the downstream flow from leading edge separation. However, an envelope of separation [80] is related to the horseshoe vortex, which can isolate the flow around MVG from the flow of neighbors. Despite its geometrical simplicity of MVG, it still has the extremely complex flow topology as proposed by a detailed analysis based on high-order LES [73, 79, 92], as well as on the surface and off-surface flow visualization [75]. This complexity may not be unexpected since geometries in three-dimensional have rich topologies. Li and Liu [73] first time gave the high order LES results in their study that eight detailed features of surface and off-surface topological structures around and downstream of the MVG are found, *i.e.*, (1) surface spiral points connected to vortex filaments,

(2) surface separation topology, (3) five pairs of vortex tubes around MVG, (4) the source of momentum deficit in the downstream flow, (5) an inflection surface that gives rise to a (6) Kelvin–Helmholz instability and (7) the generation of ring vortices, and (8) a ring vortex/shock interaction and a proposed new mechanism for separation reduction [73, 74, 79]. These features will be described in the following. The dominant one of vortex pair generated by MVG is not the above-mentioned horseshoe vortex system, but arises from the flow separating off the slant sides. From Figs. (**1.22** and **1.23**) in side view together with video clips of the experimental visualizations, high and low shear regions on the surface of MVG and flat plate surface can be clearly observed. These regions can be the suggestions of a large primary vortex structures. As can be seen from the surface streamline patterns shown in the figure, the horseshoe vortex hinders the primary vortex pair to develop downstream and then clash behind the trailing edge. Additionally, it is revealed that the singularities along the leading edge of MVG have not been observed if at all experimentally [73], see Fig.(**1.23b**). (Above description looks like a direct copy of Prof. Lu's publication; if being, it suggests some change to void copyright violation)

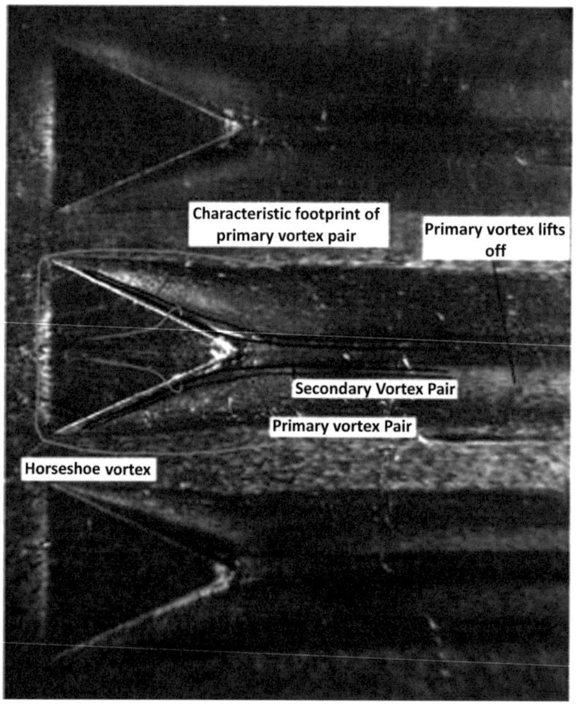

(a) From Herges et al. [56] ($M_\infty = 1.4$).

Fig. 1.22 contd.....

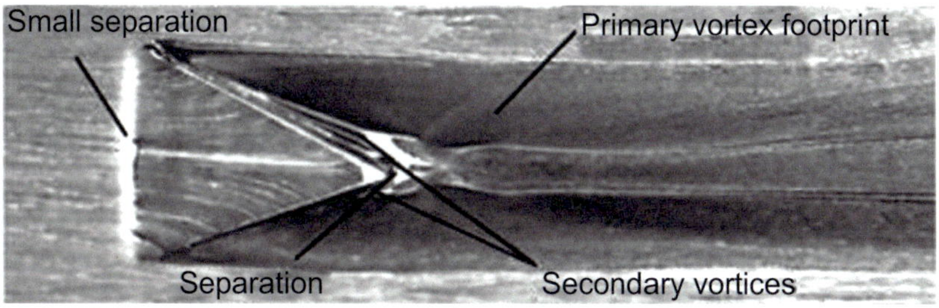

(b) From Babinsky et al. [51] ($M_\infty = 2.5$).

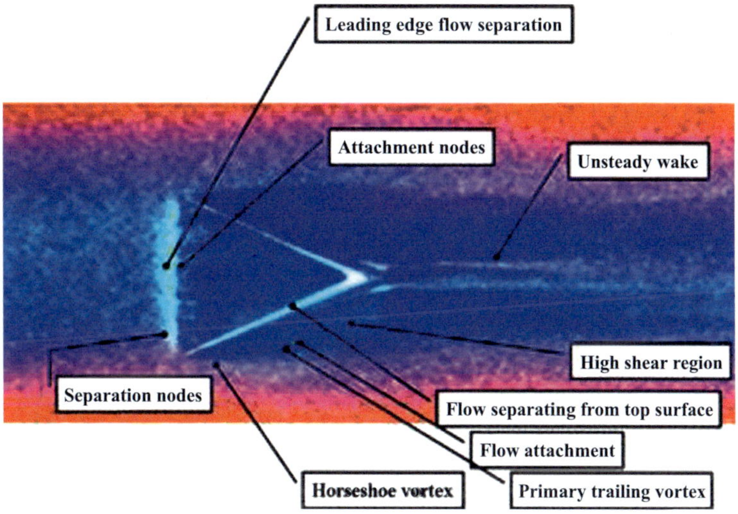

(c) From Lu et al. [93] ($M_\infty = 2.5$).

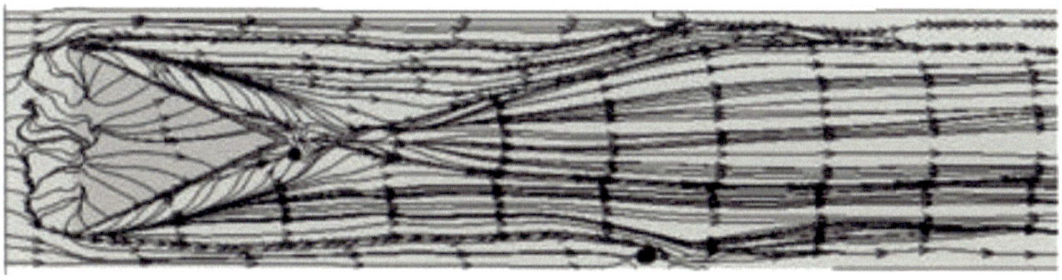

(d) From Li and Liu [73] ($M_\infty = 2.5$).

Fig. (1.22). Experimental surface flow visualization and numerical surface streamlines from investigators.

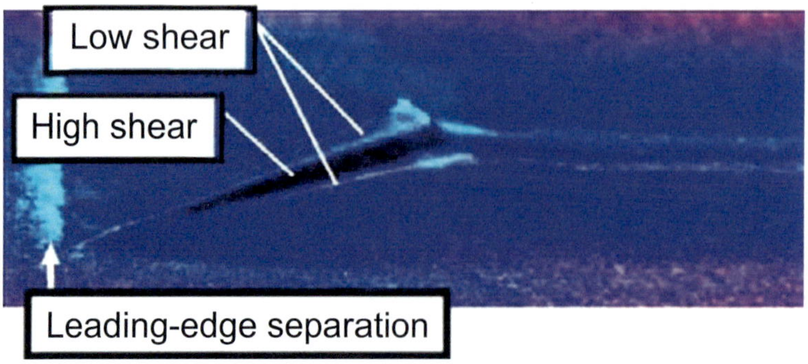

(a) Experimental surface oil-flow visualization [75].

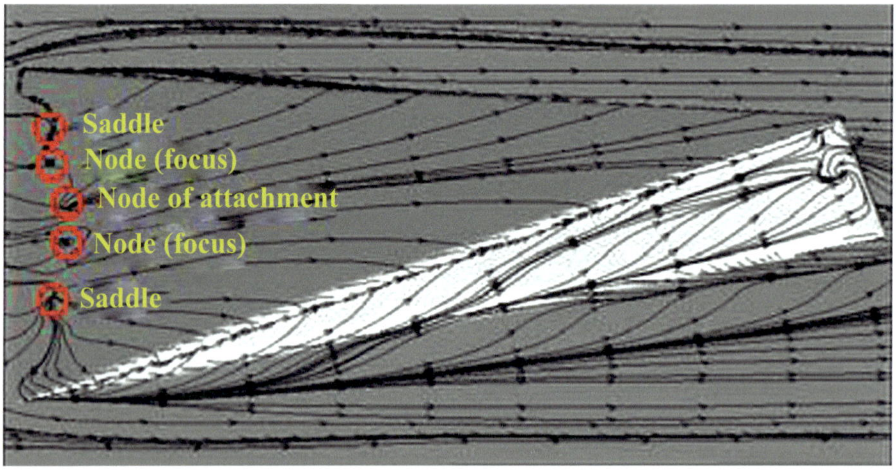

(b) Numerical surface streamlines [73, 74, 79].

Fig. (1.23). Side views.

Surface pressure distributions are another footprint to reveal flow features around MVG. A region with high-pressure lying ahead of the MVG is shown in Fig. (**1.24a**) for time-averaged, surface pressure distributions from high-order LES [73]. This is a result of the regional SBLI induced by leading edge of MVG. A slightly higher pressure also appears on both side of the MVG, while extremely low pressures exist nearest to the MVG junction on the flat plate. This low pressure can be explained as the result of the high-speed flow in this region related with an open three-dimensional separation. An asymmetry downstream especially behind the trailing edge is shown in the time-averaged result of LES as well as the experimental result, see Fig. (**1.22**). While the asymmetry may be an artifact from the simulation or the data processing, which indicating an evidence of unsteadiness. A pair of swept high-pressure regions generated in downstream of MVG. A simple construction of the inviscid wave system reveals that the high pressure regions induced by a recompression shock as the flow changed its direction from the slant sides to an axial direction. The experimental results of Herges *et al.* [56] clearly display the similar features of a high-pressure region at the MVG leading edge (see Fig. **1.24b**). The low-pressure regions caused by the rapid flow past the MVG and the high-pressure regions downstream induced by the recompression shock are also obvious in the figure. The presence of these secondary vortices follows topological rules [80]. Further, some results indicate that when either of the opposing primary or the secondary vortex pairs impinges each other, symmetry breaking occurs as the contribution to the unsteadiness [94 - 97]. The interactions between vortices enlarge the number of surface singularities which is shown by high-order LES result in Fig. (**1.25**). Some of these singularities have subsequently been discovered in the experiments. For instance, a pair of spiral points in both of the numerical LES and experimental results [98] is shown in Fig. (**1.26**). The experimental result was obtained by using the image processing of surface flow visualizations on raw videos. According to past experience, a synergistic approach of combining simulations and experiments gives a great benefit to the investigation. Therefore, analysis in detail of the numerical results with experimental support reveal a five-pair vortex model for the near-field topology in Fig. (**1.27**) [92, 98].

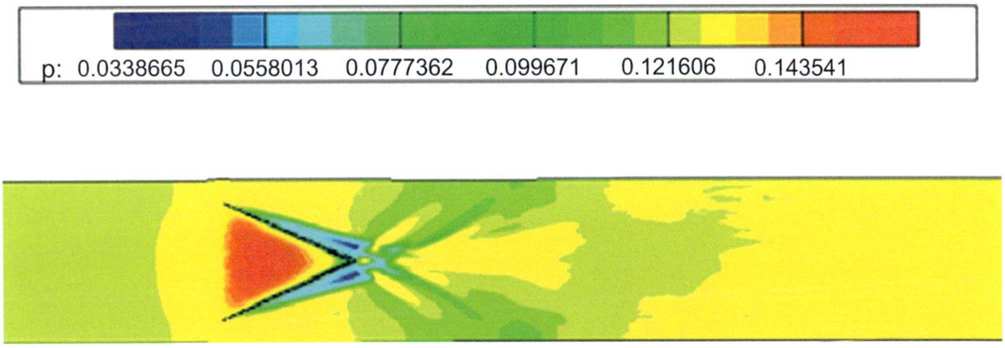

(a) Surface pressure distribution [73, 79].

Fig. 1.24 contd.....

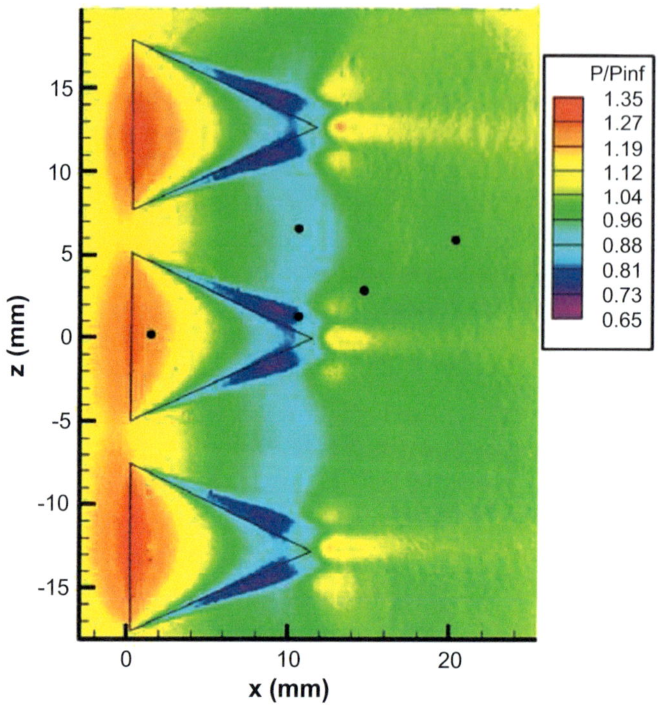

(b) From Herges et al. [56]

Fig. (1.24). Surface pressure distribution around an MVG.

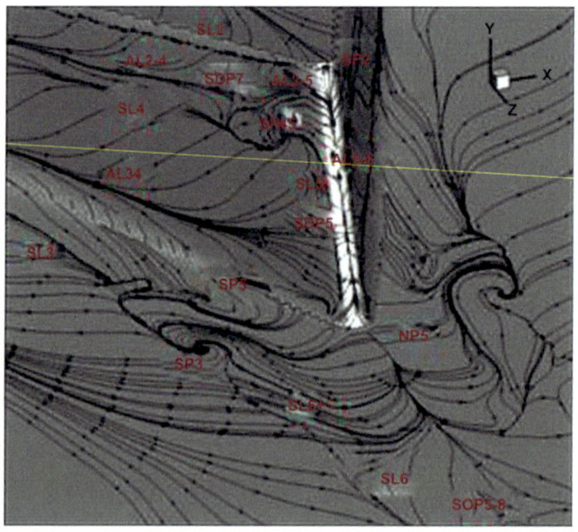

Fig. (1.25). Numerical visualization showing singularities around MVG trailing edge; AL = attachment line, SDP = saddle, SL = separation line, SP = spiral [98].

Moreover, a new technique of visualization reveals a relation between the spiral point SP5 in Fig. (**1.25**) and a flow feature in Fig. (**1.28**) which is tentatively treated as a tornado-like vortex filament. Numerical simulations in Fig. (**1.29**) confirm this feature that the streamlines originating from SP5 at the same position as the experimental "tornado" are shown. Another possible vortex filaments emanating from the top trailing edge of the MVG are given in Fig. (**1.28**). However, the flow feature near the top could be treated as a combination of a secondary vortices shed from the top slant edge of MVG and the vortex filament. This feature is possibly not just an expansion fan which would be found in schlieren imaging, but also appears on the top of the MVG trailing edge which is not always clearly seen in seeded flows.

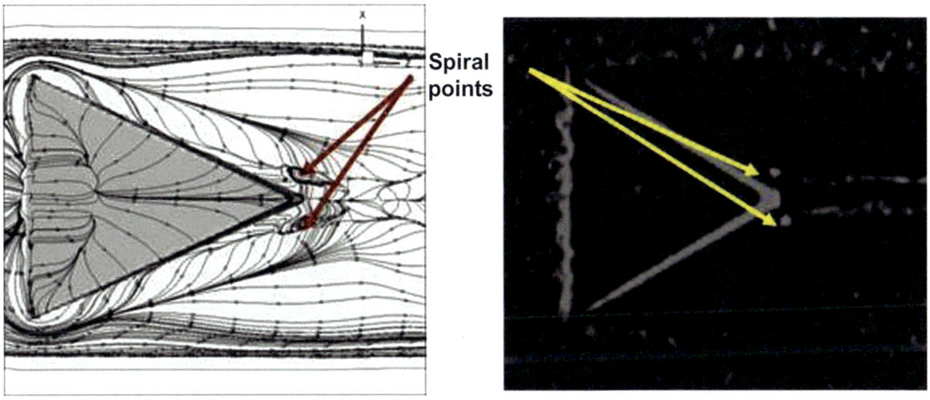

Fig. (1.26). LES and experimental visualizations showing a pair of spiral points on the flat plate at the MVG trailing edge [98]. (This figure is very similar to Fig. (**1.6**)).

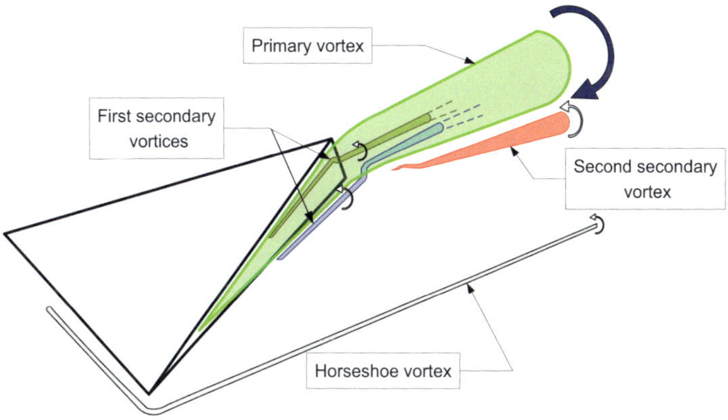

Fig. (1.27). Five-pair vortex model of Li and Liu [73]. (This is a repetition of Fig. (**1.7a**)).

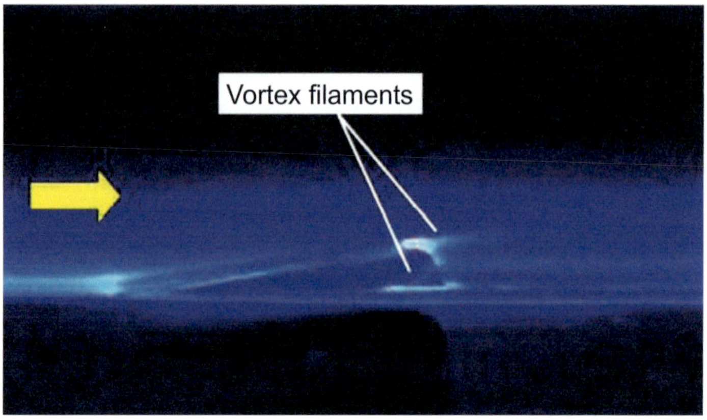

Fig. (1.28). Panoramic visualization showing vortex filaments [99].

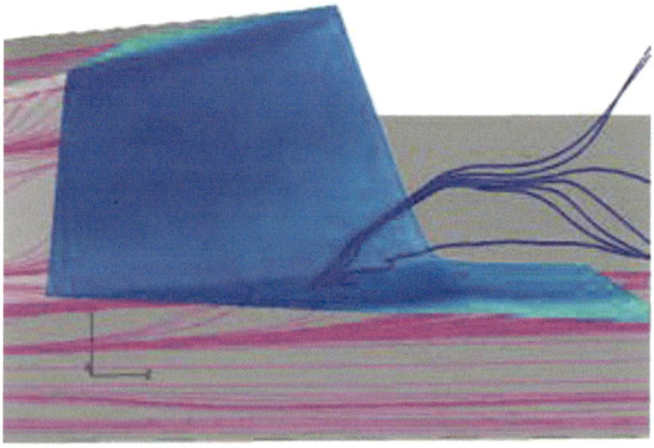

Fig. (1.29). Streamlines emanating from SP5 [98].

According to these results, Lu *et al.* [100] proposed a detailed topology of flow field. However, vortex filaments related with the surface spiral points are also included to demonstrate the complex flow topology. The complex flow features are also evident in the region near wake of the MVG. Time-averaged data, either through RANS [55, 62] or PIV/LDV [51, 56, 101] show a momentum deficit in streamwise direction near the centerline of the wake and regions of higher momentum further outboard on either side of the MVG as well as nearest to the surface, beneath the region with low momentum in Fig. (**1.30**). In Fig. (**1.31**) from Bur *et al.* [102], the laser light sheet visualizations of counter-rotating vane-typed MVGs show two spots which can be identified as trailing vortices. However, it is not known if these are low momentum regions like that discussed above. It is proposed that the primary vortices near trailing edge can entrain high momentum fluid from high-speed freestream into the low-momentum boundary layer and

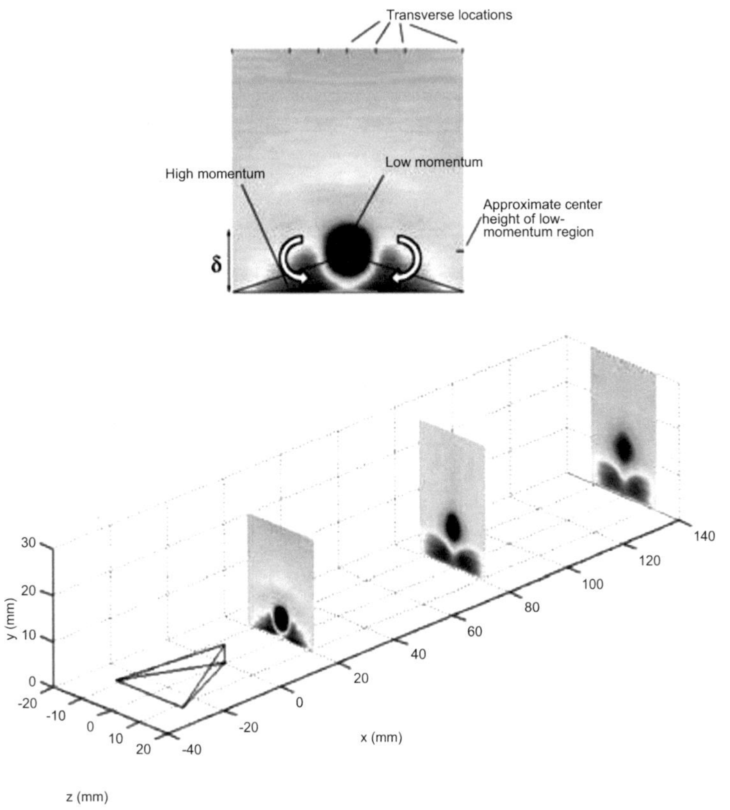

Fig. (1.30). Streamwise momentum deficit and surplus in the wake of an MVG [51].

finally lead the momentum mixing and redistribution. This experimental result was also confirmed by numerical simulation results [55, 62, 92] shown in Fig. (**1.32**). Li and Liu [73] heuristically discovered through streamline tracking that the flow in the region of momentum deficit comes from top of the MVG. This observation can be further reinforced by noting that the flow over the top edge of the MVG has suffered momentum loss in passing over the leading edge shock [55].

For further investigating the wake of MVG, Blinde *et al.* [52] interpreted the detailed stereo PIV maps. A schematic is shown in Fig. (**1.33**). The high-speed regions outboard of each MVG that serve to generate a chain of hairpin vortices directly downstream of the MVG are found in the figure. In addition, the conceptual sketch shown in Fig. (**1.33**) is instantaneous so that the flow field can be treated as unsteady flow. Blinde *et al.* stated that there are no trailing vortices apparent in instantaneous snapshots. However, large structures such as pairs of counter-rotating vortices are convected in the boundary layer.

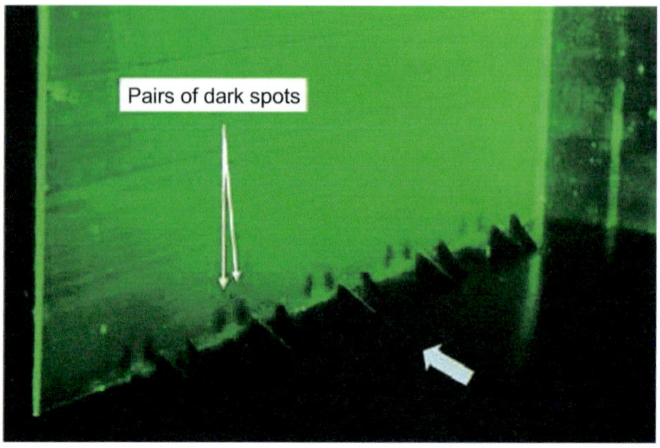

Fig. (1.31). Pairs of dark spots identified as counter-rotating vortex pairs from an array of micro-vane vortex generators [102].

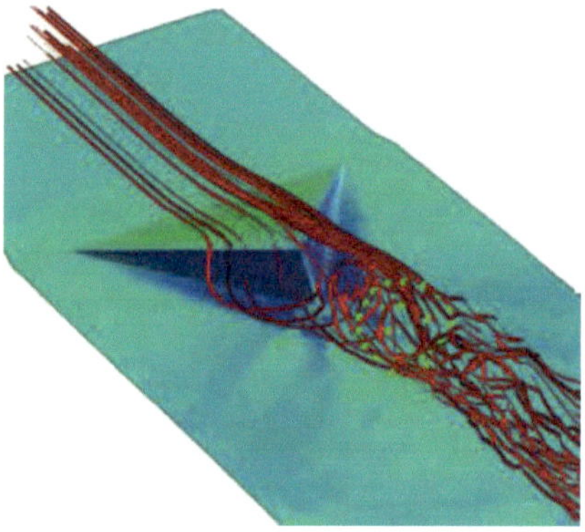

Fig. (1.32). Tracing the momentum deficit in the MVG wake to flow over the top.

These authors further stated that the observations of their studies have an agreement with observations of the wake of protuberances at low speeds in both experiment and computational DNS result [102, 103, 70]. It should be note that while Blinde *et al.* identified outboard high-speed regions, different vortex structures for entraining the flow from freestream fluid was also suggested. The vortex shedding mechanism in sub-boundary layer and its unsteadiness have been studied recently in low-speed flow [68, 70, 72]. Specifically, it is suggested by Angele and Grewe [69] that the unsteadiness contributes to maximum Reynolds stresses around the averaged vortex centers which should not be the case if the vortices of the flow remained steady. Li and Liu [104] suggested that the mean velocity profile of the MVG downstream displays an "inflection surface" that is unstable, thus causing the generation of the K-H instability [67]. The previous studies suggested that such a mechanism is the cause of turbulent vortex rings [105]. The further study where the wake behind MVG was reconstructed through using tomographic PIV is shown in Fig. (**1.34**). They identified the presence of a pair of counter-rotating vortices. In addition, the iso-surface of the crosswise vorticity magnitude $\sqrt{\omega_y^2 + \omega_x^2}$ shows the hairpin or "arch" vortices wrapping over the primary counter-rotating vortex pair. These are more complicated structures than that described in Fig. (**1.33**). Regardless of the nomenclature, such ring vortices around the trailing counter-rotating vortex pair are also shown by high-order LES result [74]. Fig. (**1.35**) gives the iso-surface of a small negative value of λ_2, which is a numerical visualization of vortices [106]. These ring vortices can be associated with the hairpin or arch vortices observed in experiment. There are similar structures between the ring and the hairpin vortices. In experimental result, the region at bottom is inaccessible to PIV instrumentation, thus is properly not observed, which is different from the high-order LES data. Additionally, large intermittent density structures can be seen from the instantaneous numerical schlieren in Fig. (**1.36**). These structures from time-averaged result give the impression of a thickened boundary layer.

The results of numerical visualizations are confirmed and supported by experiment [107]. A global visualization where the freestream flow is seeded with calcium carbonate particles with a median size of 0.7 μm is shown in Fig. (**1.37a**). In this seeding technique, the MVG wake appears dark due to lack of scattering particles. The ragged interface between the MVG wake and the freestream is similar with the numerical schlieren in Fig. (**1.36**). Large billowing structures are shown in Fig. (**1.37b**) by using a regional light sheet technique with acetone fog injected from an upstream pressure tap, which acts as a complementation of the global light sheet visualization. These structures in Fig. (**1.37c**) are attributed to the MVG, since they are not present in the light sheet visualization of the flow in boundary layer passing the flat plate.

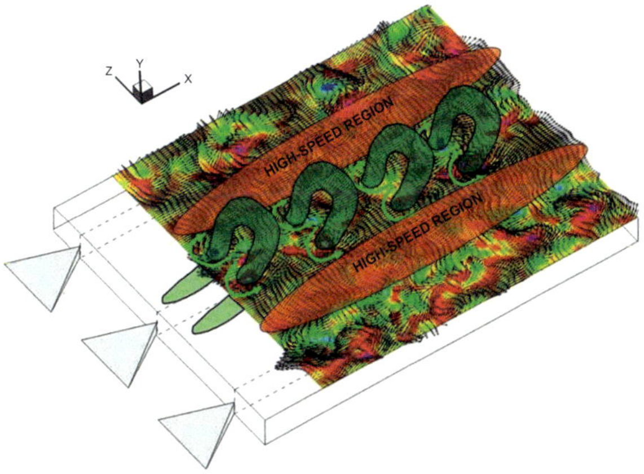

Fig. (1.33). Postulated hairpin vortex train in the wake of an MVG [52].

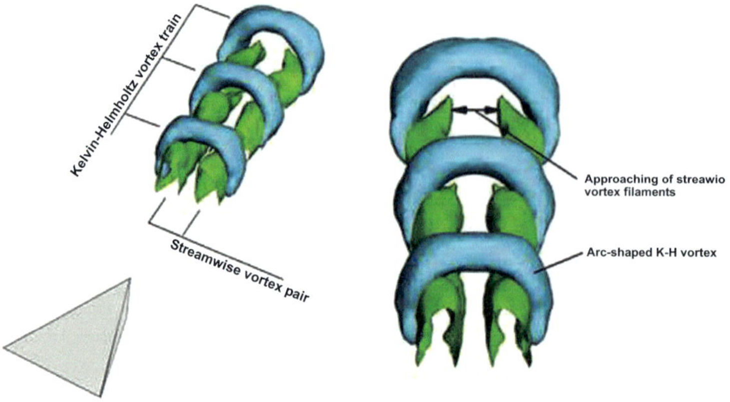

Fig. (1.34). Visualization from tomographic PIV of conditionally averaged vorticity distribution in the wake of a micro-ramp. Vorticity isosurfaces of streamwise component ωx (green) and spanwise-wall normal component $\sqrt{\omega_x^2 + \omega_y^2}$ (light blue) [101].

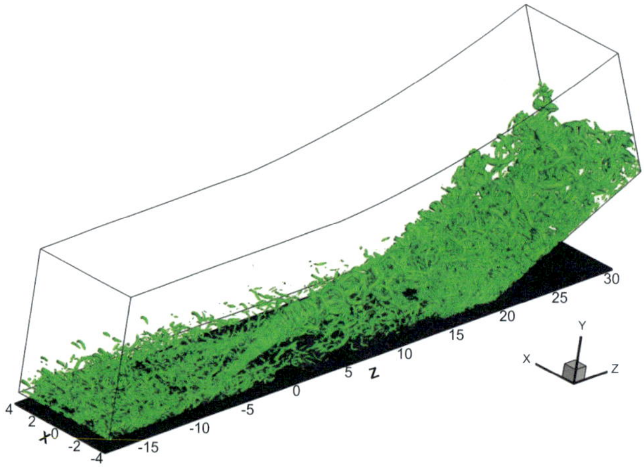

Fig. (1.35). Vortex rings shown by $\lambda 2$ iso-surface. Figure also shows vortex ring breakdown as the vortex train interacts with the leading shock [73].

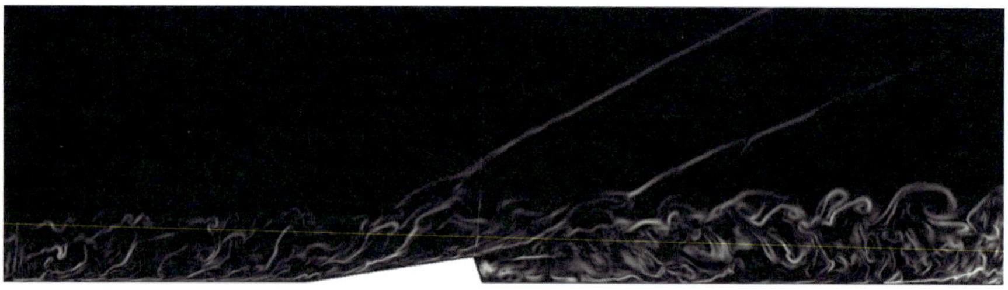

Fig. (1.36). Instantaneous numerical schlieren of centerplane showing large flow structures.

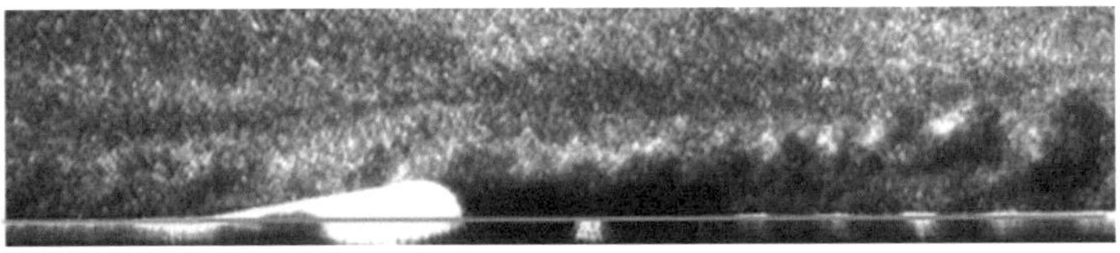

(a) Global light sheet visualization with light sheet aligned along MVG axis. (The figure has been appeared as Figure 16(b))

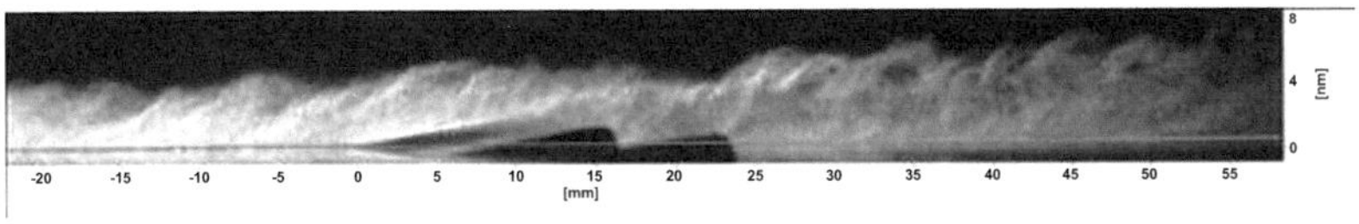

(b) With MVG.

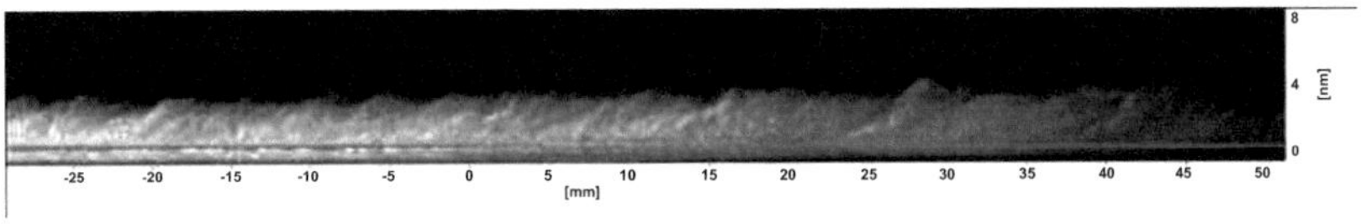

(c) Flat plate, without MVG.

Fig. (1.37). Light sheet visualizations.

In a word, the flow field flowing over MVG has extremely complex structures, due to the three-dimensional characteristics. A pair of primary vortices is the major feature that trails from the sides of the MVG with associated secondary vortices and vortex filaments. Although the horseshoe vortex that wraps around the MVG is not the major topological feature, it is crucial in generating the individual MVG wake. A momentum deficit region with a circular shape can be seen behind the MVG. A circular velocity inflection is also observed and thought to initiate a mechanism of K-H instability, which is possibly aggravated by symmetry breaking due to the two primary vortices impinge each other. The mechanism of the unsteadiness and the instability leads to the form of a chain of vortex rings, which flow into the energetic freestream and distribute it into two sides. These rings can develop a long distance downstream.

1.8. SUPERSONIC MACH NUMBERS (OBLIQUE INCIDENT SHOCK OR RAMP-INDUCED SHOCK)

According to the content discussed above, the main difference between transonic and supersonic studies is that the Mach number of the incoming flow has sufficiently past unity and that the flow behind the incident shock keeps supersonic. Anderson *et al.* [47], who raised the guidelines on geometries of

MVG by using a large test matrix, gave the recent studies. Experimental studies thus far involved incident shocks [51, 52]. Babinsky *et al.* [51] showed substantial disruptions to nominal two-dimensionality based on their surface flow visualizations, which is expected to be similar to the observations in the transonic regime. They gave the conclusion that there is a disruption instead of a separation reduction, since the reduction of separation requires further substantiation. They also suggested that MVGs needed to be smaller so that they could perform better in generation of lower wave drag, a conclusion reached separately by Bur *et al.* [102]. Lin *et al.* [36] stated in their studies that MVGs should be placed near to the interaction region at low speeds due to their small sizes. These two conclusions were also confirmed by Lee *et al.* [62], who performed a Mach 3 configuration based on MILES. Ghosh *et al.* [55] proposed a simulation according to the experiments of Babinsky *et al.* [51].

The numerical results of Lu *et al.* [78], Panaras [108] and Garrison *et al.* [109] gave the suggestion that the data from experiments might be influenced by tunnel wall interference, which was regarded to be a glancing SBLI. Blinde *et al.* [52] adopted detailed stereo particle image velocimetry mapping in their studies. They pointed out the importance of determining the size of the reversed flow region, which could be a primary method to indicate separation. They also discovered that the MVGs could have effect on the spanwise distribution of the reversed flow, and then make it beak up into isolated clumps. Their conclusions had good agreement with the findings of Babinsky *et al.* [51] as well as those reviewed by Bur *et al.* [102] earlier. Two spanwise rows of MVGs next to one another were also studied by Blinde *et al.* In their study, two rows have better effect on disrupting the separation induced by shock. Lee *et al.* [62] reported a joint experimental–computational study, with emphasis on computations. They performed a Mach 3 flow within an isolated MVG interacting with an impinging shock, which was generated by an 8 degree wedge and could cause the separation of turbulent boundary layer. They stated that the counter-rotating vortices from the trailing edge of MVG were dominant in the streamwise vorticity near the shock interaction, and also that the overall recovery of the structures was improved by MVG. These authors also suggested that a MVG with smaller size could not protrude too deeply into the boundary layer so that it would be beneficial. Also smaller MVG could generate smaller streamwise vortices which would be able to stay closer to the wall; smaller MVG near the shock interaction could reduce the displacement thickness as well as the separation area. These results could attribute to a decrease in wave drag and the case that the primary vortex pair was close proximity to the surface. A high order LES simulation reported by Li and Liu [73] indicated that a chain of vortex rings were generated and then shed from a MVG. The wake of a MVG including vortex rings "piggyback" on the counter-rotating vortex pair has been discussed at length previously. They adopted a new configuration of MVG with a trailing edge inclined at either 45 or 70 degree from the flat plate for better grid generation, also due to the smaller inclination, the vortex train could be closer to the wall for better controlling the flow; Additionally, the flow features around MVG and downstream of MVG were not expected to be different. They also found that the vortex rings could lead to strong distortion of the separation bubble by interacting with the leading shock. This finding probably can be the first study, which indicates that the interaction between the shock and the vortex rings is a crucial phenomenon in terms of the separation bubble reduction. For instance, Fig. (**1.38**) shows the numerical instantaneous visualizations of the corner-induced SBLI schlieren, which can be seen that the leading shock is absent and large vortex structures billow further outward. Li and Liu also suggested that the extent of the separation zone reduction might be related to the disappearance of the leading shock, or separation shock which is contained in lambda-foot structure. This finding is significant since the classical understanding of boundary layer separation induced by shock needs the presence of a compression wave system that generally merges together with the leading shock [90], in spite of whether the interaction is two or three dimensional and whether the incoming boundary layer is laminar or turbulent [110]. The disappearance of the leading shock does not necessarily mean there is no separation, due to the instantaneous characteristics of the numerical visualization. However, this phenomenon may indicate an extremely unsteady interaction which is induced by large and unsteady events.

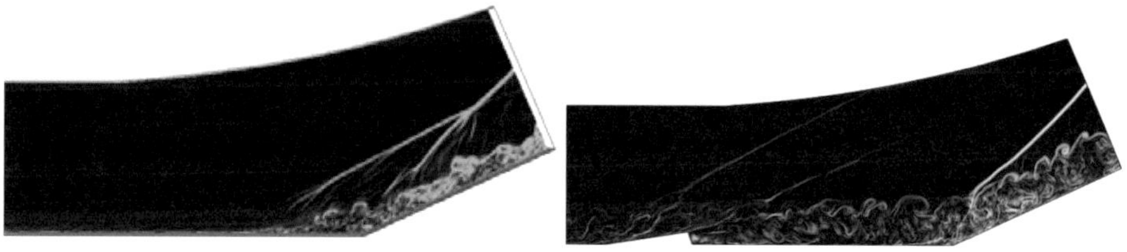

(a) No MVG. **(b) MVG upstream.**

Fig. (1.38). Instantaneous numerical schlieren [73].

The authors stated in the above that the leading shock distortions could appear from either axial, counter-rotating trailing vortex pair or from vortex rings. It is further pointed out that the interaction between vortex rings plays a positive role in the massive disruption of separation induced by SBLI. Fig. (**1.39**) gives visualizations that a vortex ring passes through the leading and rear legs of a lambda-foot shock. In Fig. (**1.39a**), the structure of lambda-foot is seen to be associated with shock boundary layer interaction, and part of the large vortical structures impinging the leading shock can also be found in the visualization. Fig. (**1.39b**) shows the leading shock disruption which is also shown in Fig. (**1.38b**).

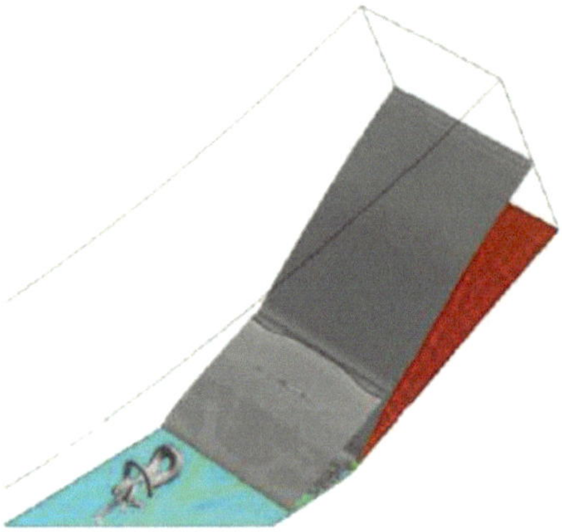

(a) Lambda-foot shock structure.

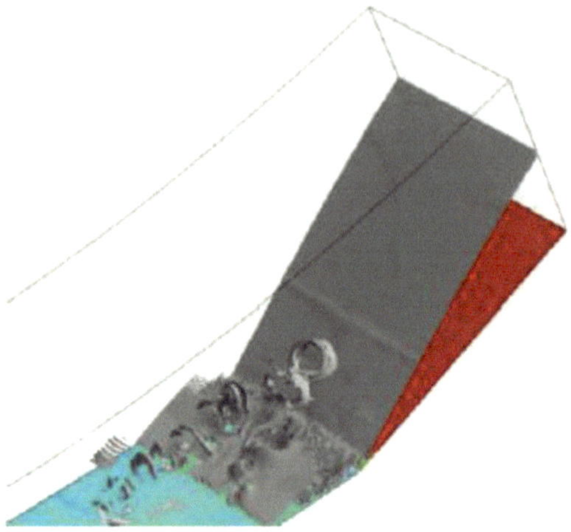

(b) Vortex ring disruption of the separation shock.

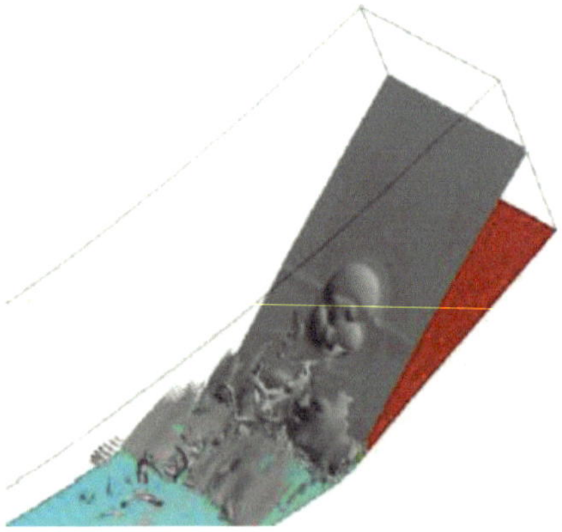

(c) Vortex ring penetrating the separation bubble and disrupting the reflection shock.

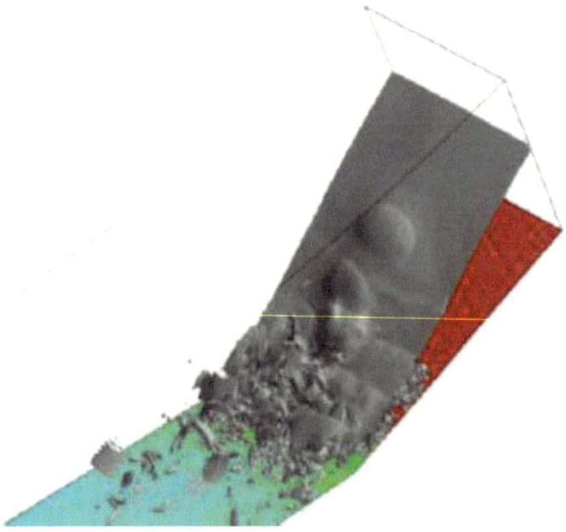

(d) Disruption of the separation.

Fig. (1.39). Sequence of frames showing vortex ring interaction with the leading and rear legs of the lambda-foot shock in the context of SBLI separation control [79].

An interpretation can be obtained from these two visualizations, which is that the leading shock is extremely broken. In Fig. (**1.39c**), the further disruption of separation or rear shock can be found. A collapse of the separation bubble is discovered in Fig. (**1.39d**). Similar consequences have been obtained from the study of SBLI where holes appear to be punched into the shock proposed by Ogawa *et al.* [63] and from the interaction between shock and vortex rings [111, 112] particularly suggested in their study that the primary consequence of the SBLI was a distortion of both shock and vortex ring. They also pointed out that the vortex ring was stiff enough and was not broken up by the shock in spite of the kinetic energy loss. Back to the topic under consideration, the wake of the MVG at least includes vortex rings wrapped over the counter-rotating vortex pair [51, 92, 113]. The above-described interaction between the shock system and the vortex rings was superposed upon the counter-rotating streamwise vortex pair interacted with the shock system. The further discussion about the interaction between counter-rotating streamwise vortex pair and the shock system was not shown in their study, however, it has been well described in the result from Reda and Murphy [114]. Unlike the former, a point regarding this latter interaction is that their study has described vortex break down. Particularly, a vortex with a wake-like velocity profile, as in vortices generated by MVG, is likely to be broken up. Therefore, the recent evidence indicates an extremely complex interaction between vortices in different types since they encounter the shock system in shock boundary layer interaction. The complex interaction indicates a disruption of the trailing vortices and the shock system; however, the vortex rings are allowed to propagate through fairly intact. The interaction is extremely unsteady, due to the passage of vortex rings through the shock.

1.9. SOME RELATED CONCEPTS

Except MVGs, there are a number of related passive and active concepts for the control of boundary-layer in high-speed flow which to be a recent interest, some of them have potential in shock induced separation alleviation. A hybrid concept that incorporates a high pressure micro air jet with jet pressure ratios of 1–3% and mass flow ratios of 0.1–0.3% of the inlet flow [115]. This type of jet was placed to energize the vortex system which is generated by MVG. It is note that such a vortex system could generate jet-like vortex, which behaved differently in breakdown compared with the "wake-like vortex" discussed above [114]. From Anderson *et al.*'s study [48], it is concluded that this type of jet-like vortex system might be adopted in modifying the boundary layer on the surface of the vehicle forebody and in controlling the flow from a serpentine or S-duct transonic inlet simultaneously. The requirements of the micro-jet flow have a tenfold decrease than ones of conventional bleeding system. The effect of air jet vortex generators (AJVGs) on alleviating separation induced by SBLI was described in detail by

Souverein and Debieve [116]. The air jet vortex generators in high speed crossflow could generate a persistent, counter-rotating vortex pair [117]. It is also discovered by Souverein and Debieve that the length as well as height of the separation bubble was decreased by the jets. Moreover, the energetic bandwidth of the shock oscillations was increased by around 50%; this phenomenon has a good agreement with the size reduction of the separation bubble. As is known, active devices like AJVGs can be modulated to specific control needs in different conditions. These control methods should be evaluated against conventional active ones so that their potential performance could be determined in practical applications.

CONCLUSION AND OUTLOOK

The flowfield around a MVG immersed in a supersonic boundary layer and the subsequent interaction downstream with separated flow induced by shock have been reviewed. A primary vortex pair that trail from the MVG sides dominates the flow field around the MVG. And two pairs of secondary vortices are also included in this complex topology. Two pairs of vortex filaments emanating from the trailing edge of the MVG are also discovered. There are at least two different views in terms of the MVG wake. Some people hold the view that the wake of the MVG produces a pair of counter-rotating streamwise vortices, which could entrain high-speed freestream fluid into boundary layer. The low momentum fluid exists near the centerline which may form a non-uniform region in spanwise, while higher momentum fluid exists further outboard. To summarize, the higher momentum can alleviate the subsequent separation through energizing the boundary layer.

However, there is a different voice. Some researchers thought that the dominant instability mechanism is to generate a chain of vortex rings superimposed on the counter-rotating streamwise vortex pair. This train of vortex rings draws in high-momentum (high-speed) freestream fluid, as in the first model. However, the vortex rings have a different interaction with the shock, compared with the counter-rotating streamwise vortex pair interacting with the shock. The vortex rings have enough robustness and are stiff enough to pass through the shock, and then cause serious structure distortions of the shock lambda-foot. Particularly, the leading shock may be further weakened. It seems that the vortex ring interaction causes the substantial unsteadiness at the region upstream of the separation zone. This unsteadiness may produce the impression of the separation size reduction. As is known in some experimental studies, tunnel sidewall interference may be the difficulty to obtain the consequences. Therefore, it is recommended that attempts are made to eliminate or remove this interference. For example, the fences should be considered as a possibility to avoid a two-dimensional interaction. However, in some computational studies, numerical methods with high order schemes are necessary to capture the fine scales present in the flow field. Moreover, appropriate vortex identification and detailed process of shock-vortex interaction may be best handled by numerical simulations integrated with experiment. Therefore, performance metrics should be further developed and studied to identify the effectiveness of MVGs.

CHAPTER 2

Governing Equations

Abstract: In this chapter, the dimensional and non-dimensional 3-D time-dependent Navier-Stokes equations in curvilinear coordinates are given in details. The flux splitting scheme and non-reflecting boundary conditions are discussed and provided.

Keywords: Curvilinear coordinates, Flux splitting, Navier-Stokes equations, Non-reflecting boundary conditions.

2.1. THE NAVIER-STOKES EQUATIONS

2.1.1. Conservation of Mass (Continuity)

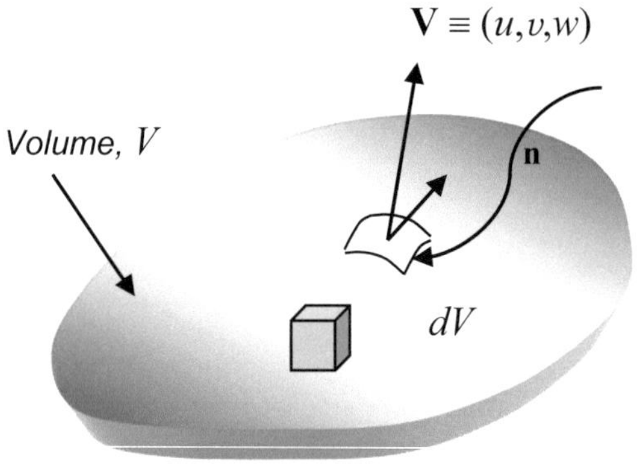

Fig. (2.1). Finite control volume fixed in space.

Let us consider a closed surface S whose position is fixed in relation to the coordinate axes and encloses a volume V completely filled with fluid (Fig. **2.1**). Given the density of the fluid ρ at a position $\boldsymbol{x}$ and at time t, the mass of the fluid enclosed by the surface at any instant is given by

$$\int_V \rho \, dV$$

Chaoqun Liu, Qin Li, Yonghua Yan, Yong Yang, Guang Yang, Xiangrui Dong

and the net rate of which the mass flows outwards across the surface is

$$\int \rho \mathbf{u} \cdot \mathbf{n}\, dS$$

where $\boldsymbol{n}$is the unit outward normal of the surface S, and dV and dS are respectively elements of the enclosed volume and of the area of the surrounding surface.

The conservation of mass of the fluid requires that

$$\frac{\partial}{\partial t}\int \rho\, dV = -\int \rho \mathbf{u} \cdot \mathbf{n}\, dS.$$

Then, since the volume V is fixed in space, the differentiation under the integral sign, and the transformation of the surface integral (by the Gaussian divergence theorem) gives

$$\int \frac{\partial \rho}{\partial t} dV + \int \nabla \cdot (\rho \mathbf{u}) dV = 0$$

or

$$\int \left[\frac{\partial \rho}{\partial t} + \nabla \cdot (\rho \boldsymbol{u}) \right] dV = 0.$$

This relation is valid for all choices of volume V that lies entirely in the fluid, and therefore, if the integrand is continuous in $\boldsymbol{x}$, it must be identically zero everywhere in the fluid. Hence, we obtain

$$\frac{\partial \rho}{\partial t} + \nabla \cdot (\rho \boldsymbol{u}) = 0 \qquad \textbf{(2.1)}$$

2.1.2. Conservation of Momentum (Equation of Motion)

For the conservation of momentum in a control volume, the changes of momentum in this volume must be equal to what is gained or lost through the surface that encloses this volume and what is created or consumed by sources and sinks inside the control volume.

Consider a volume of fluid V enclosed by a surface S, fixed with relation to the coordinate axes (Fig. **2.1**). For this body of fluid, the momentum is given by

$$\int \rho \boldsymbol{u}\, dV,$$

and its rate of change

$$\frac{\partial}{\partial t}\int \rho \boldsymbol{u}\, dV = \int \frac{\partial(\rho \boldsymbol{u})}{\partial t}\, dV.$$

Now, similar to the conservation of mass, the net rate of what is gained or lost through the surface S is given by

$$\int \rho \mathbf{u}(\mathbf{u}\cdot\mathbf{n})\, dS = \int \nabla\cdot(\rho \mathbf{u}\otimes\mathbf{u})\, dV,$$

using the divergence theorem, where ⊗ represents the outer product, *i.e.*

$$\rho \boldsymbol{u}\otimes\boldsymbol{u} = \rho \boldsymbol{u}\boldsymbol{u}^T,$$

that is a tensor.

If we represent the sources and sinks inside the control volume as $\boldsymbol{b}$, then the conservation of momentum inside the volume V is given by

$$\int \frac{\partial(\rho \boldsymbol{u})}{\partial t}\, dV = -\int \nabla\cdot(\rho \boldsymbol{u}\otimes\boldsymbol{u})\, dV + \int \rho \boldsymbol{b}\, dV,$$

or

$$\int \left[\frac{\partial(\rho \boldsymbol{u})}{\partial t} + \nabla\cdot(\rho \boldsymbol{u}\otimes\boldsymbol{u}) - \rho \boldsymbol{b}\right] dV = 0.$$

Since the volume V is arbitrary inside the fluid, we must have

$$\frac{\partial(\rho \boldsymbol{u})}{\partial t} + \nabla\cdot(\rho \boldsymbol{u}\otimes\boldsymbol{u}) - \rho \boldsymbol{b} = 0,$$

or

$$\frac{\partial(\rho \boldsymbol{u})}{\partial t}+\nabla\cdot(\rho \boldsymbol{u}\otimes \boldsymbol{u})=\rho \boldsymbol{b} \qquad \textbf{(2.2)}$$

Let us examine more carefully what is ***b***.

2.1.3. Forces Acting on the Fluid

The forces ***b*** may be separated into two types: the stress (surface) forces and body forces, such as gravity. Then, we can write that

$$\rho\mathbf{b}=\rho\mathbf{f}+(\nabla\cdot\sigma),$$

where ***f*** are the body forces and σ is the stress force tensor.

The stress force is a measure of the intensity of the total internal forces acting within a body across imaginary internal surfaces, as a reaction to external applied forces and body forces. It can be divided into normal and shear components.

If we assume a Newtonian fluid, that is, there is a linear relation between the stress and the rate of strain of the fluid, then by Stokes [118],

$$\sigma=-(p-\lambda\nabla\cdot\mathbf{u})\mathbf{I}+\mu\left[\nabla\mathbf{u}+(\nabla\mathbf{u})^T\right],$$

where p is the pressure, λ is the second viscosity. The dynamic viscosity μ and the second viscosity are related to the bulk viscosity κ

$$\kappa=\lambda+\frac{2}{3}\mu,$$

and according to Stokes' hypothesis, λ is taken to make $\kappa = 0$ and then $\lambda=-\frac{2}{3}\mu$

Therefore,

$$\sigma=-p\boldsymbol{I}+\lambda(\nabla\cdot\boldsymbol{u})\boldsymbol{I}+\mu\left[\nabla\boldsymbol{u}+(\nabla\boldsymbol{u})^T\right]=-\left[p+\frac{2}{3}\mu(\nabla\cdot\boldsymbol{u})\right]\boldsymbol{I}+\mu\left[\nabla\boldsymbol{u}+(\nabla\boldsymbol{u})^T\right] \qquad \textbf{(2.3)}$$

Hence, the conservation of momentum equations may be written as

$$\frac{\partial(\rho\mathbf{u})}{\partial t}+\nabla\cdot(\rho\mathbf{u}\otimes\mathbf{u})=\rho\mathbf{f}-\nabla p-\frac{2}{3}\nabla[\mu(\nabla\cdot\mathbf{u})]+\nabla\cdot\left[\mu\left(\nabla\mathbf{u}+(\nabla\mathbf{u})^T\right)\right] \quad \textbf{(2.4)}$$

2.1.4. Conservation of Energy

The conservation of energy for a fluid of volume V contained within a surface S can be determined by analyzing the work being done on this mass of fluid by both volume and surface forces, and also by the heat gained through transfer across the boundary and other sources inside the volume, satisfying the first law of Thermodynamics. The conserved quantity is the *total energyE*, defined as the sum of its internal energy and its kinetic energy per unit mass, or

$$E=e+\frac{\boldsymbol{u}\cdot\boldsymbol{u}}{2},$$

where e is the internal energy per unit mass of the fluid.

The rate of change of the total energy inside the volume V contained within a surface S is given by

$$\frac{\partial}{\partial t}\int\rho E\,dV=\int\frac{\partial(\rho E)}{\partial t}\,dV,$$

while the net rate of what is gained or lost through the surface is given by

$$\int(\rho E)\boldsymbol{u}\cdot\boldsymbol{n}\,dS=\int\nabla\cdot(\rho E)\boldsymbol{u}\,dV.$$

At the same time, heat may be transferred to the fluid in the volume by molecular conduction through the surface S, giving

$$\int(k\nabla T)\cdot\boldsymbol{n}\,dS=\int\nabla\cdot(k\nabla T)\,dV,$$

where T is the absolute temperature and k is the thermal conductivity coefficient of the fluid.

We will now analyze the work being done on the fluid by forces; we can separate them into volume and surface sources.

The volume sources include the volume forces $\boldsymbol{f}$, and heat sources q_H other than conduction, such as radiation or heat released by chemical reactions. This gives, for the volume V,

$$\int \left(\rho \boldsymbol{f} \cdot \boldsymbol{u} + q_H\right) dV .$$

The surface sources are the result of the work done on the fluid by the internal shear stresses σ acting on the surface of the volume considering that there are no external surface heat sources, giving

$$\int (\sigma \cdot \boldsymbol{u}) \cdot \boldsymbol{n}\, dS = \int \nabla \cdot (\sigma \cdot \boldsymbol{u}) dV .$$

Then, grouping all terms, the energy conservation equation, in integral form, becomes

$$\int \frac{\partial(\rho E)}{\partial t} dV + \int \nabla \cdot (\rho E) \boldsymbol{u}\, dV = \int \nabla \cdot (k \nabla T) dV + \int \left(\rho \boldsymbol{f} \cdot \boldsymbol{u} + q_H\right) dV + \int \nabla \cdot (\sigma \cdot \boldsymbol{u}) dV ,$$

or

$$\frac{\partial(\rho E)}{\partial t} + \nabla \cdot (\rho E) \boldsymbol{u} - \nabla \cdot (k \nabla T) - \nabla \cdot (\sigma \cdot \boldsymbol{u}) = \rho \boldsymbol{f} \cdot \boldsymbol{u} + q_H \tag{2.5}$$

with $\sigma = -\left[p + \frac{2}{3} \mu (\nabla \cdot \boldsymbol{u}) \right] \boldsymbol{I} + \mu \left[\nabla \boldsymbol{u} + (\nabla \boldsymbol{u})^T \right]$.

2.1.5. Summary of Equations

In the previous sections, we have derived the equations that satisfy the conservation of mass, the conservation of momentum, and the conservation of energy in a volume of the fluid. Together, these equations form the system known as the Navier-Stokes equations:

$$\frac{\partial \rho}{\partial t} + \nabla \cdot (\rho \boldsymbol{u}) = 0$$

$$\frac{\partial(\rho \boldsymbol{u})}{\partial t} + \nabla \cdot (\rho \boldsymbol{u} \otimes \boldsymbol{u}) = \rho \boldsymbol{f} + (\nabla \cdot \sigma),$$

$$\frac{\partial(\rho E)}{\partial t}+\nabla\cdot(\rho E)\boldsymbol{u}-\nabla\cdot(k\nabla T)-\nabla\cdot(\sigma\cdot\boldsymbol{u})=\rho\boldsymbol{f}\cdot\boldsymbol{u}+q_H$$

where

$$E=e+\frac{\boldsymbol{u}\cdot\boldsymbol{u}}{2}$$

and

$$\sigma=-\left[p+\frac{2}{3}\mu(\nabla\cdot\boldsymbol{u})\right]\boldsymbol{I}+\mu\left[\nabla\boldsymbol{u}+(\nabla\boldsymbol{u})^T\right].$$

If we assume that there are no body forces being applied to the volume of fluid and that there are no heat sources in the volume, then the equations can be rewritten as

$$\begin{aligned}\frac{\partial\rho}{\partial t}+\nabla\cdot(\rho\mathbf{u})&=0\\ \frac{\partial(\rho\mathbf{u})}{\partial t}+\nabla\cdot(\rho\mathbf{u}\otimes\mathbf{u})&=(\nabla\cdot\sigma)\\ \frac{\partial(\rho E)}{\partial t}+\nabla\cdot(\rho E)\mathbf{u}-\nabla\cdot(k\nabla T)-\nabla\cdot(\sigma\cdot\mathbf{u})&=0\end{aligned} \quad \mathbf{(2.6)}$$

with

$$E=e+\frac{\boldsymbol{u}\cdot\boldsymbol{u}}{2}$$

and

$$\sigma=-\left[p+\frac{2}{3}\mu(\nabla\cdot\boldsymbol{u})\right]\boldsymbol{I}+\mu\left[\nabla\boldsymbol{u}+(\nabla\boldsymbol{u})^T\right].$$

In three dimensions, the system above contains five equations (the conservation of momentum equation becomes three separate equations), while in two dimensions, the system has four equations. An extra equation is needed to solve the system for the unknown variables ρ, u, p, E, and T. This equation is the equation of state, for a thermally perfect gas,

$$p = \rho RT,$$

where R is a gas constant.

2.1.6. Non-dimensional Form

The equations presented in the previous section need to be reduced to a non-dimensional form. This can be achieved by dividing each variable by an appropriate dimensional reference variable. These reference variables correspond to some relevant and easily measurable quantity in the flow.

Let us define these reference variables:

- L is the characteristic length (for example, the chord length of an airfoil);
- V_∞ is the reference speed;
- ρ_∞ is the characteristic density;
- p_∞ is the reference pressure;
- T_∞ is the reference temperature;
- μ_∞ is the characteristic dynamic viscosity;
- k_∞ is the reference thermal conductivity.

With these reference variables, the non-dimensional variables are given by

$$t = \frac{t^*}{L/V_\infty}, \quad \mathbf{x} = \frac{\mathbf{x}^*}{L}, \qquad \mathbf{u} = \frac{\mathbf{u}^*}{V_\infty}, \qquad p = \frac{p^*}{\rho_\infty V_\infty^2}, \qquad \boldsymbol{\sigma} = \frac{\boldsymbol{\sigma}^*}{\rho_\infty V_\infty^2},$$

where * represents the dimensional variables.

The other non-dimensional variables assume that

$$p_\infty = \rho_\infty R T_\infty,$$

giving

$$\rho = \frac{\rho^*}{\rho_\infty}, \quad T = \frac{T^*}{T_\infty}, \qquad \mu = \mu(T) = \frac{\mu^*}{\mu_\infty}, \qquad k = k(T) = \frac{k^*}{k_\infty}.$$

The dynamic viscosity $\mu(T)$ for an ideal gas is given by Sutherland's law [119]:

$$\mu^*(T^*) = \mu_\infty \left(\frac{T^*}{T_\infty}\right)^{3/2} \frac{T_\infty + C^*}{T^* + C^*},$$

or in non-dimensional form,

$$\mu(T) = T^{3/2} \frac{1+C}{T+C}$$

with

$$C = \frac{C^*}{T_\infty}$$

and $C^* = 110.33\text{K}$ for air.

The total energy E^* can be non-dimensionalized in a form consistent with a thermally perfect gas, with

$$e^* = c_v T^*$$

and

$$E^* = e^* + \frac{\boldsymbol{u}^* \cdot \boldsymbol{u}^*}{2},$$

where c_v is a constant, defined as the specific heat. Then

$$e = \frac{e^*}{V_\infty^2},$$

and so

$$E = \frac{E^*}{V_\infty^2}.$$

Note that

$$\mathbf{x}^* = \mathbf{x}L, \mathbf{u}^* = \mathbf{u}V_\infty, t^* = t\frac{L}{V_\infty}, \rho^* = \rho\rho_\infty, p^* = p\rho_\infty V_\infty^2, \mu^* = \mu_\infty\mu, k^* = kk_\infty, \sigma^* = \sigma\frac{\mu_\infty V_\infty}{L}, \nabla^* = \frac{\nabla}{L}$$

where the superscript "*" represents the dimensional quantity and the variable without superscript represents non-dimensional quantity.

Having defined the reference variables, we can rewrite the equations of section 2.3.5 to non-dimensional form, obtaining:

$$\frac{\rho_\infty V_\infty}{L}\frac{\partial\rho}{\partial t}+\frac{\rho_\infty V_\infty}{L}\nabla\cdot\left(\rho\mathbf{u}\right)=0$$

$$\frac{\rho_\infty V_\infty^2}{L}\frac{\partial\left(\rho\mathbf{u}\right)}{\partial t}+\frac{\rho_\infty V_\infty^2}{L}\left[\nabla\cdot\left(\rho\mathbf{u}\otimes\mathbf{u}\right)\right]=\frac{\rho_\infty V_\infty^2}{L}(\nabla\cdot\sigma)$$

$$\frac{\rho_\infty V_\infty^3}{L}\frac{\partial\left(\rho E\right)}{\partial t}+\frac{\rho_\infty V_\infty^3}{L}\nabla\cdot\left(\rho E\right)\mathbf{u}-\frac{k_\infty T_\infty}{L^2}\nabla\cdot\left(k\nabla T\right)-\frac{\mu_\infty V_\infty^2}{L^2}\nabla\cdot\left(\sigma\cdot\mathbf{u}\right)=0$$

$$Non-dimensional\ \sigma=\left[-p+[\mu'(T)-\frac{2}{3}\mu(T)](\nabla\cdot\mathbf{u})\right]\mathbf{I}+\mu(T)\left[\nabla\mathbf{u}+\left(\nabla\mathbf{u}\right)^T\right]$$

$$\frac{\rho_\infty V_\infty}{L}\frac{\partial\rho}{\partial t}+\frac{\rho_\infty V_\infty}{L}\nabla\cdot(\rho\mathbf{u})=\mathbf{0}$$

$$\frac{\rho_\infty V_\infty^2}{L}\frac{\partial(\rho\mathbf{u})}{\partial t}+\frac{\rho_\infty V_\infty^2}{L}[\nabla\cdot(\rho\mathbf{u}\otimes\mathbf{u})]-\frac{\rho_\infty V_\infty^2}{L}(\nabla\cdot\sigma)=0$$

$$\frac{\rho_\infty V_\infty^3}{L}\frac{\partial(\rho E)}{\partial t}+\frac{\rho_\infty V_\infty^3}{L}\nabla\cdot(\rho E)\mathbf{u}-\frac{k_\infty T_\infty}{L^2}\nabla\cdot(k\nabla T)-\frac{\mu_\infty V_\infty^2}{L}\nabla\cdot(\sigma\cdot\mathbf{u})=0$$

or, simplifying,

$$\frac{\partial\rho}{\partial t}+\nabla\cdot(\rho\mathbf{u})=\mathbf{0}$$

$$\frac{\partial(\rho\mathbf{u})}{\partial t}+\nabla\cdot(\rho\mathbf{u}\otimes\mathbf{u})-\nabla\cdot\sigma=0$$

$$\frac{\partial(\rho E)}{\partial t}+\nabla\cdot(\rho E)\mathbf{u}-\frac{1}{\rho_\infty V_\infty^3}\frac{k_\infty T_\infty}{L}\nabla\cdot(k\nabla T)-\frac{\mu_\infty}{\rho_\infty V_\infty L}\nabla\cdot(\sigma\cdot\mathbf{u})=0$$

$$\sigma=-\left[p\frac{\rho_\infty V_\infty^2}{\mu_\infty V_\infty/L}+\frac{2}{3}\mu(T)(\nabla\cdot\mathbf{u})\right]\mathbf{I}+\mu(T)\left[\nabla\mathbf{u}+(\nabla\mathbf{u})^T\right]$$

We may define constants for the terms that did not simplify.

The Reynolds number is defined as

$$\text{Re} = \frac{\rho_\infty V_\infty L}{\mu_\infty},$$

while the Prandtl number evaluated at the reference conditions is given by

$$\text{Pr} = \frac{c_p \mu_\infty}{k_\infty}$$

and the Mach number is defined as

$M_\infty = \frac{V_\infty}{C} = \frac{V_\infty}{\sqrt{\gamma R T_\infty}}$, where C is the sound speed.

With the thermodynamic relations for the specific heats c_p and c_v, given by

$c_p = \gamma c_v$ and $c_p - c_v = R$,

we obtain

$c_v = \frac{R}{\gamma - 1}$ with $\gamma = 1.4$.

Note that

$$\frac{k_\infty T_\infty}{\rho_\infty V_\infty^3 L} = \frac{k_\infty}{\gamma R \mu_\infty} \cdot \frac{\gamma R T_\infty}{V_\infty^2} \cdot \frac{\mu_\infty}{\rho_\infty V_\infty L} = \frac{k_\infty}{\frac{\gamma R}{\gamma - 1}(\gamma - 1)\mu_\infty} \cdot \frac{C^2}{V_\infty^2} \cdot \frac{1}{\rho_\infty V_\infty L / \mu_\infty}$$

$$= \frac{k_\infty}{(\gamma - 1) C_p \mu_\infty} \cdot \frac{1}{V_\infty^2 / C^2} \cdot \frac{1}{\rho_\infty V_\infty L / \mu_\infty} = \frac{1}{(\gamma - 1)\,\mathbf{Pr}\, M_\infty^2\, \mathbf{Re}}$$

The Navier-Stokes equations in non-dimensional form can be written as

$$\begin{aligned}
&\frac{\partial \rho}{\partial t}+\nabla\cdot(\rho\mathbf{u})=\mathbf{0}\\
&\frac{\partial(\rho\mathbf{u})}{\partial t}+\nabla\cdot(\rho\mathbf{u}\otimes\mathbf{u})-\nabla\cdot\sigma=0\\
&\frac{\partial(\rho E)}{\partial t}+\nabla\cdot(\rho E)\mathbf{u}-\frac{1}{(\gamma-1)\mathbf{Pr}\,M_\infty^2\,\mathbf{Re}}\nabla\cdot(k\nabla T)-\frac{1}{\mathbf{Re}}\nabla\cdot(\sigma\cdot\mathbf{u})=0 \qquad \textbf{(2.7)}\\
&\sigma=-\left[p\,\mathbf{Re}+\frac{2}{3}\mu(T)(\nabla\cdot\mathbf{u})\right]\mathbf{I}+\mu(T)\left[\nabla\mathbf{u}+(\nabla\mathbf{u})^T\right]
\end{aligned}$$

We can also define the thermal conductivity $k(T)$ by Sutherland's law:

$$k(T)=T^{3/2}\frac{1+\hat{C}}{T+\hat{C}}$$

If we assume that $\hat{C}=C$ (from dynamic viscosity formula), then we are assuming a constant Prandtl number.

Finally, we non-dimensionalize the equation of state:

$$p^*=\rho^*RT^*,$$

obtaining

$$\rho_\infty V_\infty^2 p=\rho_\infty\rho RTT_\infty=\rho_\infty\rho\gamma RT_\infty T/\gamma=\rho_\infty\rho T\frac{C^2}{\gamma}$$

or

$$p=\rho T\frac{1}{\gamma M_\infty^2}\quad\text{or}\quad T=\frac{\gamma M_\infty^2 p}{\rho}$$

Following is the non-dimensional Navier-Stokes Equation in Cartesian Coordinates

$$\frac{\partial \rho}{\partial t}+\frac{\partial \rho u}{\partial x}+\frac{\partial \rho v}{\partial y}+\frac{\partial \rho w}{\partial z}=0$$

$$\frac{\partial \rho u}{\partial t}+\frac{\partial(\rho uu)}{\partial x}+\frac{\partial(\rho uv)}{\partial y}+\frac{\partial(\rho uw)}{\partial z}+\frac{\partial p}{\partial x}-\frac{1}{\mathrm{Re}}\left(\frac{\sigma_{xx}}{\partial x}+\frac{\partial \sigma_{xy}}{\partial y}+\frac{\partial \sigma_{xz}}{\partial z}\right)=0$$

$$\frac{\partial \rho v}{\partial t}+\frac{\partial(\rho uv)}{\partial x}+\frac{\partial(\rho vv)}{\partial y}+\frac{\partial(\rho wv)}{\partial z}+\frac{\partial p}{\partial y}-\frac{1}{\mathrm{Re}}\left(\frac{\sigma_{xy}}{\partial x}+\frac{\partial \sigma_{yy}}{\partial y}+\frac{\partial \sigma_{yz}}{\partial z}\right)=0$$

$$\frac{\partial \rho w}{\partial t}+\frac{\partial(\rho uw)}{\partial x}+\frac{\partial(\rho vw)}{\partial y}+\frac{\partial(\rho ww)}{\partial z}+\frac{\partial p}{\partial z}-\frac{1}{\mathrm{Re}}\left(\frac{\sigma_{xz}}{\partial x}+\frac{\partial \sigma_{yz}}{\partial y}+\frac{\partial \sigma_{zz}}{\partial z}\right)=0 \qquad \textbf{(2.8)}$$

$$\frac{\partial \rho E}{\partial t}+\frac{\partial(\rho uE)}{\partial x}+\frac{\partial(\rho vE)}{\partial y}+\frac{\partial(\rho wE)}{\partial z}+\frac{\partial pu}{\partial x}+\frac{\partial pv}{\partial y}+\frac{\partial pw}{\partial z}$$
$$-\frac{1}{(\gamma-1)\mathbf{Pr}\,M_\infty^2\,\mathbf{Re}}\left\{\frac{\partial}{\partial x}\left[k(T)\frac{\partial T}{\partial x}\right]+\frac{\partial}{\partial y}\left[k(T)\frac{\partial T}{\partial y}\right]+\frac{\partial}{\partial z}\left[k(T)\frac{\partial T}{\partial z}\right]\right\}$$
$$-\frac{1}{\mathbf{Re}}\left[\frac{u\sigma_{xx}+v\sigma_{xy}+w\sigma_{xz}}{\partial x}+\frac{u\sigma_{xy}+v\sigma_{yy}+w\sigma_{yz}}{\partial y}+\frac{u\sigma_{xz}+v\sigma_{yz}+w\sigma_{zz}}{\partial z}\right]=0$$

$$T=\frac{\gamma M_\infty^2 p}{\rho}$$

where,

$$\sigma_{xx}=\frac{2}{3}\mu(T)\left(2\frac{\partial u}{\partial x}-\frac{\partial v}{\partial y}-\frac{\partial w}{\partial z}\right);$$

$$\sigma_{yy}=\frac{2}{3}\mu(T)\left(-\frac{\partial u}{\partial x}+2\frac{\partial v}{\partial y}-\frac{\partial w}{\partial z}\right);$$

$$\sigma_{zz}=\frac{2}{3}\mu(T)\left(-\frac{\partial u}{\partial x}-\frac{\partial v}{\partial y}+2\frac{\partial w}{\partial z}\right);$$

$$\sigma_{xy} = \mu(T)\left(\frac{\partial u}{\partial y} + \frac{\partial v}{\partial x}\right);$$

$$\sigma_{xz} = \mu(T)\left(\frac{\partial u}{\partial z} + \frac{\partial w}{\partial x}\right);$$

$$\sigma_{yz} = \mu(T)\left(\frac{\partial w}{\partial y} + \frac{\partial v}{\partial z}\right).$$

$$k(T) = T^{3/2}\frac{1+\hat{C}}{T+\hat{C}} \text{ and } \mu(T) = T^{3/2}\frac{1+C}{T+C}, \text{ assume } \hat{C} = C$$

$$C = \frac{C^*}{T_\infty}$$

where $C^* = 110.33\text{K}$ and Prandtl number $\text{Pr} = 0.72$ for air.

2.1.7. Expansion in Curvilinear Coordinates

We will now expand the non-dimensionalized Navier-Stokes equations (from the previous section) to curvilinear coordinates.

For this, it is necessary to expand first the Navier-Stokes equations in 3D Cartesian coordinates x, y, and z.

In vector form, we may write the equations as

$$\frac{\partial \mathbf{q}}{\partial t} + \frac{\partial \mathbf{F}}{\partial x} + \frac{\partial \mathbf{G}}{\partial y} + \frac{\partial \mathbf{H}}{\partial z} = \frac{1}{\text{Re}}\left(\frac{\partial \mathbf{F}_v}{\partial x} + \frac{\partial \mathbf{G}_v}{\partial y} + \frac{\partial \mathbf{H}_v}{\partial z}\right) \tag{2.9}$$

where

$$\mathbf{q}=\begin{bmatrix} \rho & \rho u & \rho v & \rho w & \rho E \end{bmatrix}^T;$$

$$\mathbf{F}=\begin{bmatrix} \rho u & \left(\rho u^2+p\right) & \rho uv & \rho uw & \left(\rho E+p\right)u \end{bmatrix}^T;$$

$$\mathbf{G}=\begin{bmatrix} \rho v & \rho uv & \left(\rho v^2+p\right) & \rho vw & \left(\rho E+p\right)v \end{bmatrix}^T;$$

$$\mathbf{H}=\begin{bmatrix} \rho w & \rho uw & \rho vw & \left(\rho w^2+p\right) & \left(\rho E+p\right)w \end{bmatrix}^T;$$

$$\mathbf{F}_{\mathbf{v}}=\begin{bmatrix} 0 & \sigma_{xx} & \sigma_{xy} & \sigma_{xz} & \left(u\sigma_{xx}+v\sigma_{xy}+w\sigma_{xz}+\frac{1}{(\gamma-1)\Pr M_\infty^2}k(T)\frac{\partial T}{\partial x}\right) \end{bmatrix}^T;$$

$$\mathbf{G}_{\mathbf{v}}=\begin{bmatrix} 0 & \sigma_{xy} & \sigma_{yy} & \sigma_{yz} & \left(u\sigma_{xy}+v\sigma_{yy}+w\sigma_{yz}+\frac{1}{(\gamma-1)\Pr M_\infty^2}k(T)\frac{\partial T}{\partial y}\right) \end{bmatrix}^T;$$

$$\mathbf{H}_{\mathbf{v}}=\begin{bmatrix} 0 & \sigma_{xz} & \sigma_{yz} & \sigma_{zz} & \left(u\sigma_{xz}+v\sigma_{yz}+w\sigma_{zz}+\frac{1}{(\gamma-1)\Pr M_\infty^2}k(T)\frac{\partial T}{\partial z}\right) \end{bmatrix}^T;$$

$$\sigma_{xx}=\frac{2}{3}\mu(T)\left(2\frac{\partial u}{\partial x}-\frac{\partial v}{\partial y}-\frac{\partial w}{\partial z}\right);$$

$$\sigma_{yy}=\frac{2}{3}\mu(T)\left(-\frac{\partial u}{\partial x}+2\frac{\partial v}{\partial y}-\frac{\partial w}{\partial z}\right);$$

$$\sigma_{zz}=\frac{2}{3}\mu(T)\left(-\frac{\partial u}{\partial x}-\frac{\partial v}{\partial y}+2\frac{\partial w}{\partial z}\right);$$

$$\sigma_{xy}=\mu(T)\left(\frac{\partial u}{\partial y}+\frac{\partial v}{\partial x}\right);$$

$$\sigma_{xz}=\mu(T)\left(\frac{\partial u}{\partial z}+\frac{\partial w}{\partial x}\right);$$

$$\sigma_{yz}=\mu(T)\left(\frac{\partial w}{\partial y}+\frac{\partial v}{\partial z}\right).$$

2.1.7.1. Curvilinear Coordinate Transformation

Let us assume that the position frame of reference is fixed in time, that is, the generalized coordinates do not change with time. Then, we can define the curvilinear coordinates in relation to the Cartesian coordinates as

$$\begin{aligned}
&\tau = t;\\
&\xi = \xi(x, y, z);\\
&\eta = \eta(x, y, z);\\
&\zeta = \zeta(x, y, z),
\end{aligned}$$

such that

$$\begin{aligned}
&\frac{\partial}{\partial t} = \frac{\partial}{\partial \tau};\\
&\frac{\partial}{\partial x} = \xi_x \frac{\partial}{\partial \xi} + \eta_x \frac{\partial}{\partial \eta} + \zeta_x \frac{\partial}{\partial \zeta};\\
&\frac{\partial}{\partial y} = \xi_y \frac{\partial}{\partial \xi} + \eta_y \frac{\partial}{\partial \eta} + \zeta_y \frac{\partial}{\partial \zeta};\\
&\frac{\partial}{\partial z} = \xi_z \frac{\partial}{\partial \xi} + \eta_z \frac{\partial}{\partial \eta} + \zeta_z \frac{\partial}{\partial \zeta}.
\end{aligned}$$

But

$$\begin{aligned}
&d\tau = dt;\\
&d\xi = \xi_x dx + \xi_y dy + \xi_z dz;\\
&d\eta = \eta_x dx + \eta_y dy + \eta_z dz;\\
&d\zeta = \zeta_x dx + \zeta_y dy + \zeta_z dz,
\end{aligned}$$

or

$$\begin{bmatrix} d\tau \\ d\xi \\ d\eta \\ d\zeta \end{bmatrix} = \begin{bmatrix} 1 & 0 & 0 & 0 \\ 0 & \xi_x & \xi_y & \xi_z \\ 0 & \eta_x & \eta_y & \eta_z \\ 0 & \zeta_x & \zeta_y & \zeta_z \end{bmatrix} \begin{bmatrix} dt \\ dx \\ dy \\ dz \end{bmatrix},$$

and

$$\begin{aligned}
&dt = d\tau;\\
&dx = x_\xi d\xi + x_\eta d\eta + x_\zeta d\zeta;\\
&dy = y_\xi d\xi + y_\eta d\eta + y_\zeta d\zeta;\\
&dz = z_\xi d\xi + z_\eta d\eta + z_\zeta d\zeta,
\end{aligned}$$

or

$$\begin{bmatrix} dt \\ dx \\ dy \\ dz \end{bmatrix} = \begin{bmatrix} 1 & 0 & 0 & 0 \\ 0 & x_\xi & x_\eta & x_\zeta \\ 0 & y_\xi & y_\eta & y_\zeta \\ 0 & z_\xi & z_\eta & z_\zeta \end{bmatrix} \begin{bmatrix} d\tau \\ d\xi \\ d\eta \\ d\zeta \end{bmatrix}.$$

Then,

$$\begin{bmatrix} 1 & 0 & 0 & 0 \\ 0 & \xi_x & \xi_y & \xi_z \\ 0 & \eta_x & \eta_y & \eta_z \\ 0 & \zeta_x & \zeta_y & \zeta_z \end{bmatrix} = \begin{bmatrix} 1 & 0 & 0 & 0 \\ 0 & x_\xi & x_\eta & x_\zeta \\ 0 & y_\xi & y_\eta & y_\zeta \\ 0 & z_\xi & z_\eta & z_\zeta \end{bmatrix}^{-1}.$$

Let us define

$$J = \begin{vmatrix} 1 & 0 & 0 & 0 \\ 0 & \xi_x & \xi_y & \xi_z \\ 0 & \eta_x & \eta_y & \eta_z \\ 0 & \zeta_x & \zeta_y & \zeta_z \end{vmatrix} = \frac{1}{\begin{vmatrix} 1 & 0 & 0 & 0 \\ 0 & x_\xi & x_\eta & x_\zeta \\ 0 & y_\xi & y_\eta & y_\zeta \\ 0 & z_\xi & z_\eta & z_\zeta \end{vmatrix}} = .$$

$$= 1/\left[x_\xi\left(y_\eta z_\zeta - y_\zeta z_\eta\right) - x_\eta\left(y_\xi z_\zeta - y_\zeta z_\xi\right) + x_\zeta\left(y_\xi z_\eta - y_\eta z_\xi\right)\right]$$

Therefore,

$$\begin{bmatrix} \xi_x & \xi_y & \xi_z \\ \eta_x & \eta_y & \eta_z \\ \zeta_x & \zeta_y & \zeta_z \end{bmatrix} = J \begin{bmatrix} \left(y_\eta z_\zeta - y_\zeta z_\eta\right) & \left(x_\zeta z_\eta - x_\eta z_\zeta\right) & \left(x_\eta y_\zeta - x_\zeta y_\eta\right) \\ \left(y_\zeta z_\xi - y_\xi z_\zeta\right) & \left(x_\xi z_\zeta - x_\zeta z_\xi\right) & \left(x_\zeta y_\xi - x_\xi y_\zeta\right) \\ \left(y_\xi z_\eta - y_\eta z_\xi\right) & \left(x_\eta z_\xi - x_\xi z_\eta\right) & \left(x_\xi y_\eta - x_\eta y_\xi\right) \end{bmatrix} \quad \textbf{(2.10)}$$

2.1.7.2. Governing Equations in General Coordinates

In vector form, we can write the governing equations in general coordinates as

$$\begin{aligned} &\frac{\partial \mathbf{q}}{\partial \tau} + \xi_x \frac{\partial \mathbf{F}}{\partial \xi} + \eta_x \frac{\partial \mathbf{F}}{\partial \eta} + \zeta_x \frac{\partial \mathbf{F}}{\partial \zeta} + \xi_y \frac{\partial \mathbf{G}}{\partial \xi} + \eta_y \frac{\partial \mathbf{G}}{\partial \eta} + \zeta_y \frac{\partial \mathbf{G}}{\partial \zeta} \\ &+ \xi_z \frac{\partial \mathbf{H}}{\partial \xi} + \eta_z \frac{\partial \mathbf{H}}{\partial \eta} + \zeta_z \frac{\partial \mathbf{H}}{\partial \zeta} = \\ &= \frac{1}{\mathrm{Re}}\left(\xi_x \frac{\partial \mathbf{F_v}}{\partial \xi} + \eta_x \frac{\partial \mathbf{F_v}}{\partial \eta} + \zeta_x \frac{\partial \mathbf{F_v}}{\partial \zeta} + \xi_y \frac{\partial \mathbf{G_v}}{\partial \xi} + \eta_y \frac{\partial \mathbf{G_v}}{\partial \eta} + \zeta_y \frac{\partial \mathbf{G_v}}{\partial \zeta} + \xi_z \frac{\partial \mathbf{H_v}}{\partial \xi} + \eta_z \frac{\partial \mathbf{H_v}}{\partial \eta} + \zeta_z \frac{\partial \mathbf{H_v}}{\partial \zeta} \right) \end{aligned} \quad \textbf{(2.11)}$$

Dividing by J and using metric identities this equation may be written in the following form:

$$
\begin{aligned}
&\frac{\partial}{\partial \tau}\left(\frac{\mathbf{q}}{J}\right)+\frac{\partial}{\partial \xi}\left(\frac{\backslash \xi_x \mathbf{F}+\xi_y \mathbf{G}+\xi_z \mathbf{H}}{J}\right)+\frac{\partial}{\partial \eta}\left(\frac{\eta_x \mathbf{F}+\eta_y \mathbf{G}+\eta_z \mathbf{H}}{J}\right)+\\
&+\frac{\partial}{\partial \zeta}\left(\frac{\zeta_x \mathbf{F}+\zeta_y \mathbf{G}+\zeta_z \mathbf{H}}{J}\right)-\mathbf{F}\left[\left(\frac{\xi_x}{J}\right)_\xi+\left(\frac{\eta_x}{J}\right)_\eta+\left(\frac{\zeta_x}{J}\right)_\zeta\right]+\\
&-\mathbf{G}\left[\left(\frac{\xi_y}{J}\right)_\xi+\left(\frac{\eta_y}{J}\right)_\eta+\left(\frac{\zeta_y}{J}\right)_\zeta\right]-\mathbf{H}\left[\left(\frac{\xi_z}{J}\right)_\xi+\left(\frac{\eta_z}{J}\right)_\eta+\left(\frac{\zeta_z}{J}\right)_\zeta\right]=\\
&=\frac{1}{\mathrm{Re}}\left\{\frac{\partial}{\partial \xi}\left(\frac{\xi_x \mathbf{F}_v+\xi_y \mathbf{G}_v+\xi_z \mathbf{H}_v}{J}\right)+\frac{\partial}{\partial \eta}\left(\frac{\eta_x \mathbf{F}_v+\eta_y \mathbf{G}_v+\eta_z \mathbf{H}_v}{J}\right)+\frac{\partial}{\partial \zeta}\left(\frac{\zeta_x \mathbf{F}_v+\zeta_y \mathbf{G}_v+\zeta_z \mathbf{H}_v}{J}\right)\right\}+\\
&-\frac{1}{\mathrm{Re}}\left\{\mathbf{F}_v\left[\left(\frac{\xi_x}{J}\right)_\xi+\left(\frac{\eta_x}{J}\right)_\eta+\left(\frac{\zeta_x}{J}\right)_\zeta\right]+\mathbf{G}_v\left[\left(\frac{\xi_y}{J}\right)_\xi+\left(\frac{\eta_y}{J}\right)_\eta+\left(\frac{\zeta_y}{J}\right)_\zeta\right]+\mathbf{H}_v\left[\left(\frac{\xi_y}{J}\right)_\xi+\left(\frac{\eta_y}{J}\right)_\eta+\left(\frac{\zeta_y}{J}\right)_\zeta\right]\right\}
\end{aligned}
$$

Let us analyze the terms inside [.]. By using the definitions given at the end of the previous section, and considering that $a_{\alpha\beta} = a_{\beta\alpha}$, for $a = x, y, z$ and $\alpha, \beta = \tau, \xi, \eta, \zeta$, we obtain:

$$
\begin{aligned}
&\left(\frac{\xi_x}{J}\right)_\xi+\left(\frac{\eta_x}{J}\right)_\eta+\left(\frac{\zeta_x}{J}\right)_\zeta=(y_\eta z_\zeta-y_\zeta z_\eta)_\xi+(y_\zeta z_\xi-y_\xi z_\zeta)_\eta+(y_\xi z_\eta-y_\eta z_\xi)_\zeta=\\
&=\left(y_{\eta\xi}z_\zeta+y_\eta z_{\zeta\xi}-y_{\zeta\xi}z_\eta-y_\zeta z_{\eta\xi}\right)+\left(y_{\zeta\eta}z_\xi+y_\zeta z_{\xi\eta}-y_{\xi\eta}z_\zeta-y_\xi z_{\zeta\eta}\right)+\left(y_{\xi\zeta}z_\eta-y_{\eta\zeta}z_\xi+y_\xi z_{\eta\zeta}-y_\eta z_{\xi\zeta}\right)=\\
&=0;
\end{aligned}
$$

$$
\begin{aligned}
&\left(\frac{\xi_y}{J}\right)_\xi+\left(\frac{\eta_y}{J}\right)_\eta+\left(\frac{\zeta_y}{J}\right)_\zeta=(x_\zeta z_\eta-x_\eta z_\zeta)_\xi+(x_\xi z_\zeta-x_\zeta z_\xi)_\eta+(x_\eta z_\xi-x_\xi z_\eta)_\zeta=\\
&=\left(x_{\zeta\xi}z_\eta+x_\zeta z_{\eta\xi}-x_{\eta\xi}z_\zeta-x_\eta z_{\zeta\xi}\right)+\left(x_{\xi\eta}z_\zeta+x_\xi z_{\zeta\eta}-x_{\zeta\eta}z_\xi-x_\zeta z_{\xi\eta}\right)+\left(x_{\eta\zeta}z_\xi+x_\eta z_{\xi\zeta}-x_{\xi\zeta}z_\eta-x_\xi z_{\eta\zeta}\right)=\\
&=0;
\end{aligned}
$$

$$
\begin{aligned}
&\left(\frac{\xi_z}{J}\right)_\xi+\left(\frac{\eta_z}{J}\right)_\eta+\left(\frac{\zeta_z}{J}\right)_\zeta=(x_\eta y_\zeta-x_\zeta y_\eta)_\xi+(x_\zeta y_\xi-x_\xi y_\zeta)_\eta+(x_\xi y_\eta-x_\eta y_\xi)_\zeta=\\
&=\left(x_{\eta\xi}y_\zeta+x_\eta y_{\zeta\xi}-x_{\zeta\xi}y_\eta-x_\zeta y_{\eta\xi}\right)+\left(x_{\zeta\eta}y_\xi+x_\zeta y_{\xi\eta}-x_{\xi\eta}y_\zeta-x_\xi y_{\zeta\eta}\right)+\left(x_{\xi\zeta}y_\eta+x_\xi y_{\eta\zeta}-x_{\eta\zeta}y_\xi-x_\eta y_{\xi\zeta}\right)=\\
&=0.
\end{aligned}
$$

Then, we may write the governing equations as

$$\frac{\partial \hat{\boldsymbol{q}}}{\partial \tau}+\frac{\partial \hat{\boldsymbol{F}}}{\partial \xi}+\frac{\partial \hat{\boldsymbol{G}}}{\partial \eta}+\frac{\partial \hat{\boldsymbol{H}}}{\partial \zeta}=\frac{1}{\mathrm{Re}}\left[\frac{\partial \hat{\boldsymbol{F}}_v}{\partial \xi}+\frac{\partial \hat{\boldsymbol{G}}_v}{\partial \eta}+\frac{\partial \hat{\boldsymbol{H}}_v}{\partial \zeta}\right] \tag{2.12}$$

where

$$\begin{aligned}
\hat{\boldsymbol{q}} &= \frac{\boldsymbol{q}}{J}\\
\hat{\boldsymbol{F}} &= \frac{\xi_x \boldsymbol{F}+\xi_y \boldsymbol{G}+\xi_z \boldsymbol{H}}{J}\\
\hat{\boldsymbol{G}} &= \frac{\eta_x \boldsymbol{F}+\eta_y \boldsymbol{G}+\eta_z \boldsymbol{H}}{J}\\
\hat{\boldsymbol{H}} &= \frac{\zeta_x \boldsymbol{F}+\zeta_y \boldsymbol{G}+\zeta_z \boldsymbol{H}}{J}\\
\hat{\boldsymbol{F}}_v &= \frac{\xi_x \boldsymbol{F}_v+\xi_y \boldsymbol{G}_v+\xi_z \boldsymbol{H}_v}{J}\\
\hat{\boldsymbol{G}}_v &= \frac{\eta_x \boldsymbol{F}_v+\eta_y \boldsymbol{G}_v+\eta_z \boldsymbol{H}_v}{J}\\
\hat{\boldsymbol{H}}_v &= \frac{\zeta_x \boldsymbol{F}_v+\zeta_y \boldsymbol{G}_v+\zeta_z \boldsymbol{H}_v}{J}
\end{aligned}.$$

Now, we can expand the flux vectors and the viscous terms. The flux vectors become

$$\hat{\boldsymbol{F}}=\frac{1}{J}\begin{bmatrix}
\xi_t\rho+\xi_x\rho u+\xi_y\rho v+\xi_z\rho w\\
\xi_t\rho u+\xi_x\left(\rho u^2+p\right)+\xi_y\rho uv+\xi_z\rho uw\\
\xi_t\rho v+\xi_x\rho uv+\xi_y\left(\rho v^2+p\right)+\xi_z\rho vw\\
\xi_t\rho w+\xi_x\rho uw+\xi_y\rho vw+\xi_z\left(\rho w^2+p\right)\\
\xi_t\rho E+\xi_x(\rho E+p)u+\xi_y(\rho E+p)v+\xi_z(\rho E+p)w
\end{bmatrix};$$

$$\hat{\boldsymbol{G}}=\frac{1}{J}\begin{bmatrix}
\eta_t\rho+\eta_x\rho u+\eta_y\rho v+\eta_z\rho w\\
\eta_t\rho u+\eta_x\left(\rho u^2+p\right)+\eta_y\rho uv+\eta_z\rho uw\\
\eta_t\rho v+\eta_x\rho uv+\eta_y\left(\rho v^2+p\right)+\eta_z\rho vw\\
\eta_t\rho w+\eta_x\rho uw+\eta_y\rho vw+\eta_z\left(\rho w^2+p\right)\\
\eta_t\rho E+\eta_x(\rho E+p)u+\eta_y(\rho E+p)v+\eta_z(\rho E+p)w
\end{bmatrix};$$

$$\hat{\boldsymbol{H}}=\frac{1}{J}\begin{bmatrix}
\zeta_t\rho+\zeta_x\rho u+\zeta_y\rho v+\zeta_z\rho w\\
\zeta_t\rho u+\zeta_x\left(\rho u^2+p\right)+\zeta_y\rho uv+\zeta_z\rho uw\\
\zeta_t\rho v+\zeta_x\rho uv+\zeta_y\left(\rho v^2+p\right)+\zeta_z\rho vw\\
\zeta_t\rho w+\zeta_x\rho uw+\zeta_y\rho vw+\zeta_z\left(\rho w^2+p\right)\\
\zeta_t\rho E+\zeta_x(\rho E+p)u+\zeta_y(\rho E+p)v+\zeta_z(\rho E+p)w
\end{bmatrix},$$

while for the viscous terms, we must first determine the expansion of the stress tensor terms in generalized coordinates:

$$\sigma_{xx} = \mu(T)\left[2\left(\xi_x \frac{\partial u}{\partial \xi} + \eta_x \frac{\partial u}{\partial \eta} + \zeta_x \frac{\partial u}{\partial \zeta}\right) - \frac{2}{3}\left(\xi_x \frac{\partial u}{\partial \xi} + \xi_y \frac{\partial v}{\partial \xi} + \xi_z \frac{\partial w}{\partial \xi} + \eta_x \frac{\partial u}{\partial \eta} + \eta_y \frac{\partial v}{\partial \eta} + \eta_z \frac{\partial w}{\partial \eta}\right.\right.$$
$$\left.\left. + \zeta_x \frac{\partial u}{\partial \zeta} + \zeta_y \frac{\partial v}{\partial \zeta} + \zeta_z \frac{\partial w}{\partial \zeta}\right)\right];$$

$$\sigma_{yy} = \mu(T)\left[2\left(\xi_y \frac{\partial v}{\partial \xi} + \eta_y \frac{\partial v}{\partial \eta} + \zeta_y \frac{\partial v}{\partial \zeta}\right) - \frac{2}{3}\left(\xi_x \frac{\partial u}{\partial \xi} + \xi_y \frac{\partial v}{\partial \xi} + \xi_z \frac{\partial w}{\partial \xi} + \eta_x \frac{\partial u}{\partial \eta} + \eta_y \frac{\partial v}{\partial \eta} + \eta_z \frac{\partial w}{\partial \eta}\right.\right.$$
$$\left.\left. + \zeta_x \frac{\partial u}{\partial \zeta} + \zeta_y \frac{\partial v}{\partial \zeta} + \zeta_z \frac{\partial w}{\partial \zeta}\right)\right];$$

$$\sigma_{zz} = \mu(T)\left[2\left(\xi_z \frac{\partial w}{\partial \xi} + \eta_z \frac{\partial w}{\partial \eta} + \zeta_z \frac{\partial w}{\partial \zeta}\right) - \frac{2}{3}\left(\xi_x \frac{\partial u}{\partial \xi} + \xi_y \frac{\partial v}{\partial \xi} + \xi_z \frac{\partial w}{\partial \xi} + \eta_x \frac{\partial u}{\partial \eta} + \eta_y \frac{\partial v}{\partial \eta} + \eta_z \frac{\partial w}{\partial \eta}\right.\right.$$
$$\left.\left. + \zeta_x \frac{\partial u}{\partial \zeta} + \zeta_y \frac{\partial v}{\partial \zeta} + \zeta_z \frac{\partial w}{\partial \zeta}\right)\right];$$

$$\sigma_{xy} = \mu(T)\left(\xi_x \frac{\partial v}{\partial \xi} + \xi_y \frac{\partial u}{\partial \xi} + \eta_x \frac{\partial v}{\partial \eta} + \eta_y \frac{\partial u}{\partial \eta} + \zeta_x \frac{\partial v}{\partial \zeta} + \zeta_y \frac{\partial u}{\partial \zeta}\right);$$

$$\sigma_{xz} = \mu(T)\left(\xi_x \frac{\partial w}{\partial \xi} + \xi_z \frac{\partial u}{\partial \xi} + \eta_x \frac{\partial w}{\partial \eta} + \eta_z \frac{\partial u}{\partial \eta} + \zeta_x \frac{\partial w}{\partial \zeta} + \zeta_z \frac{\partial u}{\partial \zeta}\right);$$

$$\sigma_{yz} = \mu(T)\left(\xi_y \frac{\partial w}{\partial \xi} + \xi_z \frac{\partial v}{\partial \xi} + \eta_y \frac{\partial w}{\partial \eta} + \eta_z \frac{\partial v}{\partial \eta} + \zeta_y \frac{\partial w}{\partial \zeta} + \zeta_z \frac{\partial v}{\partial \zeta}\right).$$

Now, let us define

$$\phi^{(\xi)} = -\frac{2}{3}\mu(T)\left(\xi_x \frac{\partial u}{\partial \xi} + \xi_y \frac{\partial v}{\partial \xi} + \xi_z \frac{\partial w}{\partial \xi}\right);$$
$$\phi^{(\eta)} = -\frac{2}{3}\mu(T)\left(\eta_x \frac{\partial u}{\partial \eta} + \eta_y \frac{\partial v}{\partial \eta} + \eta_z \frac{\partial w}{\partial \eta}\right);$$
$$\phi^{(\zeta)} = -\frac{2}{3}\mu(T)\left(\zeta_x \frac{\partial u}{\partial \zeta} + \zeta_y \frac{\partial v}{\partial \zeta} + \zeta_z \frac{\partial w}{\partial \zeta}\right).$$

Then, to possibly simplify for the implementation of the viscous terms, they should be split as follows:

$$\hat{\boldsymbol{F}}_v = \boldsymbol{F}_v^{(\xi)} + \boldsymbol{F}_v^{(\eta)} + \boldsymbol{F}_v^{(\zeta)}$$
$$\hat{\boldsymbol{G}}_v = \boldsymbol{G}_v^{(\xi)} + \boldsymbol{G}_v^{(\eta)} + \boldsymbol{G}_v^{(\zeta)}$$
$$\hat{\boldsymbol{H}}_v = \boldsymbol{H}_v^{(\xi)} + \boldsymbol{H}_v^{(\eta)} + \boldsymbol{H}_v^{(\zeta)}$$

such that

$$\frac{1}{\mathrm{Re}}\left[\frac{\partial \hat{\boldsymbol{F}}_v}{\partial \xi} + \frac{\partial \hat{\boldsymbol{G}}_v}{\partial \eta} + \frac{\partial \hat{\boldsymbol{H}}_v}{\partial \zeta}\right] = \frac{1}{\mathrm{Re}}\left[\frac{\partial}{\partial \xi}(\boldsymbol{v}_1 + \boldsymbol{w}_1) + \frac{\partial}{\partial \eta}(\boldsymbol{v}_2 + \boldsymbol{w}_2) + \frac{\partial}{\partial \zeta}(\boldsymbol{v}_3 + \boldsymbol{w}_3)\right]$$

where

$$\boldsymbol{v}_1 = \frac{\xi_x \boldsymbol{F}_v^{(\xi)} + \xi_y \boldsymbol{G}_v^{(\xi)} + \xi_z \boldsymbol{H}_v^{(\xi)}}{J};$$
$$\boldsymbol{v}_2 = \frac{\eta_x \boldsymbol{F}_v^{(\eta)} + \eta_y \boldsymbol{G}_v^{(\eta)} + \eta_z \boldsymbol{H}_v^{(\eta)}}{J};$$
$$\boldsymbol{v}_3 = \frac{\zeta_x \boldsymbol{F}_v^{(\zeta)} + \zeta_y \boldsymbol{G}_v^{(\zeta)} + \zeta_z \boldsymbol{H}_v^{(\zeta)}}{J};$$
$$\boldsymbol{w}_1 = \frac{\xi_x\left(\boldsymbol{F}_v^{(\eta)} + \boldsymbol{F}_v^{(\zeta)}\right) + \xi_y\left(\boldsymbol{G}_v^{(\eta)} + \boldsymbol{G}_v^{(\zeta)}\right) + \xi_z\left(\boldsymbol{H}_v^{(\eta)} + \boldsymbol{H}_v^{(\zeta)}\right)}{J};$$
$$\boldsymbol{w}_2 = \frac{\eta_x\left(\boldsymbol{F}_v^{(\xi)} + \boldsymbol{F}_v^{(\zeta)}\right) + \eta_y\left(\boldsymbol{G}_v^{(\xi)} + \boldsymbol{G}_v^{(\zeta)}\right) + \eta_z\left(\boldsymbol{H}_v^{(\xi)} + \boldsymbol{H}_v^{(\zeta)}\right)}{J};$$
$$\boldsymbol{w}_3 = \frac{\zeta_x\left(\boldsymbol{F}_v^{(\xi)} + \boldsymbol{F}_v^{(\eta)}\right) + \zeta_y\left(\boldsymbol{G}_v^{(\xi)} + \boldsymbol{G}_v^{(\eta)}\right) + \zeta_z\left(\boldsymbol{H}_v^{(\xi)} + \boldsymbol{H}_v^{(\eta)}\right)}{J}.$$

and

$$\boldsymbol{F}_v^{(\xi)} = \begin{bmatrix} 0 \\ 2\mu(T)\xi_x \dfrac{\partial u}{\partial \xi} + \phi^{(\xi)} \\ \mu(T)\left(\xi_x \dfrac{\partial v}{\partial \xi} + \xi_y \dfrac{\partial u}{\partial \xi}\right) \\ \mu(T)\left(\xi_z \dfrac{\partial u}{\partial \xi} + \xi_x \dfrac{\partial w}{\partial \xi}\right) \\ u f_2^{(\xi)} + v f_3^{(\xi)} + w f_4^{(\xi)} - \dfrac{k\xi_x}{(\gamma-1)\mathrm{Pr} M_\infty^2}\dfrac{\partial T}{\partial \xi} \end{bmatrix} = \begin{bmatrix} f_1^{(\xi)} \\ f_2^{(\xi)} \\ f_3^{(\xi)} \\ f_4^{(\xi)} \\ f_5^{(\xi)} \end{bmatrix};$$

$$\boldsymbol{F_v}^{(\eta)}=\begin{bmatrix}0\\2\mu(T)\eta_x\dfrac{\partial u}{\partial\eta}+\phi^{(\eta)}\\\mu(T)\left(\eta_x\dfrac{\partial v}{\partial\eta}+\eta_y\dfrac{\partial u}{\partial\eta}\right)\\\mu(T)\left(\eta_z\dfrac{\partial u}{\partial\eta}+\eta_x\dfrac{\partial w}{\partial\eta}\right)\\uf_2^{(\eta)}+vf_3^{(\eta)}+wf_4^{(\eta)}-\dfrac{k\eta_x}{(\gamma-1)\Pr M_\infty^2}\dfrac{\partial T}{\partial\eta}\end{bmatrix};\quad \boldsymbol{F_v}^{(\zeta)}=\begin{bmatrix}0\\2\mu(T)\zeta_x\dfrac{\partial u}{\partial\zeta}+\phi^{(\zeta)}\\\mu(T)\left(\zeta_x\dfrac{\partial v}{\partial\zeta}+\zeta_y\dfrac{\partial u}{\partial\zeta}\right)\\\mu(T)\left(\zeta_z\dfrac{\partial u}{\partial\zeta}+\zeta_x\dfrac{\partial w}{\partial\zeta}\right)\\uf_2^{(\zeta)}+vf_3^{(\zeta)}+wf_4^{(\zeta)}-\dfrac{k\zeta_x}{(\gamma-1)\Pr M_\infty^2}\dfrac{\partial T}{\partial\zeta}\end{bmatrix};$$

$$\boldsymbol{G_v}^{(\xi)}=\begin{bmatrix}0\\\mu(T)\left(\xi_x\dfrac{\partial v}{\partial\xi}+\xi_y\dfrac{\partial u}{\partial\xi}\right)\\2\mu(T)\xi_y\dfrac{\partial v}{\partial\xi}+\phi^{(\xi)}\\\mu(T)\left(\xi_z\dfrac{\partial v}{\partial\xi}+\xi_y\dfrac{\partial w}{\partial\xi}\right)\\ug_2^{(\xi)}+vg_3^{(\xi)}+wg_4^{(\xi)}-\dfrac{k\xi_y}{(\gamma-1)\Pr M_\infty^2}\dfrac{\partial T}{\partial\xi}\end{bmatrix}=\begin{bmatrix}g_1^{(\xi)}\\g_2^{(\xi)}\\g_3^{(\xi)}\\g_4^{(\xi)}\\g_5^{(\xi)}\end{bmatrix};$$

$$\boldsymbol{G_v}^{(\eta)}=\begin{bmatrix}0\\\mu(T)\left(\eta_x\dfrac{\partial v}{\partial\eta}+\eta_y\dfrac{\partial u}{\partial\eta}\right)\\2\mu(T)\eta_y\dfrac{\partial v}{\partial\eta}+\phi^{(\eta)}\\\mu(T)\left(\eta_z\dfrac{\partial v}{\partial\eta}+\eta_y\dfrac{\partial w}{\partial\eta}\right)\\ug_2^{(\eta)}+vg_3^{(\eta)}+wg_4^{(\eta)}-\dfrac{k\eta_y}{(\gamma-1)\Pr M_\infty^2}\dfrac{\partial T}{\partial\eta}\end{bmatrix};\quad \boldsymbol{G_v}^{(\zeta)}=\begin{bmatrix}0\\\mu(T)\left(\zeta_x\dfrac{\partial v}{\partial\zeta}+\zeta_y\dfrac{\partial u}{\partial\zeta}\right)\\2\mu(T)\zeta_y\dfrac{\partial v}{\partial\zeta}+\phi^{(\zeta)}\\\mu(T)\left(\zeta_z\dfrac{\partial v}{\partial\zeta}+\zeta_y\dfrac{\partial w}{\partial\zeta}\right)\\ug_2^{(\zeta)}+vg_3^{(\zeta)}+wg_4^{(\zeta)}-\dfrac{k\zeta_y}{(\gamma-1)\Pr M_\infty^2}\dfrac{\partial T}{\partial\zeta}\end{bmatrix};$$

$$\boldsymbol{H_v}^{(\xi)}=\begin{bmatrix}0\\\mu(T)\left(\xi_x\dfrac{\partial w}{\partial\xi}+\xi_z\dfrac{\partial u}{\partial\xi}\right)\\\mu(T)\left(\xi_z\dfrac{\partial v}{\partial\xi}+\xi_y\dfrac{\partial w}{\partial\xi}\right)\\2\mu(T)\xi_z\dfrac{\partial w}{\partial\xi}+\phi^{(\xi)}\\uh_2^{(\xi)}+vh_3^{(\xi)}+wh_4^{(\xi)}-\dfrac{k\xi_z}{(\gamma-1)\Pr M_\infty^2}\dfrac{\partial T}{\partial\xi}\end{bmatrix}=\begin{bmatrix}h_1^{(\xi)}\\h_2^{(\xi)}\\h_3^{(\xi)}\\h_4^{(\xi)}\\h_5^{(\xi)}\end{bmatrix};$$

$$\boldsymbol{H}_v^{(\eta)} = \begin{bmatrix} 0 \\ \mu(T)\left(\eta_x \frac{\partial w}{\partial \eta} + \eta_z \frac{\partial u}{\partial \eta}\right) \\ \mu(T)\left(\eta_z \frac{\partial v}{\partial \eta} + \eta_y \frac{\partial w}{\partial \eta}\right) \\ 2\mu(T)\eta_z \frac{\partial w}{\partial \eta} + \phi^{(\eta)} \\ uh_2^{(\eta)} + vh_3^{(\eta)} + wh_4^{(\eta)} - \frac{k\eta_z}{(\gamma-1)\Pr M_\infty^2} \frac{\partial T}{\partial \eta} \end{bmatrix}; \quad \boldsymbol{H}_v^{(\zeta)} = \begin{bmatrix} 0 \\ \mu(T)\left(\zeta_x \frac{\partial w}{\partial \zeta} + \zeta_z \frac{\partial u}{\partial \zeta}\right) \\ \mu(T)\left(\zeta_z \frac{\partial v}{\partial \zeta} + \zeta_y \frac{\partial w}{\partial \zeta}\right) \\ 2\mu(T)\zeta_z \frac{\partial w}{\partial \zeta} + \phi^{(\zeta)} \\ uh_2^{(\zeta)} + vh_3^{(\zeta)} + wh_4^{(\zeta)} - \frac{k\zeta_z}{(\gamma-1)\Pr M_\infty^2} \frac{\partial T}{\partial \zeta} \end{bmatrix}.$$

2.2. FLUX VECTOR SPLITTING

According to Steger and Warming [120], finite-difference schemes for the conservation-law form of the gas dynamic equations are restricted to a very limited class of spatial difference approximations in subsonic flow regions. For stability in both the positive and negative characteristic speeds that are associated with the spatial flux terms in subsonic flow, it is necessary to split the flux into the positive and negative directions, and then apply one-sided spatial difference operators to each direction.

To determine the flux splitting form of the 3D Navier-Stokes equations, it is necessary to first set the equations into generalized characteristic form.

2.2.1. Transformation to Generalized Characteristic Form

In Part I, we determined that the non-dimensionalized Navier-Stokes equations in curvilinear coordinates can be written as

$$\frac{\partial \hat{\boldsymbol{q}}}{\partial \tau} + \frac{\partial \hat{\boldsymbol{F}}}{\partial \xi} + \frac{\partial \hat{\boldsymbol{G}}}{\partial \eta} + \frac{\partial \hat{\boldsymbol{H}}}{\partial \zeta} = \frac{1}{\mathrm{Re}}\left[\frac{\partial \hat{\boldsymbol{F}}_v}{\partial \xi} + \frac{\partial \hat{\boldsymbol{G}}_v}{\partial \eta} + \frac{\partial \hat{\boldsymbol{H}}_v}{\partial \zeta}\right]. \tag{2.13}$$

This equation can be transformed into a generalized characteristic form.

We can rewrite the above equation as

$$\frac{\partial \hat{\boldsymbol{q}}}{\partial \tau} + \frac{\partial \hat{\boldsymbol{F}}}{\partial \xi} + \frac{\partial \hat{\boldsymbol{G}}}{\partial \eta} + \frac{\partial \hat{\boldsymbol{H}}}{\partial \zeta} = \hat{\boldsymbol{S}}_v,$$

where

$$\hat{\boldsymbol{S}}_v = \frac{1}{\mathrm{Re}}\left[\frac{\partial \hat{\boldsymbol{F}}_v}{\partial \xi} + \frac{\partial \hat{\boldsymbol{G}}_v}{\partial \eta} + \frac{\partial \hat{\boldsymbol{H}}_v}{\partial \zeta}\right].$$

If we define a vector of primitive variables

$$\mathbf{Q} = \frac{1}{J}\begin{bmatrix}\rho & u & v & w & p\end{bmatrix}^T,$$

then we can determine the Jacobian matrix that relates the conservative variables in $\boldsymbol{q}$ with the primitive variables in $\boldsymbol{Q}$. Let

$$\rho E = \rho e + \rho\frac{\mathbf{u}\cdot\mathbf{u}}{2} = \frac{p}{\gamma-1} + \rho\frac{\mathbf{u}\cdot\mathbf{u}}{2}.$$

Then,

$$\mathbf{P} = \frac{\partial\hat{\mathbf{q}}}{\partial\mathbf{Q}} = \begin{bmatrix} 1 & 0 & 0 & 0 & 0 \\ u & \rho & 0 & 0 & 0 \\ v & 0 & \rho & 0 & 0 \\ w & 0 & 0 & \rho & 0 \\ \dfrac{u^2+v^2+w^2}{2} & \rho u & \rho v & \rho w & \dfrac{1}{\gamma-1} \end{bmatrix}$$

and

$$\mathbf{P}^{-1} = \frac{\partial\mathbf{Q}}{\partial\hat{\mathbf{q}}} = \begin{bmatrix} 1 & 0 & 0 & 0 & 0 \\ -\dfrac{u}{\rho} & \dfrac{1}{\rho} & 0 & 0 & 0 \\ -\dfrac{v}{\rho} & 0 & \dfrac{1}{\rho} & 0 & 0 \\ -\dfrac{w}{\rho} & 0 & 0 & \dfrac{1}{\rho} & 0 \\ (\gamma-1)\dfrac{u^2+v^2+w^2}{2} & -(\gamma-1)u & -(\gamma-1)v & -(\gamma-1)w & (\gamma-1) \end{bmatrix}.$$

The equation

$$\frac{\partial\hat{\mathbf{q}}}{\partial\tau} + \frac{\partial\hat{\mathbf{F}}}{\partial\xi} + \frac{\partial\hat{\mathbf{G}}}{\partial\eta} + \frac{\partial\hat{\mathbf{H}}}{\partial\zeta} = \hat{\mathbf{S}}_v$$

can then be transformed into

$$\boldsymbol{P}\frac{\partial \boldsymbol{Q}}{\partial \tau}+\frac{\partial \hat{\boldsymbol{F}}}{\partial \hat{\boldsymbol{q}}}\boldsymbol{P}\frac{\partial \boldsymbol{Q}}{\partial \xi}+\frac{\partial \hat{\boldsymbol{G}}}{\partial \hat{\boldsymbol{q}}}\boldsymbol{P}\frac{\partial \boldsymbol{Q}}{\partial \eta}+\frac{\partial \hat{\boldsymbol{H}}}{\partial \hat{\boldsymbol{q}}}\boldsymbol{P}\frac{\partial \boldsymbol{Q}}{\partial \zeta}=\hat{\boldsymbol{S}}_v \tag{2.14}$$

or

$$\boldsymbol{P}\frac{\partial \boldsymbol{Q}}{\partial \tau}+\boldsymbol{AP}\frac{\partial \boldsymbol{Q}}{\partial \xi}+\boldsymbol{BP}\frac{\partial \boldsymbol{Q}}{\partial \eta}+\boldsymbol{CP}\frac{\partial \boldsymbol{Q}}{\partial \zeta}=\hat{\boldsymbol{S}}_v,$$

where

$$\boldsymbol{A}=\frac{\partial \hat{\boldsymbol{F}}}{\partial \hat{\boldsymbol{q}}}=\begin{bmatrix} 0 & \xi_x & \xi_y & \xi_z & 0 \\ \xi_x\phi-uU & \xi_x(2-\gamma)u+U & \xi_y u-\xi_x(\gamma-1)v & \xi_z u-\xi_x(\gamma-1)w & \xi_x(\gamma-1) \\ \xi_y\phi-vU & \xi_x v-\xi_y(\gamma-1)u & \xi_y(2-\gamma)v+U & \xi_z v-\xi_y(\gamma-1)w & \xi_y(\gamma-1) \\ \xi_z\phi-wU & \xi_x w-\xi_z(\gamma-1)u & \xi_y w-\xi_z(\gamma-1)v & \xi_z(2-\gamma)w+U & \xi_z(\gamma-1) \\ [2\phi-\gamma E]U & \xi_x[\gamma E-\phi]-(\gamma-1)uU & \xi_y[\gamma E-\phi]-(\gamma-1)vU & \xi_z[\gamma E-\phi]-(\gamma-1)wU & \gamma U \end{bmatrix};$$

$$\boldsymbol{B}=\frac{\partial \hat{\boldsymbol{G}}}{\partial \hat{\boldsymbol{q}}}=\begin{bmatrix} 0 & \eta_x & \eta_y & \eta_z & 0 \\ \eta_x\phi-uV & \eta_x(2-\gamma)u+V & \eta_y u-\eta_x(\gamma-1)v & \eta_z u-\eta_x(\gamma-1)w & \eta_x(\gamma-1) \\ \eta_y\phi-vV & \eta_x v-\eta_y(\gamma-1)u & \eta_y(2-\gamma)v+V & \eta_z v-\eta_y(\gamma-1)w & \eta_y(\gamma-1) \\ \eta_z\phi-wV & \eta_x w-\eta_z(\gamma-1)u & \eta_y w-\eta_z(\gamma-1)v & \eta_z(2-\gamma)w+V & \eta_z(\gamma-1) \\ [2\phi-\gamma E]V & \eta_x[\gamma E-\phi]-(\gamma-1)uV & \eta_y[\gamma E-\phi]-(\gamma-1)vV & \eta_z[\gamma E-\phi]-(\gamma-1)wV & \gamma V \end{bmatrix};$$

$$\boldsymbol{C}=\frac{\partial \hat{\boldsymbol{H}}}{\partial \hat{\boldsymbol{q}}}=\begin{bmatrix} 0 & \zeta_x & \zeta_y & \zeta_z & 0 \\ \zeta_x\phi-uW & \zeta_x(2-\gamma)u+W & \zeta_y u-\zeta_x(\gamma-1)v & \zeta_z u-\zeta_x(\gamma-1)w & \zeta_x(\gamma-1) \\ \zeta_y\phi-vW & \zeta_x v-\zeta_y(\gamma-1)u & \zeta_y(2-\gamma)v+W & \zeta_z v-\zeta_y(\gamma-1)w & \zeta_y(\gamma-1) \\ \zeta_z\phi-wW & \zeta_x w-\zeta_z(\gamma-1)u & \zeta_y w-\zeta_z(\gamma-1)v & \zeta_z(2-\gamma)w+W & \zeta_z(\gamma-1) \\ [2\phi-\gamma E]W & \zeta_x[\gamma E-\phi]-(\gamma-1)uW & \zeta_y[\gamma E-\phi]-(\gamma-1)vW & \zeta_z[\gamma E-\phi]-(\gamma-1)wW & \gamma W \end{bmatrix},$$

with

$$\begin{aligned} \phi &= \frac{\gamma-1}{2}\left(u^2+v^2+w^2\right); \\ U &= \xi_x u+\xi_y v+\xi_z w; \\ V &= \eta_x u+\eta_y v+\eta_z w; \\ W &= \zeta_x u+\zeta_y v+\zeta_z w. \end{aligned}$$

Multiplying the equation by $\boldsymbol{P}^{-1}$ gives

$$\boldsymbol{I}\frac{\partial \boldsymbol{Q}}{\partial \tau}+\boldsymbol{P}^{-1}\boldsymbol{AP}\frac{\partial \boldsymbol{Q}}{\partial \xi}+\boldsymbol{P}^{-1}\boldsymbol{BP}\frac{\partial \boldsymbol{Q}}{\partial \eta}+\boldsymbol{P}^{-1}\boldsymbol{CP}\frac{\partial \boldsymbol{Q}}{\partial \zeta}=\boldsymbol{P}^{-1}\hat{\boldsymbol{S}}_v. \tag{2.15}$$

$\boldsymbol{P^{-1}AP}$, $\boldsymbol{P^{-1}BP}$,and $\boldsymbol{P^{-1}CP}$ have, respectively, eigenvalue diagonal matrices

$$\boldsymbol{\Lambda}_A = \begin{bmatrix} \lambda_A^1 & 0 & 0 & 0 & 0 \\ 0 & \lambda_A^2 & 0 & 0 & 0 \\ 0 & 0 & \lambda_A^3 & 0 & 0 \\ 0 & 0 & 0 & \lambda_A^4 & 0 \\ 0 & 0 & 0 & 0 & \lambda_A^5 \end{bmatrix} = \begin{bmatrix} U & 0 & 0 & 0 & 0 \\ 0 & U & 0 & 0 & 0 \\ 0 & 0 & U & 0 & 0 \\ 0 & 0 & 0 & U + c\sqrt{\xi_x^2 + \xi_y^2 + \xi_z^2} & 0 \\ 0 & 0 & 0 & 0 & U - c\sqrt{\xi_x^2 + \xi_y^2 + \xi_z^2} \end{bmatrix};$$

$$\boldsymbol{\Lambda}_B = \begin{bmatrix} \lambda_B^1 & 0 & 0 & 0 & 0 \\ 0 & \lambda_B^2 & 0 & 0 & 0 \\ 0 & 0 & \lambda_B^3 & 0 & 0 \\ 0 & 0 & 0 & \lambda_B^4 & 0 \\ 0 & 0 & 0 & 0 & \lambda_B^5 \end{bmatrix} = \begin{bmatrix} V & 0 & 0 & 0 & 0 \\ 0 & V & 0 & 0 & 0 \\ 0 & 0 & V & 0 & 0 \\ 0 & 0 & 0 & V + c\sqrt{\eta_x^2 + \eta_y^2 + \eta_z^2} & 0 \\ 0 & 0 & 0 & 0 & V - c\sqrt{\eta_x^2 + \eta_y^2 + \eta_z^2} \end{bmatrix};$$

$$\boldsymbol{\Lambda}_C = \begin{bmatrix} \lambda_C^1 & 0 & 0 & 0 & 0 \\ 0 & \lambda_C^2 & 0 & 0 & 0 \\ 0 & 0 & \lambda_C^3 & 0 & 0 \\ 0 & 0 & 0 & \lambda_C^4 & 0 \\ 0 & 0 & 0 & 0 & \lambda_C^5 \end{bmatrix} = \begin{bmatrix} W & 0 & 0 & 0 & 0 \\ 0 & W & 0 & 0 & 0 \\ 0 & 0 & W & 0 & 0 \\ 0 & 0 & 0 & W + c\sqrt{\zeta_x^2 + \zeta_y^2 + \zeta_z^2} & 0 \\ 0 & 0 & 0 & 0 & W - c\sqrt{\zeta_x^2 + \zeta_y^2 + \zeta_z^2} \end{bmatrix},$$

where c is the speed of sound, given by

$$c = \sqrt{\frac{\gamma p}{\rho}}$$

Since the matrices $\boldsymbol{P^{-1}AP}$, $\boldsymbol{P^{-1}BP}$,and $\boldsymbol{P^{-1}CP}$ are respectively similar to $\boldsymbol{A}$, $\boldsymbol{B}$, and $\boldsymbol{C}$, then they have the same eigenvalues and (right) eigenvectors, which are, respectively, the columns of the following matrices:

$$\boldsymbol{M}_A = \begin{bmatrix} \frac{\xi_x}{\sqrt{\xi_x^2+\xi_y^2+\xi_z^2}} & \frac{\xi_y}{\sqrt{\xi_x^2+\xi_y^2+\xi_z^2}} & \frac{\xi_z}{\sqrt{\xi_x^2+\xi_y^2+\xi_z^2}} & \frac{\rho}{\sqrt{2}c} & \frac{\rho}{\sqrt{2}c} \\ 0 & -\frac{\xi_z}{\sqrt{\xi_x^2+\xi_y^2+\xi_z^2}} & \frac{\xi_y}{\sqrt{\xi_x^2+\xi_y^2+\xi_z^2}} & \frac{\xi_x}{\sqrt{2}\sqrt{\xi_x^2+\xi_y^2+\xi_z^2}} & -\frac{\xi_x}{\sqrt{2}\sqrt{\xi_x^2+\xi_y^2+\xi_z^2}} \\ \frac{\xi_z}{\sqrt{\xi_x^2+\xi_y^2+\xi_z^2}} & 0 & -\frac{\xi_x}{\sqrt{\xi_x^2+\xi_y^2+\xi_z^2}} & \frac{\xi_y}{\sqrt{2}\sqrt{\xi_x^2+\xi_y^2+\xi_z^2}} & -\frac{\xi_y}{\sqrt{2}\sqrt{\xi_x^2+\xi_y^2+\xi_z^2}} \\ -\frac{\xi_y}{\sqrt{\xi_x^2+\xi_y^2+\xi_z^2}} & \frac{\xi_x}{\sqrt{\xi_x^2+\xi_y^2+\xi_z^2}} & 0 & \frac{\xi_z}{\sqrt{2}\sqrt{\xi_x^2+\xi_y^2+\xi_z^2}} & -\frac{\xi_z}{\sqrt{2}\sqrt{\xi_x^2+\xi_y^2+\xi_z^2}} \\ 0 & 0 & 0 & \frac{\rho c}{\sqrt{2}} & \frac{\rho c}{\sqrt{2}} \end{bmatrix}$$

$$\boldsymbol{M}_B = \begin{bmatrix} \frac{\eta_x}{\sqrt{\eta_x^2+\eta_y^2+\eta_z^2}} & \frac{\eta_y}{\sqrt{\eta_x^2+\eta_y^2+\eta_z^2}} & \frac{\eta_z}{\sqrt{\eta_x^2+\eta_y^2+\eta_z^2}} & \frac{\rho}{\sqrt{2}c} & \frac{\rho}{\sqrt{2}c} \\ 0 & -\frac{\eta_z}{\sqrt{\eta_x^2+\eta_y^2+\eta_z^2}} & \frac{\eta_y}{\sqrt{\eta_x^2+\eta_y^2+\eta_z^2}} & \frac{\eta_x}{\sqrt{2}\sqrt{\eta_x^2+\eta_y^2+\eta_z^2}} & -\frac{\eta_x}{\sqrt{2}\sqrt{\eta_x^2+\eta_y^2+\eta_z^2}} \\ \frac{\eta_z}{\sqrt{\eta_x^2+\eta_y^2+\eta_z^2}} & 0 & -\frac{\eta_x}{\sqrt{\eta_x^2+\eta_y^2+\eta_z^2}} & \frac{\eta_y}{\sqrt{2}\sqrt{\eta_x^2+\eta_y^2+\eta_z^2}} & -\frac{\eta_y}{\sqrt{2}\sqrt{\eta_x^2+\eta_y^2+\eta_z^2}} \\ -\frac{\eta_y}{\sqrt{\eta_x^2+\eta_y^2+\eta_z^2}} & \frac{\eta_x}{\sqrt{\eta_x^2+\eta_y^2+\eta_z^2}} & 0 & \frac{\eta_z}{\sqrt{2}\sqrt{\eta_x^2+\eta_y^2+\eta_z^2}} & -\frac{\eta_z}{\sqrt{2}\sqrt{\eta_x^2+\eta_y^2+\eta_z^2}} \\ 0 & 0 & 0 & \frac{\rho c}{\sqrt{2}} & \frac{\rho c}{\sqrt{2}} \end{bmatrix}$$

$$\boldsymbol{M}_C = \begin{bmatrix} \frac{\zeta_x}{\sqrt{\zeta_x^2+\zeta_y^2+\zeta_z^2}} & \frac{\zeta_y}{\sqrt{\zeta_x^2+\zeta_y^2+\zeta_z^2}} & \frac{\zeta_z}{\sqrt{\zeta_x^2+\zeta_y^2+\zeta_z^2}} & \frac{\rho}{\sqrt{2}c} & \frac{\rho}{\sqrt{2}c} \\ 0 & -\frac{\zeta_z}{\sqrt{\zeta_x^2+\zeta_y^2+\zeta_z^2}} & \frac{\zeta_y}{\sqrt{\zeta_x^2+\zeta_y^2+\zeta_z^2}} & \frac{\zeta_x}{\sqrt{2}\sqrt{\zeta_x^2+\zeta_y^2+\zeta_z^2}} & -\frac{\zeta_x}{\sqrt{2}\sqrt{\zeta_x^2+\zeta_y^2+\zeta_z^2}} \\ \frac{\zeta_z}{\sqrt{\zeta_x^2+\zeta_y^2+\zeta_z^2}} & 0 & -\frac{\zeta_x}{\sqrt{\zeta_x^2+\zeta_y^2+\zeta_z^2}} & \frac{\zeta_y}{\sqrt{2}\sqrt{\zeta_x^2+\zeta_y^2+\zeta_z^2}} & -\frac{\zeta_y}{\sqrt{2}\sqrt{\zeta_x^2+\zeta_y^2+\zeta_z^2}} \\ -\frac{\zeta_y}{\sqrt{\zeta_x^2+\zeta_y^2+\zeta_z^2}} & \frac{\zeta_x}{\sqrt{\zeta_x^2+\zeta_y^2+\zeta_z^2}} & 0 & \frac{\zeta_z}{\sqrt{2}\sqrt{\zeta_x^2+\zeta_y^2+\zeta_z^2}} & -\frac{\zeta_z}{\sqrt{2}\sqrt{\zeta_x^2+\zeta_y^2+\zeta_z^2}} \\ 0 & 0 & 0 & \frac{\rho c}{\sqrt{2}} & \frac{\rho c}{\sqrt{2}} \end{bmatrix}$$

Their inverses are, respectively,

$$\boldsymbol{M}_A^{-1} = \begin{bmatrix} \frac{\xi_x}{\sqrt{\xi_x^2+\xi_y^2+\xi_z^2}} & 0 & \frac{\xi_z}{\sqrt{\xi_x^2+\xi_y^2+\xi_z^2}} & -\frac{\xi_y}{\sqrt{\xi_x^2+\xi_y^2+\xi_z^2}} & -\frac{\xi_x}{c^2\sqrt{\xi_x^2+\xi_y^2+\xi_z^2}} \\ \frac{\xi_y}{\sqrt{\xi_x^2+\xi_y^2+\xi_z^2}} & -\frac{\xi_z}{\sqrt{\xi_x^2+\xi_y^2+\xi_z^2}} & 0 & \frac{\xi_x}{\sqrt{\xi_x^2+\xi_y^2+\xi_z^2}} & -\frac{\xi_y}{c^2\sqrt{\xi_x^2+\xi_y^2+\xi_z^2}} \\ \frac{\xi_z}{\sqrt{\xi_x^2+\xi_y^2+\xi_z^2}} & \frac{\xi_y}{\sqrt{\xi_x^2+\xi_y^2+\xi_z^2}} & -\frac{\xi_x}{\sqrt{\xi_x^2+\xi_y^2+\xi_z^2}} & 0 & -\frac{\xi_z}{c^2\sqrt{\xi_x^2+\xi_y^2+\xi_z^2}} \\ 0 & \frac{\xi_x}{\sqrt{2}\sqrt{\xi_x^2+\xi_y^2+\xi_z^2}} & \frac{\xi_y}{\sqrt{2}\sqrt{\xi_x^2+\xi_y^2+\xi_z^2}} & \frac{\xi_z}{\sqrt{2}\sqrt{\xi_x^2+\xi_y^2+\xi_z^2}} & \frac{1}{\sqrt{2}\rho c} \\ 0 & -\frac{\xi_x}{\sqrt{2}\sqrt{\xi_x^2+\xi_y^2+\xi_z^2}} & -\frac{\xi_y}{\sqrt{2}\sqrt{\xi_x^2+\xi_y^2+\xi_z^2}} & -\frac{\xi_z}{\sqrt{2}\sqrt{\xi_x^2+\xi_y^2+\xi_z^2}} & \frac{1}{\sqrt{2}\rho c} \end{bmatrix}$$

$$\boldsymbol{M}_B^{-1} = \begin{bmatrix} \frac{\eta_x}{\sqrt{\eta_x^2+\eta_y^2+\eta_z^2}} & 0 & \frac{\eta_z}{\sqrt{\eta_x^2+\eta_y^2+\eta_z^2}} & -\frac{\eta_y}{\sqrt{\eta_x^2+\eta_y^2+\eta_z^2}} & -\frac{\eta_x}{c^2\sqrt{\eta_x^2+\eta_y^2+\eta_z^2}} \\ \frac{\eta_y}{\sqrt{\eta_x^2+\eta_y^2+\eta_z^2}} & -\frac{\eta_z}{\sqrt{\eta_x^2+\eta_y^2+\eta_z^2}} & 0 & \frac{\eta_x}{\sqrt{\eta_x^2+\eta_y^2+\eta_z^2}} & -\frac{\eta_y}{c^2\sqrt{\eta_x^2+\eta_y^2+\eta_z^2}} \\ \frac{\eta_z}{\sqrt{\eta_x^2+\eta_y^2+\eta_z^2}} & \frac{\eta_y}{\sqrt{\eta_x^2+\eta_y^2+\eta_z^2}} & -\frac{\eta_x}{\sqrt{\eta_x^2+\eta_y^2+\eta_z^2}} & 0 & -\frac{\eta_z}{c^2\sqrt{\eta_x^2+\eta_y^2+\eta_z^2}} \\ 0 & \frac{\eta_x}{\sqrt{2}\sqrt{\eta_x^2+\eta_y^2+\eta_z^2}} & \frac{\eta_y}{\sqrt{2}\sqrt{\eta_x^2+\eta_y^2+\eta_z^2}} & \frac{\eta_z}{\sqrt{2}\sqrt{\eta_x^2+\eta_y^2+\eta_z^2}} & \frac{1}{\sqrt{2}\rho c} \\ 0 & -\frac{\eta_x}{\sqrt{2}\sqrt{\eta_x^2+\eta_y^2+\eta_z^2}} & -\frac{\eta_y}{\sqrt{2}\sqrt{\eta_x^2+\eta_y^2+\eta_z^2}} & -\frac{\eta_z}{\sqrt{2}\sqrt{\eta_x^2+\eta_y^2+\eta_z^2}} & \frac{1}{\sqrt{2}\rho c} \end{bmatrix}$$

$$\boldsymbol{M}_C^{-1} = \begin{bmatrix} \frac{\zeta_x}{\sqrt{\zeta_x^2+\zeta_y^2+\zeta_z^2}} & 0 & \frac{\zeta_z}{\sqrt{\zeta_x^2+\zeta_y^2+\zeta_z^2}} & -\frac{\zeta_y}{\sqrt{\zeta_x^2+\zeta_y^2+\zeta_z^2}} & -\frac{\zeta_x}{c^2\sqrt{\zeta_x^2+\zeta_y^2+\zeta_z^2}} \\ \frac{\zeta_y}{\sqrt{\zeta_x^2+\zeta_y^2+\zeta_z^2}} & -\frac{\zeta_z}{\sqrt{\zeta_x^2+\zeta_y^2+\zeta_z^2}} & 0 & \frac{\zeta_x}{\sqrt{\zeta_x^2+\zeta_y^2+\zeta_z^2}} & -\frac{\zeta_y}{c^2\sqrt{\zeta_x^2+\zeta_y^2+\zeta_z^2}} \\ \frac{\zeta_z}{\sqrt{\zeta_x^2+\zeta_y^2+\zeta_z^2}} & \frac{\zeta_y}{\sqrt{\zeta_x^2+\zeta_y^2+\zeta_z^2}} & -\frac{\zeta_x}{\sqrt{\zeta_x^2+\zeta_y^2+\zeta_z^2}} & 0 & -\frac{\zeta_z}{c^2\sqrt{\zeta_x^2+\zeta_y^2+\zeta_z^2}} \\ 0 & \frac{\zeta_x}{\sqrt{2}\sqrt{\zeta_x^2+\zeta_y^2+\zeta_z^2}} & \frac{\zeta_y}{\sqrt{2}\sqrt{\zeta_x^2+\zeta_y^2+\zeta_z^2}} & \frac{\zeta_z}{\sqrt{2}\sqrt{\zeta_x^2+\zeta_y^2+\zeta_z^2}} & \frac{1}{\sqrt{2}\rho c} \\ 0 & -\frac{\zeta_x}{\sqrt{2}\sqrt{\zeta_x^2+\zeta_y^2+\zeta_z^2}} & -\frac{\zeta_y}{\sqrt{2}\sqrt{\zeta_x^2+\zeta_y^2+\zeta_z^2}} & -\frac{\zeta_z}{\sqrt{2}\sqrt{\zeta_x^2+\zeta_y^2+\zeta_z^2}} & \frac{1}{\sqrt{2}\rho c} \end{bmatrix}$$

Now, we can write

$$\boldsymbol{P}^{-1}\boldsymbol{A}\boldsymbol{P} = \boldsymbol{M}_A\boldsymbol{\Lambda}_A\boldsymbol{M}_A^{-1}$$
$$\boldsymbol{P}^{-1}\boldsymbol{B}\boldsymbol{P} = \boldsymbol{M}_B\boldsymbol{\Lambda}_B\boldsymbol{M}_B^{-1}$$
$$\boldsymbol{P}^{-1}\boldsymbol{C}\boldsymbol{P} = \boldsymbol{M}_C\boldsymbol{\Lambda}_C\boldsymbol{M}_C^{-1}$$

and so the equation becomes

$$\boldsymbol{I}\frac{\partial \boldsymbol{Q}}{\partial \tau} + \boldsymbol{M}_A\boldsymbol{\Lambda}_A\boldsymbol{M}_A^{-1}\frac{\partial \boldsymbol{Q}}{\partial \xi} + \boldsymbol{M}_B\boldsymbol{\Lambda}_B\boldsymbol{M}_B^{-1}\frac{\partial \boldsymbol{Q}}{\partial \eta} + \boldsymbol{M}_C\boldsymbol{\Lambda}_C\boldsymbol{M}_C^{-1}\frac{\partial \boldsymbol{Q}}{\partial \zeta} = \boldsymbol{P}^{-1}\hat{\boldsymbol{S}}_v \tag{2.16}$$

with

$$\boldsymbol{M}_A\boldsymbol{\Lambda}_A\boldsymbol{M}_A^{-1}=\begin{bmatrix} U & \xi_x\rho & \xi_y\rho & \xi_z\rho & 0 \\ 0 & U & 0 & 0 & \frac{\xi_x}{\rho} \\ 0 & 0 & U & 0 & \frac{\xi_y}{\rho} \\ 0 & 0 & 0 & U & \frac{\xi_z}{\rho} \\ 0 & \xi_x\rho c^2 & \xi_y\rho c^2 & \xi_z\rho c^2 & U \end{bmatrix};$$

$$\boldsymbol{M}_B\boldsymbol{\Lambda}_B\boldsymbol{M}_B^{-1}=\begin{bmatrix} V & \eta_x\rho & \eta_y\rho & \eta_z\rho & 0 \\ 0 & V & 0 & 0 & \frac{\eta_x}{\rho} \\ 0 & 0 & V & 0 & \frac{\eta_y}{\rho} \\ 0 & 0 & 0 & V & \frac{\eta_z}{\rho} \\ 0 & \eta_x\rho c^2 & \eta_y\rho c^2 & \eta_z\rho c^2 & V \end{bmatrix};$$

$$\boldsymbol{M}_C\boldsymbol{\Lambda}_C\boldsymbol{M}_C^{-1}=\begin{bmatrix} W & \zeta_x\rho & \zeta_y\rho & \zeta_z\rho & 0 \\ 0 & W & 0 & 0 & \frac{\zeta_x}{\rho} \\ 0 & 0 & W & 0 & \frac{\zeta_y}{\rho} \\ 0 & 0 & 0 & W & \frac{\zeta_z}{\rho} \\ 0 & \zeta_x\rho c^2 & \zeta_y\rho c^2 & \zeta_z\rho c^2 & W \end{bmatrix}.$$

2.2.2. Flux Vector Splitting Form

The vectors $\hat{\boldsymbol{F}}$, $\hat{\boldsymbol{G}}$, and $\hat{\boldsymbol{H}}$ can be considered homogeneous functions of degree one in $\hat{\boldsymbol{q}}$. This means that they can be written as

$$\hat{\boldsymbol{F}}=\boldsymbol{A}\hat{\boldsymbol{q}};$$
$$\hat{\boldsymbol{G}}=\boldsymbol{B}\hat{\boldsymbol{q}};$$
$$\hat{\boldsymbol{H}}=\boldsymbol{C}\hat{\boldsymbol{q}}.$$

Since

$$\boldsymbol{P^{-1}AP} = \boldsymbol{M_A \Lambda_A M_A}^{-1}$$
$$\boldsymbol{P^{-1}BP} = \boldsymbol{M_B \Lambda_B M_B}^{-1}$$
$$\boldsymbol{P^{-1}CP} = \boldsymbol{M_C \Lambda_C M_C}^{-1}$$

then

$$\boldsymbol{A} = \boldsymbol{PM_A \Lambda_A M_A}^{-1}\boldsymbol{P^{-1}}$$
$$\boldsymbol{B} = \boldsymbol{PM_B \Lambda_B M_B}^{-1}\boldsymbol{P^{-1}}$$
$$\boldsymbol{C} = \boldsymbol{PM_C \Lambda_C M_C}^{-1}\boldsymbol{P^{-1}}$$

so, if we define

$$\boldsymbol{T_A} = \boldsymbol{PM_A} = \begin{bmatrix}
\frac{\xi_x}{\sqrt{\xi_x^2+\xi_y^2+\xi_z^2}} & \frac{\xi_y}{\sqrt{\xi_x^2+\xi_y^2+\xi_z^2}} & \frac{\xi_z}{\sqrt{\xi_x^2+\xi_y^2+\xi_z^2}} & \frac{\rho}{\sqrt{2}c} & \frac{\rho}{\sqrt{2}c} \\
\frac{u\xi_x}{\sqrt{\xi_x^2+\xi_y^2+\xi_z^2}} & \frac{u\xi_y-\rho\xi_z}{\sqrt{\xi_x^2+\xi_y^2+\xi_z^2}} & \frac{u\xi_z+\rho\xi_y}{\sqrt{\xi_x^2+\xi_y^2+\xi_z^2}} & \frac{\rho}{\sqrt{2}c}\left(u+\frac{c\xi_x}{\sqrt{\xi_x^2+\xi_y^2+\xi_z^2}}\right) & \frac{\rho}{\sqrt{2}c}\left(u-\frac{c\xi_x}{\sqrt{\xi_x^2+\xi_y^2+\xi_z^2}}\right) \\
\frac{v\xi_x+\rho\xi_z}{\sqrt{\xi_x^2+\xi_y^2+\xi_z^2}} & \frac{v\xi_y}{\sqrt{\xi_x^2+\xi_y^2+\xi_z^2}} & \frac{v\xi_z-\rho\xi_x}{\sqrt{\xi_x^2+\xi_y^2+\xi_z^2}} & \frac{\rho}{\sqrt{2}c}\left(v+\frac{c\xi_y}{\sqrt{\xi_x^2+\xi_y^2+\xi_z^2}}\right) & \frac{\rho}{\sqrt{2}c}\left(v-\frac{c\xi_y}{\sqrt{\xi_x^2+\xi_y^2+\xi_z^2}}\right) \\
\frac{w\xi_x-\rho\xi_y}{\sqrt{\xi_x^2+\xi_y^2+\xi_z^2}} & \frac{w\xi_y+\rho\xi_x}{\sqrt{\xi_x^2+\xi_y^2+\xi_z^2}} & \frac{w\xi_z}{\sqrt{\xi_x^2+\xi_y^2+\xi_z^2}} & \frac{\rho}{\sqrt{2}c}\left(w+\frac{c\xi_z}{\sqrt{\xi_x^2+\xi_y^2+\xi_z^2}}\right) & \frac{\rho}{\sqrt{2}c}\left(w-\frac{c\xi_z}{\sqrt{\xi_x^2+\xi_y^2+\xi_z^2}}\right) \\
\frac{\frac{\phi}{\gamma-1}\xi_x+\rho(v\xi_z-w\xi_y)}{\sqrt{\xi_x^2+\xi_y^2+\xi_z^2}} & \frac{\frac{\phi}{\gamma-1}\xi_y+\rho(w\xi_x-u\xi_z)}{\sqrt{\xi_x^2+\xi_y^2+\xi_z^2}} & \frac{\frac{\phi}{\gamma-1}\xi_z+\rho(u\xi_y-v\xi_x)}{\sqrt{\xi_x^2+\xi_y^2+\xi_z^2}} & \frac{\rho}{\sqrt{2}c}\left(\frac{\phi+c^2}{\gamma-1}+\frac{cU}{\sqrt{\xi_x^2+\xi_y^2+\xi_z^2}}\right) & \frac{\rho}{\sqrt{2}c}\left(\frac{\phi+c^2}{\gamma-1}-\frac{cU}{\sqrt{\xi_x^2+\xi_y^2+\xi_z^2}}\right)
\end{bmatrix};$$

$$
\boldsymbol{T}_B = \boldsymbol{PM}_B = \begin{bmatrix}
\frac{\eta_x}{\sqrt{\eta_x^2+\eta_y^2+\eta_z^2}} & \frac{\eta_y}{\sqrt{\eta_x^2+\eta_y^2+\eta_z^2}} & \frac{\eta_z}{\sqrt{\eta_x^2+\eta_y^2+\eta_z^2}} & \frac{\rho}{\sqrt{2}c} & \frac{\rho}{\sqrt{2}c} \\
\frac{u\eta_x}{\sqrt{\eta_x^2+\eta_y^2+\eta_z^2}} & \frac{u\eta_y-\rho\eta_z}{\sqrt{\eta_x^2+\eta_y^2+\eta_z^2}} & \frac{u\eta_z+\rho\eta_y}{\sqrt{\eta_x^2+\eta_y^2+\eta_z^2}} & \frac{\rho}{\sqrt{2}c}\left(u+\frac{c\eta_x}{\sqrt{\eta_x^2+\eta_y^2+\eta_z^2}}\right) & \frac{\rho}{\sqrt{2}c}\left(u-\frac{c\eta_x}{\sqrt{\eta_x^2+\eta_y^2+\eta_z^2}}\right) \\
\frac{v\eta_x+\rho\eta_z}{\sqrt{\eta_x^2+\eta_y^2+\eta_z^2}} & \frac{v\eta_y}{\sqrt{\eta_x^2+\eta_y^2+\eta_z^2}} & \frac{v\eta_z-\rho\eta_x}{\sqrt{\eta_x^2+\eta_y^2+\eta_z^2}} & \frac{\rho}{\sqrt{2}c}\left(v+\frac{c\eta_y}{\sqrt{\eta_x^2+\eta_y^2+\eta_z^2}}\right) & \frac{\rho}{\sqrt{2}c}\left(v-\frac{c\eta_y}{\sqrt{\eta_x^2+\eta_y^2+\eta_z^2}}\right) \\
\frac{w\eta_x-\rho\eta_y}{\sqrt{\eta_x^2+\eta_y^2+\eta_z^2}} & \frac{w\eta_y+\rho\eta_x}{\sqrt{\eta_x^2+\eta_y^2+\eta_z^2}} & \frac{w\eta_z}{\sqrt{\eta_x^2+\eta_y^2+\eta_z^2}} & \frac{\rho}{\sqrt{2}c}\left(w+\frac{c\eta_z}{\sqrt{\eta_x^2+\eta_y^2+\eta_z^2}}\right) & \frac{\rho}{\sqrt{2}c}\left(w-\frac{c\eta_z}{\sqrt{\eta_x^2+\eta_y^2+\eta_z^2}}\right) \\
\frac{\frac{\phi}{\gamma-1}\eta_x+\rho(v\eta_z-w\eta_y)}{\sqrt{\eta_x^2+\eta_y^2+\eta_z^2}} & \frac{\frac{\phi}{\gamma-1}\eta_y+\rho(w\eta_x-u\eta_z)}{\sqrt{\eta_x^2+\eta_y^2+\eta_z^2}} & \frac{\frac{\phi}{\gamma-1}\eta_z+\rho(u\eta_y-v\eta_x)}{\sqrt{\eta_x^2+\eta_y^2+\eta_z^2}} & \frac{\rho}{\sqrt{2}c}\left(\frac{\phi+c^2}{\gamma-1}+\frac{cV}{\sqrt{\eta_x^2+\eta_y^2+\eta_z^2}}\right) & \frac{\rho}{\sqrt{2}c}\left(\frac{\phi+c^2}{\gamma-1}-\frac{cV}{\sqrt{\eta_x^2+\eta_y^2+\eta_z^2}}\right)
\end{bmatrix};
$$

$$
\boldsymbol{T}_C = \boldsymbol{P}\boldsymbol{M}_C = \begin{bmatrix}
\frac{\zeta_x}{\sqrt{\zeta_x^2+\zeta_y^2+\zeta_z^2}} & \frac{\zeta_y}{\sqrt{\zeta_x^2+\zeta_y^2+\zeta_z^2}} & \frac{\zeta_z}{\sqrt{\zeta_x^2+\zeta_y^2+\zeta_z^2}} \\
\frac{u\zeta_x}{\sqrt{\zeta_x^2+\zeta_y^2+\zeta_z^2}} & \frac{u\zeta_y-\rho\zeta_z}{\sqrt{\zeta_x^2+\zeta_y^2+\zeta_z^2}} & \frac{u\zeta_z+\rho\zeta_y}{\sqrt{\zeta_x^2+\zeta_y^2+\zeta_z^2}} \\
\frac{v\zeta_x+\rho\zeta_z}{\sqrt{\zeta_x^2+\zeta_y^2+\zeta_z^2}} & \frac{v\zeta_y}{\sqrt{\zeta_x^2+\zeta_y^2+\zeta_z^2}} & \frac{v\zeta_z-\rho\zeta_x}{\sqrt{\zeta_x^2+\zeta_y^2+\zeta_z^2}} \\
\frac{w\zeta_x-\rho\zeta_y}{\sqrt{\zeta_x^2+\zeta_y^2+\zeta_z^2}} & \frac{w\zeta_y+\rho\zeta_x}{\sqrt{\zeta_x^2+\zeta_y^2+\zeta_z^2}} & \frac{w\zeta_z}{\sqrt{\zeta_x^2+\zeta_y^2+\zeta_z^2}} \\
\frac{\frac{\phi}{\gamma-1}\zeta_x+\rho(v\zeta_z-w\zeta_y)}{\sqrt{\zeta_x^2+\zeta_y^2+\zeta_z^2}} & \frac{\frac{\phi}{\gamma-1}\zeta_y+\rho(w\zeta_x-u\zeta_z)}{\sqrt{\zeta_x^2+\zeta_y^2+\zeta_z^2}} & \frac{\frac{\phi}{\gamma-1}\zeta_z+\rho(u\zeta_y-v\zeta_x)}{\sqrt{\zeta_x^2+\zeta_y^2+\zeta_z^2}}
\end{bmatrix.}
$$

$$
\begin{bmatrix.}
\frac{\rho}{\sqrt{2}c} & \frac{\rho}{\sqrt{2}c} \\
\frac{\rho}{\sqrt{2}c}\left(u+\frac{c\zeta_x}{\sqrt{\zeta_x^2+\zeta_y^2+\zeta_z^2}}\right) & \frac{\rho}{\sqrt{2}c}\left(u-\frac{c\zeta_x}{\sqrt{\zeta_x^2+\zeta_y^2+\zeta_z^2}}\right) \\
\frac{\rho}{\sqrt{2}c}\left(v+\frac{c\zeta_y}{\sqrt{\zeta_x^2+\zeta_y^2+\zeta_z^2}}\right) & \frac{\rho}{\sqrt{2}c}\left(v-\frac{c\zeta_y}{\sqrt{\zeta_x^2+\zeta_y^2+\zeta_z^2}}\right) \\
\frac{\rho}{\sqrt{2}c}\left(w+\frac{c\zeta_z}{\sqrt{\zeta_x^2+\zeta_y^2+\zeta_z^2}}\right) & \frac{\rho}{\sqrt{2}c}\left(w-\frac{c\zeta_z}{\sqrt{\zeta_x^2+\zeta_y^2+\zeta_z^2}}\right) \\
\frac{\rho}{\sqrt{2}c}\left(\frac{\phi+c^2}{\gamma-1}+\frac{cW}{\sqrt{\zeta_x^2+\zeta_y^2+\zeta_z^2}}\right) & \frac{\rho}{\sqrt{2}c}\left(\frac{\phi+c^2}{\gamma-1}-\frac{cW}{\sqrt{\zeta_x^2+\zeta_y^2+\zeta_z^2}}\right)
\end{bmatrix},
$$

with

$$
\begin{aligned}
\phi &= \frac{\gamma-1}{2}\left(u^2+v^2+w^2\right) \\
U &= \xi_x u+\xi_y v+\xi_z w \\
V &= \eta_x u+\eta_y v+\eta_z w \\
W &= \zeta_x u+\zeta_y v+\zeta_z w
\end{aligned}
$$

then

$$\boldsymbol{A} = \boldsymbol{T}_A \boldsymbol{\Lambda}_A \boldsymbol{T}_A^{-1}$$
$$\boldsymbol{B} = \boldsymbol{T}_B \boldsymbol{\Lambda}_B \boldsymbol{T}_B^{-1}$$
$$\boldsymbol{C} = \boldsymbol{T}_C \boldsymbol{\Lambda}_C \boldsymbol{T}_C^{-1}$$

and thus

$$\hat{\boldsymbol{F}} = \boldsymbol{T}_A \boldsymbol{\Lambda}_A \boldsymbol{T}_A^{-1} \hat{\boldsymbol{q}}$$
$$\hat{\boldsymbol{G}} = \boldsymbol{T}_B \boldsymbol{\Lambda}_B \boldsymbol{T}_B^{-1} \hat{\boldsymbol{q}}$$
$$\hat{\boldsymbol{H}} = \boldsymbol{T}_C \boldsymbol{\Lambda}_C \boldsymbol{T}_C^{-1} \hat{\boldsymbol{q}}$$

If we split each of the eigenvalue matrices $\boldsymbol{\Lambda}_A$, $\boldsymbol{\Lambda}_B$, and $\boldsymbol{\Lambda}_C$ into three parts, as such:

$$\boldsymbol{\Lambda}_A = \boldsymbol{\Lambda}_A^1 + \boldsymbol{\Lambda}_A^4 + \boldsymbol{\Lambda}_A^5 = \begin{bmatrix} U & 0 & 0 & 0 & 0 \\ 0 & U & 0 & 0 & 0 \\ 0 & 0 & U & 0 & 0 \\ 0 & 0 & 0 & 0 & 0 \\ 0 & 0 & 0 & 0 & 0 \end{bmatrix} + \begin{bmatrix} 0 & 0 & 0 & 0 & 0 \\ 0 & 0 & 0 & 0 & 0 \\ 0 & 0 & 0 & 0 & 0 \\ 0 & 0 & 0 & U + c\sqrt{\xi_x^2 + \xi_y^2 + \xi_z^2} & 0 \\ 0 & 0 & 0 & 0 & 0 \end{bmatrix} + \begin{bmatrix} 0 & 0 & 0 & 0 & 0 \\ 0 & 0 & 0 & 0 & 0 \\ 0 & 0 & 0 & 0 & 0 \\ 0 & 0 & 0 & 0 & 0 \\ 0 & 0 & 0 & 0 & U - c\sqrt{\xi_x^2 + \xi_y^2 + \xi_z^2} \end{bmatrix}$$

$$\boldsymbol{\Lambda}_B = \boldsymbol{\Lambda}_B^1 + \boldsymbol{\Lambda}_B^4 + \boldsymbol{\Lambda}_B^5 = \begin{bmatrix} V & 0 & 0 & 0 & 0 \\ 0 & V & 0 & 0 & 0 \\ 0 & 0 & V & 0 & 0 \\ 0 & 0 & 0 & 0 & 0 \\ 0 & 0 & 0 & 0 & 0 \end{bmatrix} + \begin{bmatrix} 0 & 0 & 0 & 0 & 0 \\ 0 & 0 & 0 & 0 & 0 \\ 0 & 0 & 0 & 0 & 0 \\ 0 & 0 & 0 & V + c\sqrt{\eta_x^2 + \eta_y^2 + \eta_z^2} & 0 \\ 0 & 0 & 0 & 0 & 0 \end{bmatrix} + \begin{bmatrix} 0 & 0 & 0 & 0 & 0 \\ 0 & 0 & 0 & 0 & 0 \\ 0 & 0 & 0 & 0 & 0 \\ 0 & 0 & 0 & 0 & 0 \\ 0 & 0 & 0 & 0 & V - c\sqrt{\eta_x^2 + \eta_y^2 + \eta_z^2} \end{bmatrix}$$

$$\boldsymbol{\Lambda}_C = \boldsymbol{\Lambda}_C^1 + \boldsymbol{\Lambda}_C^4 + \boldsymbol{\Lambda}_C^5 = \begin{bmatrix} W & 0 & 0 & 0 & 0 \\ 0 & W & 0 & 0 & 0 \\ 0 & 0 & W & 0 & 0 \\ 0 & 0 & 0 & 0 & 0 \\ 0 & 0 & 0 & 0 & 0 \end{bmatrix} + \begin{bmatrix} 0 & 0 & 0 & 0 & 0 \\ 0 & 0 & 0 & 0 & 0 \\ 0 & 0 & 0 & 0 & 0 \\ 0 & 0 & 0 & W + c\sqrt{\zeta_x^2 + \zeta_y^2 + \zeta_z^2} & 0 \\ 0 & 0 & 0 & 0 & 0 \end{bmatrix} + \begin{bmatrix} 0 & 0 & 0 & 0 & 0 \\ 0 & 0 & 0 & 0 & 0 \\ 0 & 0 & 0 & 0 & 0 \\ 0 & 0 & 0 & 0 & 0 \\ 0 & 0 & 0 & 0 & W - c\sqrt{\zeta_x^2 + \zeta_y^2 + \zeta_z^2} \end{bmatrix}$$

then the flux vectors may also be separated into three parts, according to their eigenvalues:

$$\hat{\boldsymbol{F}} = \hat{\boldsymbol{F}}_1 + \hat{\boldsymbol{F}}_4 + \hat{\boldsymbol{F}}_5 = \boldsymbol{T}_A \boldsymbol{\Lambda}_A^1 \boldsymbol{T}_A^{-1} \hat{\boldsymbol{q}} + \boldsymbol{T}_A \boldsymbol{\Lambda}_A^4 \boldsymbol{T}_A^{-1} \hat{\boldsymbol{q}} + \boldsymbol{T}_A \boldsymbol{\Lambda}_A^5 \boldsymbol{T}_A^{-1} \hat{\boldsymbol{q}}$$
$$\hat{\boldsymbol{G}} = \hat{\boldsymbol{G}}_1 + \hat{\boldsymbol{G}}_4 + \hat{\boldsymbol{G}}_5 = \boldsymbol{T}_B \boldsymbol{\Lambda}_B^1 \boldsymbol{T}_B^{-1} \hat{\boldsymbol{q}} + \boldsymbol{T}_B \boldsymbol{\Lambda}_B^4 \boldsymbol{T}_B^{-1} \hat{\boldsymbol{q}} + \boldsymbol{T}_B \boldsymbol{\Lambda}_B^5 \boldsymbol{T}_B^{-1} \hat{\boldsymbol{q}}$$
$$\hat{\boldsymbol{H}} = \hat{\boldsymbol{H}}_1 + \hat{\boldsymbol{H}}_4 + \hat{\boldsymbol{H}}_5 = \boldsymbol{T}_C \boldsymbol{\Lambda}_C^1 \boldsymbol{T}_C^{-1} \hat{\boldsymbol{q}} + \boldsymbol{T}_C \boldsymbol{\Lambda}_C^4 \boldsymbol{T}_C^{-1} \hat{\boldsymbol{q}} + \boldsymbol{T}_C \boldsymbol{\Lambda}_C^5 \boldsymbol{T}_C^{-1} \hat{\boldsymbol{q}}$$

or, expanding,

$$\hat{\boldsymbol{F}} = \hat{\boldsymbol{F}}_1 + \hat{\boldsymbol{F}}_4 + \hat{\boldsymbol{F}}_5 =$$

$$= \frac{\lambda_A^1}{J}\frac{\gamma-1}{\gamma}\begin{bmatrix} \rho \\ \rho u \\ \rho v \\ \rho w \\ \frac{\rho}{2}\left(u^2+v^2+w^2\right) \end{bmatrix} + \frac{\lambda_A^4}{J}\frac{1}{2\gamma}\begin{bmatrix} \rho \\ \rho\left(u+\frac{c\xi_x}{\sqrt{\xi_x{}^2+\xi_y{}^2+\xi_z{}^2}}\right) \\ \rho\left(v+\frac{c\xi_y}{\sqrt{\xi_x{}^2+\xi_y{}^2+\xi_z{}^2}}\right) \\ \rho\left(w+\frac{c\xi_z}{\sqrt{\xi_x{}^2+\xi_y{}^2+\xi_z{}^2}}\right) \\ \frac{\rho c^2}{(\gamma-1)}+\frac{\rho}{2}\left(u^2+v^2+w^2\right)+\frac{\rho c U}{\sqrt{\xi_x{}^2+\xi_y{}^2+\xi_z{}^2}} \end{bmatrix} +$$

$$+ \frac{\lambda_A^5}{J}\frac{1}{2\gamma}\begin{bmatrix} \rho \\ \rho\left(u-\frac{c\xi_x}{\sqrt{\xi_x{}^2+\xi_y{}^2+\xi_z{}^2}}\right) \\ \rho\left(v-\frac{c\xi_y}{\sqrt{\xi_x{}^2+\xi_y{}^2+\xi_z{}^2}}\right) \\ \rho\left(w-\frac{c\xi_z}{\sqrt{\xi_x{}^2+\xi_y{}^2+\xi_z{}^2}}\right) \\ \frac{\rho c^2}{(\gamma-1)}+\frac{\rho}{2}\left(u^2+v^2+w^2\right)-\frac{\rho c U}{\sqrt{\xi_x{}^2+\xi_y{}^2+\xi_z{}^2}} \end{bmatrix};$$

$$\hat{\boldsymbol{G}} = \hat{\boldsymbol{G}}_1 + \hat{\boldsymbol{G}}_4 + \hat{\boldsymbol{G}}_5 =$$

$$= \frac{\lambda_B^1}{J}\frac{\gamma-1}{\gamma}\begin{bmatrix} \rho \\ \rho u \\ \rho v \\ \rho w \\ \frac{\rho}{2}\left(u^2+v^2+w^2\right) \end{bmatrix} + \frac{\lambda_B^4}{J}\frac{1}{2\gamma}\begin{bmatrix} \rho \\ \rho\left(u+\frac{c\eta_x}{\sqrt{\eta_x^{\ 2}+\eta_y^{\ 2}+\eta_z^{\ 2}}}\right) \\ \rho\left(v+\frac{c\eta_y}{\sqrt{\eta_x^{\ 2}+\eta_y^{\ 2}+\eta_z^{\ 2}}}\right) \\ \rho\left(w+\frac{c\eta_z}{\sqrt{\eta_x^{\ 2}+\eta_y^{\ 2}+\eta_z^{\ 2}}}\right) \\ \frac{\rho c^2}{(\gamma-1)}+\frac{\rho}{2}\left(u^2+v^2+w^2\right)+\frac{\rho c V}{\sqrt{\eta_x^{\ 2}+\eta_y^{\ 2}+\eta_z^{\ 2}}} \end{bmatrix} +$$

$$+ \frac{\lambda_B^5}{J}\frac{1}{2\gamma}\begin{bmatrix} \rho \\ \rho\left(u-\frac{c\eta_x}{\sqrt{\eta_x^{\ 2}+\eta_y^{\ 2}+\eta_z^{\ 2}}}\right) \\ \rho\left(v-\frac{c\eta_y}{\sqrt{\eta_x^{\ 2}+\eta_y^{\ 2}+\eta_z^{\ 2}}}\right) \\ \rho\left(w-\frac{c\eta_z}{\sqrt{\eta_x^{\ 2}+\eta_y^{\ 2}+\eta_z^{\ 2}}}\right) \\ \frac{\rho c^2}{(\gamma-1)}+\frac{\rho}{2}\left(u^2+v^2+w^2\right)-\frac{\rho c V}{\sqrt{\eta_x^{\ 2}+\eta_y^{\ 2}+\eta_z^{\ 2}}} \end{bmatrix};$$

$$\hat{\boldsymbol{H}} = \hat{\boldsymbol{H}}_1 + \hat{\boldsymbol{H}}_4 + \hat{\boldsymbol{H}}_5 =$$

$$= \frac{\lambda_C^1}{J}\frac{\gamma-1}{\gamma}\begin{bmatrix} \rho \\ \rho u \\ \rho v \\ \rho w \\ \frac{\rho}{2}\left(u^2+v^2+w^2\right) \end{bmatrix} + \frac{\lambda_C^4}{J}\frac{1}{2\gamma}\begin{bmatrix} \rho \\ \rho\left(u + \frac{c\zeta_x}{\sqrt{\zeta_x^2+\zeta_y^2+\zeta_z^2}}\right) \\ \rho\left(v + \frac{c\zeta_y}{\sqrt{\zeta_x^2+\zeta_y^2+\zeta_z^2}}\right) \\ \rho\left(w + \frac{c\zeta_z}{\sqrt{\zeta_x^2+\zeta_y^2+\zeta_z^2}}\right) \\ \frac{\rho c^2}{(\gamma-1)} + \frac{\rho}{2}\left(u^2+v^2+w^2\right) + \frac{\rho c V}{\sqrt{\zeta_x^2+\zeta_y^2+\zeta_z^2}} \end{bmatrix} +$$

$$+ \frac{\lambda_C^5}{J}\frac{1}{2\gamma}\begin{bmatrix} \rho \\ \rho\left(u - \frac{c\zeta_x}{\sqrt{\zeta_x^2+\zeta_y^2+\zeta_z^2}}\right) \\ \rho\left(v - \frac{c\zeta_y}{\sqrt{\zeta_x^2+\zeta_y^2+\zeta_z^2}}\right) \\ \rho\left(w - \frac{c\zeta_z}{\sqrt{\zeta_x^2+\zeta_y^2+\zeta_z^2}}\right) \\ \frac{\rho c^2}{(\gamma-1)} + \frac{\rho}{2}\left(u^2+v^2+w^2\right) - \frac{\rho c V}{\sqrt{\zeta_x^2+\zeta_y^2+\zeta_z^2}} \end{bmatrix}.$$

With this information, we can split the flux vectors into positive and negative directions. This splitting is determined by the positive or negative values of the eigenvalues. If we define

$$\lambda_{k\pm}^i = \frac{\lambda_k^i \pm \left|\lambda_k^i\right|}{2}, \quad i = 1,4,5; \; k = A,B,C,$$

then

$$\begin{aligned} \lambda_k^i &= \lambda_{k+}^i + \lambda_{k-}^i \\ \left|\lambda_k^i\right| &= \lambda_{k+}^i - \lambda_{k-}^i \end{aligned}, \quad i = 1,4,5; \; k = A,B,C,$$

and so

$$\begin{aligned}\hat{\boldsymbol{F}} &= \hat{\boldsymbol{F}}_1 + \hat{\boldsymbol{F}}_4 + \hat{\boldsymbol{F}}_5 = \left(\hat{\boldsymbol{F}}_1^{\ +} + \hat{\boldsymbol{F}}_4^{\ +} + \hat{\boldsymbol{F}}_5^{\ +}\right) + \left(\hat{\boldsymbol{F}}_1^{\ -} + \hat{\boldsymbol{F}}_4^{\ -} + \hat{\boldsymbol{F}}_5^{\ -}\right) = \hat{\boldsymbol{F}}^{+} + \hat{\boldsymbol{F}}^{-};\\ \hat{\boldsymbol{G}} &= \hat{\boldsymbol{G}}_1 + \hat{\boldsymbol{G}}_4 + \hat{\boldsymbol{G}}_5 = \left(\hat{\boldsymbol{G}}_1^{\ +} + \hat{\boldsymbol{G}}_4^{\ +} + \hat{\boldsymbol{G}}_5^{\ +}\right) + \left(\hat{\boldsymbol{G}}_1^{\ -} + \hat{\boldsymbol{G}}_4^{\ -} + \hat{\boldsymbol{G}}_5^{\ -}\right) = \hat{\boldsymbol{G}}^{+} + \hat{\boldsymbol{G}}^{-};\\ \hat{\boldsymbol{H}} &= \hat{\boldsymbol{H}}_1 + \hat{\boldsymbol{H}}_4 + \hat{\boldsymbol{H}}_5 = \left(\hat{\boldsymbol{H}}_1^{\ +} + \hat{\boldsymbol{H}}_4^{\ +} + \hat{\boldsymbol{H}}_5^{\ +}\right) + \left(\hat{\boldsymbol{H}}_1^{\ -} + \hat{\boldsymbol{H}}_4^{\ -} + \hat{\boldsymbol{H}}_5^{\ -}\right) = \hat{\boldsymbol{H}}^{+} + \hat{\boldsymbol{H}}^{-},\end{aligned}$$

where

$$\hat{\boldsymbol{F}}^{\pm} = \frac{\rho}{2J\gamma}\begin{bmatrix} 2(\gamma-1)\lambda_{A\pm}^{1} + \lambda_{A\pm}^{4} + \lambda_{A\pm}^{5} \\ \left(2(\gamma-1)\lambda_{A\pm}^{1} + \lambda_{A\pm}^{4} + \lambda_{A\pm}^{5}\right)u + \dfrac{\left(\lambda_{A\pm}^{4} - \lambda_{A\pm}^{5}\right)c\xi_x}{\sqrt{\xi_x^{\ 2} + \xi_y^{\ 2} + \xi_z^{\ 2}}} \\ \left(2(\gamma-1)\lambda_{A\pm}^{1} + \lambda_{A\pm}^{4} + \lambda_{A\pm}^{5}\right)v + \dfrac{\left(\lambda_{A\pm}^{4} - \lambda_{A\pm}^{5}\right)c\xi_y}{\sqrt{\xi_x^{\ 2} + \xi_y^{\ 2} + \xi_z^{\ 2}}} \\ \left(2(\gamma-1)\lambda_{A\pm}^{1} + \lambda_{A\pm}^{4} + \lambda_{A\pm}^{5}\right)w + \dfrac{\left(\lambda_{A\pm}^{4} - \lambda_{A\pm}^{5}\right)c\xi_z}{\sqrt{\xi_x^{\ 2} + \xi_y^{\ 2} + \xi_z^{\ 2}}} \\ \dfrac{c^2}{\gamma-1}\left(\lambda_{A\pm}^{4} + \lambda_{A\pm}^{5}\right) + \left(2(\gamma-1)\lambda_{A\pm}^{1} + \lambda_{A\pm}^{4} + \lambda_{A\pm}^{5}\right)\dfrac{\left(u^2+v^2+w^2\right)}{2} + \dfrac{\left(\lambda_{A\pm}^{4} - \lambda_{A\pm}^{5}\right)cU}{\sqrt{\xi_x^{\ 2} + \xi_y^{\ 2} + \xi_z^{\ 2}}} \end{bmatrix};$$

$$\hat{\boldsymbol{G}}^{\pm} = \frac{\rho}{2J\gamma}\begin{bmatrix} 2(\gamma-1)\lambda_{B\pm}^{1} + \lambda_{B\pm}^{4} + \lambda_{B\pm}^{5} \\ \left(2(\gamma-1)\lambda_{B\pm}^{1} + \lambda_{B\pm}^{4} + \lambda_{B\pm}^{5}\right)u + \dfrac{\left(\lambda_{B\pm}^{4} - \lambda_{B\pm}^{5}\right)c\eta_x}{\sqrt{\eta_x^{\ 2} + \eta_y^{\ 2} + \eta_z^{\ 2}}} \\ \left(2(\gamma-1)\lambda_{B\pm}^{1} + \lambda_{B\pm}^{4} + \lambda_{B\pm}^{5}\right)v + \dfrac{\left(\lambda_{B\pm}^{4} - \lambda_{B\pm}^{5}\right)c\eta_y}{\sqrt{\eta_x^{\ 2} + \eta_y^{\ 2} + \eta_z^{\ 2}}} \\ \left(2(\gamma-1)\lambda_{B\pm}^{1} + \lambda_{B\pm}^{4} + \lambda_{B\pm}^{5}\right)w + \dfrac{\left(\lambda_{B\pm}^{4} - \lambda_{B\pm}^{5}\right)c\eta_z}{\sqrt{\eta_x^{\ 2} + \eta_y^{\ 2} + \eta_z^{\ 2}}} \\ \dfrac{c^2}{\gamma-1}\left(\lambda_{B\pm}^{4} + \lambda_{B\pm}^{5}\right) + \left(2(\gamma-1)\lambda_{B\pm}^{1} + \lambda_{B\pm}^{4} + \lambda_{B\pm}^{5}\right)\dfrac{\left(u^2+v^2+w^2\right)}{2} + \dfrac{\left(\lambda_{B\pm}^{4} - \lambda_{B\pm}^{5}\right)cV}{\sqrt{\eta_x^{\ 2} + \eta_y^{\ 2} + \eta_z^{\ 2}}} \end{bmatrix};$$

$$\hat{\boldsymbol{H}}^{\pm} = \frac{\rho}{2J\gamma}\begin{bmatrix} 2(\gamma-1)\lambda_{C\pm}^{1} + \lambda_{C\pm}^{4} + \lambda_{C\pm}^{5} \\ \left(2(\gamma-1)\lambda_{C\pm}^{1} + \lambda_{C\pm}^{4} + \lambda_{C\pm}^{5}\right)u + \frac{\left(\lambda_{C\pm}^{4} - \lambda_{C\pm}^{5}\right)c\zeta_x}{\sqrt{\zeta_x^2 + \zeta_y^2 + \zeta_z^2}} \\ \left(2(\gamma-1)\lambda_{C\pm}^{1} + \lambda_{C\pm}^{4} + \lambda_{C\pm}^{5}\right)v + \frac{\left(\lambda_{C\pm}^{4} - \lambda_{C\pm}^{5}\right)c\zeta_y}{\sqrt{\zeta_x^2 + \zeta_y^2 + \zeta_z^2}} \\ \left(2(\gamma-1)\lambda_{C\pm}^{1} + \lambda_{C\pm}^{4} + \lambda_{C\pm}^{5}\right)w + \frac{\left(\lambda_{C\pm}^{4} - \lambda_{C\pm}^{5}\right)c\zeta_z}{\sqrt{\zeta_x^2 + \zeta_y^2 + \zeta_z^2}} \\ \frac{c^2}{\gamma-1}\left(\lambda_{C\pm}^{4} + \lambda_{C\pm}^{5}\right) + \left(2(\gamma-1)\lambda_{C\pm}^{1} + \lambda_{C\pm}^{4} + \lambda_{C\pm}^{5}\right)\frac{\left(u^2+v^2+w^2\right)}{2} + \frac{\left(\lambda_{C\pm}^{4} - \lambda_{C\pm}^{5}\right)cW}{\sqrt{\zeta_x^2 + \zeta_y^2 + \zeta_z^2}} \end{bmatrix},$$

with

$$\begin{aligned} &\lambda_A^1 = U; \quad \lambda_A^4 = U + c\sqrt{\xi_x^2 + \xi_y^2 + \xi_z^2}; \quad \lambda_A^5 = U - c\sqrt{\xi_x^2 + \xi_y^2 + \xi_z^2}; \\ &\lambda_B^1 = V; \quad \lambda_B^4 = V + c\sqrt{\eta_x^2 + \eta_y^2 + \eta_z^2}; \quad \lambda_B^5 = V - c\sqrt{\eta_x^2 + \eta_y^2 + \eta_z^2}; \\ &\lambda_C^1 = W; \quad \lambda_C^4 = W + c\sqrt{\zeta_x^2 + \zeta_y^2 + \zeta_z^2}; \quad \lambda_C^5 = W - c\sqrt{\zeta_x^2 + \zeta_y^2 + \zeta_z^2}. \end{aligned}$$

Hence, the Navier-Stokes equations to be solved become

$$\frac{\partial \hat{\boldsymbol{q}}}{\partial \tau} + \left(\frac{\partial \hat{\boldsymbol{F}}^{+}}{\partial \xi} + \frac{\partial \hat{\boldsymbol{F}}^{-}}{\partial \xi}\right) + \left(\frac{\partial \hat{\boldsymbol{G}}^{+}}{\partial \eta} + \frac{\partial \hat{\boldsymbol{G}}^{-}}{\partial \eta}\right) + \left(\frac{\partial \hat{\boldsymbol{H}}^{+}}{\partial \zeta} + \frac{\partial \hat{\boldsymbol{H}}^{-}}{\partial \zeta}\right) = \hat{\boldsymbol{S}}_v \tag{2.17}$$

Other forms of flux vector splitting are described in the paper by Steger and Warming [120].

2.3. BOUNDARY CONDITIONS

Having finished deriving and expanding the Navier-Stokes equations in curvilinear coordinates, and determining the flux vector splitting, let us derive the characteristic variable boundary conditions that can be applied. For this, we must modify the characteristic form to reflect only the direction normal to the boundary.

2.3.1. Characteristic Variable Boundary Conditions

Let us consider that the direction normal to the boundary is the ξ direction (the boundary is where ξ keeps a constant value). Boundaries in other directions may be considered by taking the terms in the corresponding direction.

Taking the equation

$$\boldsymbol{I}\frac{\partial \boldsymbol{Q}}{\partial \tau}+\boldsymbol{M}_A\boldsymbol{\Lambda}_A\boldsymbol{M}_A^{-1}\frac{\partial \boldsymbol{Q}}{\partial \xi}+\boldsymbol{M}_B\boldsymbol{\Lambda}_B\boldsymbol{M}_B^{-1}\frac{\partial \boldsymbol{Q}}{\partial \eta}+\boldsymbol{M}_C\boldsymbol{\Lambda}_C\boldsymbol{M}_C^{-1}\frac{\partial \boldsymbol{Q}}{\partial \zeta}=\boldsymbol{P}^{-1}\hat{\boldsymbol{S}}_v \tag{2.18}$$

from Part II, by moving the last two terms from the left hand side to the right hand side, we obtain

$$\boldsymbol{I}\frac{\partial \boldsymbol{Q}}{\partial \tau}+\boldsymbol{M}_A\boldsymbol{\Lambda}_A\boldsymbol{M}_A^{-1}\frac{\partial \boldsymbol{Q}}{\partial \xi}=\hat{\boldsymbol{S}}_C,$$

where

$$\hat{\boldsymbol{S}}_C=\boldsymbol{P}^{-1}\hat{\boldsymbol{S}}_v-\boldsymbol{M}_B\boldsymbol{\Lambda}_B\boldsymbol{M}_B^{-1}\frac{\partial \boldsymbol{Q}}{\partial \eta}-\boldsymbol{M}_C\boldsymbol{\Lambda}_C\boldsymbol{M}_C^{-1}\frac{\partial \boldsymbol{Q}}{\partial \zeta}.$$

The matrices $\boldsymbol{\Lambda}_A$, $\boldsymbol{M}_A$ and $\boldsymbol{M}_A^{-1}$ (eigenvalue and eigenvector matrices) can be redefined as

$$\boldsymbol{\Lambda}_A=\begin{bmatrix}\lambda_A^1&0&0&0&0\\0&\lambda_A^2&0&0&0\\0&0&\lambda_A^3&0&0\\0&0&0&\lambda_A^4&0\\0&0&0&0&\lambda_A^5\end{bmatrix}=\begin{bmatrix}U-c\sqrt{\xi_x^2+\xi_y^2+\xi_z^2}&0&0&0&0\\0&U&0&0&0\\0&0&U&0&0\\0&0&0&U&0\\0&0&0&0&U+c\sqrt{\xi_x^2+\xi_y^2+\xi_z^2}\end{bmatrix};$$

$$\boldsymbol{M}_A=\begin{bmatrix}\frac{1}{2c^2}&\frac{1}{c^2}&0&0&\frac{1}{2c^2}\\-\frac{\xi_x}{2\rho c\sqrt{\xi_x^2+\xi_y^2+\xi_z^2}}&0&-\frac{\xi_y}{\xi_x^2+\xi_y^2+\xi_z^2}&-\frac{\xi_z}{\xi_x^2+\xi_y^2+\xi_z^2}&\frac{\xi_x}{2\rho c\sqrt{\xi_x^2+\xi_y^2+\xi_z^2}}\\-\frac{\xi_y}{2\rho c\sqrt{\xi_x^2+\xi_y^2+\xi_z^2}}&0&\frac{\xi_x^2+\xi_z^2}{\xi_x\left(\xi_x^2+\xi_y^2+\xi_z^2\right)}&-\frac{\xi_y\xi_z}{\xi_x\left(\xi_x^2+\xi_y^2+\xi_z^2\right)}&\frac{\xi_y}{2\rho c\sqrt{\xi_x^2+\xi_y^2+\xi_z^2}}\\-\frac{\xi_z}{2\rho c\sqrt{\xi_x^2+\xi_y^2+\xi_z^2}}&0&-\frac{\xi_y\xi_z}{\xi_x\left(\xi_x^2+\xi_y^2+\xi_z^2\right)}&\frac{\xi_x^2+\xi_y^2}{\xi_x\left(\xi_x^2+\xi_y^2+\xi_z^2\right)}&\frac{\xi_z}{2\rho c\sqrt{\xi_x^2+\xi_y^2+\xi_z^2}}\\\frac{1}{2}&0&0&0&\frac{1}{2}\end{bmatrix};$$

$$\boldsymbol{M}_A^{-1} = \begin{bmatrix} 0 & -\dfrac{\rho c \xi_x}{\sqrt{\xi_x^2+\xi_y^2+\xi_z^2}} & -\dfrac{\rho c \xi_y}{\sqrt{\xi_x^2+\xi_y^2+\xi_z^2}} & -\dfrac{\rho c \xi_z}{\sqrt{\xi_x^2+\xi_y^2+\xi_z^2}} & 1 \\ c^2 & 0 & 0 & 0 & -1 \\ 0 & -\xi_y & \xi_x & 0 & 0 \\ 0 & -\xi_z & 0 & \xi_x & 0 \\ 0 & \dfrac{\rho c \xi_x}{\sqrt{\xi_x^2+\xi_y^2+\xi_z^2}} & \dfrac{\rho c \xi_y}{\sqrt{\xi_x^2+\xi_y^2+\xi_z^2}} & \dfrac{\rho c \xi_z}{\sqrt{\xi_x^2+\xi_y^2+\xi_z^2}} & 1 \end{bmatrix}.$$

Also, let us define

$$\frac{\partial \boldsymbol{Q}}{\partial \xi} = \boldsymbol{M}_A \boldsymbol{\Lambda}_A^{-1} \boldsymbol{L}$$

or

$$\boldsymbol{L} = \boldsymbol{\Lambda}_A \boldsymbol{M}_A^{-1} \frac{\partial \boldsymbol{Q}}{\partial \xi}$$

Then,

$$\frac{\partial \boldsymbol{Q}}{\partial \tau} + \boldsymbol{M}_A \boldsymbol{L} = \hat{\boldsymbol{S}}_C$$

$$\boldsymbol{L} = \begin{bmatrix} L_1 \\ L_2 \\ L_3 \\ L_4 \\ L_5 \end{bmatrix} = \frac{1}{J} \begin{bmatrix} \left(U - c\sqrt{\xi_x^2+\xi_y^2+\xi_z^2}\right)\left(-\dfrac{\rho c}{\sqrt{\xi_x^2+\xi_y^2+\xi_z^2}}\left(\xi_x \dfrac{\partial u}{\partial \xi} + \xi_y \dfrac{\partial v}{\partial \xi} + \xi_z \dfrac{\partial w}{\partial \xi}\right) + \dfrac{\partial p}{\partial \xi}\right) \\ U\left(c^2 \dfrac{\partial \rho}{\partial \xi} - \dfrac{\partial p}{\partial \xi}\right) \\ U\left(-\xi_y \dfrac{\partial u}{\partial \xi} + \xi_x \dfrac{\partial v}{\partial \xi}\right) \\ U\left(-\xi_z \dfrac{\partial u}{\partial \xi} + \xi_x \dfrac{\partial w}{\partial \xi}\right) \\ \left(U + c\sqrt{\xi_x^2+\xi_y^2+\xi_z^2}\right)\left(\dfrac{\rho c}{\sqrt{\xi_x^2+\xi_y^2+\xi_z^2}}\left(\xi_x \dfrac{\partial u}{\partial \xi} + \xi_y \dfrac{\partial v}{\partial \xi} + \xi_z \dfrac{\partial w}{\partial \xi}\right) + \dfrac{\partial p}{\partial \xi}\right) \end{bmatrix},$$

and

$$\frac{\partial \mathbf{Q}}{\partial \xi} = \mathbf{M}_A \mathbf{\Lambda}_A^{-1} \mathbf{L} =$$

$$= \begin{bmatrix} \frac{1}{c^2}\left(\frac{1}{2}\left(\frac{L_1}{U - c\sqrt{\xi_x^2+\xi_y^2+\xi_z^2}} + \frac{L_5}{U + c\sqrt{\xi_x^2+\xi_y^2+\xi_z^2}}\right) + \frac{L_2}{U}\right) \\ \frac{\xi_x}{2\rho c\sqrt{\xi_x^2+\xi_y^2+\xi_z^2}}\left(\frac{L_5}{U + c\sqrt{\xi_x^2+\xi_y^2+\xi_z^2}} - \frac{L_1}{U - c\sqrt{\xi_x^2+\xi_y^2+\xi_z^2}}\right) - \frac{\xi_y L_3 + \xi_z L_4}{U\left(\xi_x^2+\xi_y^2+\xi_z^2\right)} \\ \frac{\xi_y}{2\rho c\sqrt{\xi_x^2+\xi_y^2+\xi_z^2}}\left(\frac{L_5}{U + c\sqrt{\xi_x^2+\xi_y^2+\xi_z^2}} - \frac{L_1}{U - c\sqrt{\xi_x^2+\xi_y^2+\xi_z^2}}\right) + \frac{\left(\xi_x^2+\xi_z^2\right)L_3 - \xi_y\xi_z L_4}{\xi_x U\left(\xi_x^2+\xi_y^2+\xi_z^2\right)} \\ \frac{\xi_z}{2\rho c\sqrt{\xi_x^2+\xi_y^2+\xi_z^2}}\left(\frac{L_5}{U + c\sqrt{\xi_x^2+\xi_y^2+\xi_z^2}} - \frac{L_1}{U - c\sqrt{\xi_x^2+\xi_y^2+\xi_z^2}}\right) - \frac{\xi_y\xi_z L_3 - \left(\xi_x^2+\xi_y^2\right)L_4}{\xi_x U\left(\xi_x^2+\xi_y^2+\xi_z^2\right)} \\ \frac{1}{2}\left(\frac{L_1}{U - c\sqrt{\xi_x^2+\xi_y^2+\xi_z^2}} + \frac{L_5}{U + c\sqrt{\xi_x^2+\xi_y^2+\xi_z^2}}\right) \end{bmatrix}.$$

From this we can determine

$$\mathbf{d} = \mathbf{M}_A \mathbf{L} = \begin{bmatrix} d_1 \\ d_2 \\ d_3 \\ d_4 \\ d_5 \end{bmatrix} = \frac{1}{J}\begin{bmatrix} \frac{1}{c^2}\left(\frac{L_1+L_5}{2} + L_2\right) \\ \frac{\xi_x}{2\rho c\sqrt{\xi_x^2+\xi_y^2+\xi_z^2}}\left(L_5 - L_1\right) - \frac{1}{\xi_x^2+\xi_y^2+\xi_z^2}\left(\xi_y L_3 + \xi_z L_4\right) \\ \frac{\xi_y}{2\rho c\sqrt{\xi_x^2+\xi_y^2+\xi_z^2}}\left(L_5 - L_1\right) + \frac{1}{\xi_x\left(\xi_x^2+\xi_y^2+\xi_z^2\right)}\left(\left(\xi_x^2+\xi_z^2\right)L_3 - \xi_y\xi_z L_4\right) \\ \frac{\xi_z}{2\rho c\sqrt{\xi_x^2+\xi_y^2+\xi_z^2}}\left(L_5 - L_1\right) - \frac{1}{\xi_x\left(\xi_x^2+\xi_y^2+\xi_z^2\right)}\left(\xi_y\xi_z L_3 - \left(\xi_x^2+\xi_y^2\right)L_4\right) \\ \frac{1}{2}\left(L_1 + L_5\right) \end{bmatrix}.$$

Therefore,

$$\frac{\partial \mathbf{Q}}{\partial \tau} + \mathbf{d} = \hat{\mathbf{S}}_C \tag{2.19}$$

or

$$\frac{\partial\rho}{\partial\tau}+\frac{1}{c^2}\left(\frac{L_1+L_5}{2}+L_2\right)=JS_{C1}$$

$$\frac{\partial u}{\partial\tau}+\frac{\xi_x}{2\rho c\sqrt{\xi_x^2+\xi_y^2+\xi_z^2}}(L_5-L_1)-\frac{1}{\xi_x^2+\xi_y^2+\xi_z^2}(\xi_y L_3+\xi_z L_4)=JS_{C2}$$

$$\frac{\partial v}{\partial\tau}+\frac{\xi_y}{2\rho c\sqrt{\xi_x^2+\xi_y^2+\xi_z^2}}(L_5-L_1)+\frac{1}{\xi_x\left(\xi_x^2+\xi_y^2+\xi_z^2\right)}\left(\left(\xi_x^2+\xi_z^2\right)L_3-\xi_y\xi_z L_4\right)=JS_{C3}$$

$$\frac{\partial w}{\partial\tau}+\frac{\xi_z}{2\rho c\sqrt{\xi_x^2+\xi_y^2+\xi_z^2}}(L_5-L_1)-\frac{1}{\xi_x\left(\xi_x^2+\xi_y^2+\xi_z^2\right)}\left(\xi_y\xi_z L_3-\left(\xi_x^2+\xi_y^2\right)L_4\right)=JS_{C4}$$

$$\frac{\partial p}{\partial\tau}+\frac{1}{2}(L_1+L_5)=JS_{C5}$$

These equations will be used next to determine the characteristic variable boundary conditions.

2.3.2. Boundary Treatment

The boundary conditions will be determined by the equations obtained above.

We will determine the values of the L_i for $i = 1,2,3,4,5$ by either assigning a value from the boundary conditions or determining the value from the inside information with one-sided derivative approximations, depending on the direction of propagation of waves described by the characteristic speeds. With these values, we can solve the equation

$$\frac{\partial \boldsymbol{Q}}{\partial\tau}+\boldsymbol{M}_A\boldsymbol{L}=\hat{\boldsymbol{S}}_C \tag{2.20}$$

at the boundary points.

At $\xi = \xi_{\min}$, for all cases where $\lambda_i < 0$ (outgoing waves), the L_i are calculated from the equations given above, while for $\lambda_i > 0$ (incoming waves), the values for L_i are specified from the boundary conditions. Similarly, at $\xi = \xi_{\max}$, we calculate the L_i from the inside values when $\lambda_i > 0$ (outgoing), while for $\lambda_i < 0$ (incoming), the L_i are specified from the boundary conditions.

Let us consider a few special cases for nonreflecting boundary conditions.

2.3.2.1. No-Slip Wall (Solid Boundary)

In this case, all the velocity components and their derivatives in time should be zero. Then, the initial condition data must have $u = v = w = 0$ at the boundary points being considered.

At $\xi = \xi_{\min}$, $\lambda_1 = -c\sqrt{\xi_x^2 + \xi_y^2 + \xi_z^2} < 0$, $\lambda_2 = \lambda_3 = \lambda_4 = 0$, $\lambda_5 = c\sqrt{\xi_x^2 + \xi_y^2 + \xi_z^2} > 0$. Therefore, L_5 must be specified, and the other values of L_i will be determined from the equations.

We can determine L_1 from

$$JL_1 = \rho c^2\left(\xi_x \frac{\partial u}{\partial \xi} + \xi_y \frac{\partial v}{\partial \xi} + \xi_z \frac{\partial w}{\partial \xi}\right) - c\sqrt{\xi_x^2 + \xi_y^2 + \xi_z^2}\,\frac{\partial p}{\partial \xi}.$$

To guarantee that the transverse velocities are zero at the boundary, L_3 and L_4 may be determined from the following equations

$$\frac{\partial u}{\partial \tau} + \frac{\xi_x}{2\rho c\sqrt{\xi_x^2 + \xi_y^2 + \xi_z^2}}(L_5 - L_1) - \frac{1}{\xi_x^2 + \xi_y^2 + \xi_z^2}(\xi_y L_3 + \xi_z L_4) = JS_{C2}$$

$$\frac{\partial v}{\partial \tau} + \frac{\xi_y}{2\rho c\sqrt{\xi_x^2 + \xi_y^2 + \xi_z^2}}(L_5 - L_1) + \frac{1}{\xi_x(\xi_x^2 + \xi_y^2 + \xi_z^2)}\left((\xi_x^2 + \xi_z^2)L_3 - \xi_y \xi_z L_4\right) = JS_{C3}$$

$$\frac{\partial w}{\partial \tau} + \frac{\xi_z}{2\rho c\sqrt{\xi_x^2 + \xi_y^2 + \xi_z^2}}(L_5 - L_1) - \frac{1}{\xi_x(\xi_x^2 + \xi_y^2 + \xi_z^2)}\left(\xi_y \xi_z L_3 - (\xi_x^2 + \xi_y^2)L_4\right) = JS_{C4}$$

as

$$L_3 = \xi_x JS_{C3} - \xi_y JS_{C2}$$
$$L_4 = \xi_x JS_{C4} - \xi_z JS_{C2}$$

Since the convection speeds for L_3 and L_4 are zero, then each term is also zero, $L_3 = L_4 = 0$. Therefore,

$$S_{C3} = \frac{\xi_y}{\xi_x} S_{C2}$$
$$S_{C4} = \frac{\xi_z}{\xi_x} S_{C2}$$

To determine L_5, we use the equations above and obtain

$$L_5 = L_1 + \frac{2\rho c}{\sqrt{\xi_x^2+\xi_y^2+\xi_z^2}}\left(\xi_x JS_{C2} + \xi_y JS_{C3} + \xi_z JS_{C4}\right) =$$
$$= L_1 + \frac{2\rho c\sqrt{\xi_x^2+\xi_y^2+\xi_z^2}}{\xi_x} JS_{C2}$$

Now, at $\xi = \xi_{\max}$, L_1 must be specified, and the other values of L_i will be determined from the equations. Then,

$$L_1 = L_5 - \frac{2\rho c}{\xi_x\sqrt{\xi_x^2+\xi_y^2+\xi_z^2}}\left(\xi_y L_3 + \xi_z L_4\right) - \frac{2\rho c\sqrt{\xi_x^2+\xi_y^2+\xi_z^2}}{\xi_x} JS_{C2}$$

$$L_3 = \xi_x JS_{C3} - \xi_y JS_{C2}$$
$$L_4 = \xi_x JS_{C4} - \xi_z JS_{C2}$$

$$JL_5 = \rho c^2\left(\xi_x \frac{\partial u}{\partial \xi} + \xi_y \frac{\partial v}{\partial \xi} + \xi_z \frac{\partial w}{\partial \xi}\right) + c\sqrt{\xi_x^2+\xi_y^2+\xi_z^2}\,\frac{\partial p}{\partial \xi}.$$

or, similarly to the previous case,

$$L_1 = L_5 - \frac{2\rho c\sqrt{\xi_x^2+\xi_y^2+\xi_z^2}}{\xi_x} JS_{C2}$$

$$L_3 = L_4 = 0$$

$$S_{C3} = \frac{\xi_y}{\xi_x} S_{C2}$$
$$S_{C4} = \frac{\xi_z}{\xi_x} S_{C2}$$

$$JL_5 = \rho c^2\left(\xi_x \frac{\partial u}{\partial \xi} + \xi_y \frac{\partial v}{\partial \xi} + \xi_z \frac{\partial w}{\partial \xi}\right) + c\sqrt{\xi_x^2+\xi_y^2+\xi_z^2}\,\frac{\partial p}{\partial \xi}.$$

For L_2 in both cases, we need to, initially, determine the equation for temperature.

Since

$$T = \gamma M_\infty^2 \frac{p}{\rho},$$

then taking the derivative in the normal direction gives

$$\frac{\partial T}{\partial \xi} = \gamma M_\infty^2 \frac{\partial}{\partial \xi}\left(\frac{p}{\rho}\right) =$$
$$= \gamma M_\infty^2 \frac{1}{\rho}\frac{\partial p}{\partial \xi} - \gamma M_\infty^2 \frac{p}{\rho^2}\frac{\partial \rho}{\partial \xi}$$

Substituting the derivative values, we obtain

$$\frac{\partial T}{\partial \xi} = \frac{M_\infty^2}{\rho}\left[\frac{\gamma-1}{2}\left(\frac{L_1}{U - c\sqrt{\xi_x^2+\xi_y^2+\xi_z^2}} + \frac{L_5}{U + c\sqrt{\xi_x^2+\xi_y^2+\xi_z^2}}\right) - \frac{L_2}{U}\right].$$

Now, we can combine the equations

$$\frac{\partial \rho}{\partial \tau} + \frac{1}{c^2}\left(\frac{L_1+L_5}{2} + L_2\right) = JS_{C1}$$

$$\frac{\partial p}{\partial \tau} + \frac{1}{2}(L_1 + L_5) = JS_{C5}$$

to obtain

$$\frac{\partial T}{\partial \tau} + \frac{M_\infty^2}{\rho}\left[(\gamma-1)\frac{L_1+L_5}{2} - L_2\right] = \frac{M_\infty^2}{\rho}\left(\gamma JS_{C5} - c^2 JS_{C1}\right).$$

At the no-slip solid boundary, we consider that the temperature is constant. Then,

$$L_2 = (\gamma - 1)\frac{L_1 + L_5}{2} - \left(\gamma J S_{C5} - c^2 J S_{C1}\right).$$

In the adiabatic wall case,

$$\frac{\partial T}{\partial \xi} = 0,$$

and so

$$L_2 = \frac{U}{2}(\gamma - 1)\left(\frac{L_1}{U - c\sqrt{\xi_x^2 + \xi_y^2 + \xi_z^2}} + \frac{L_5}{U + c\sqrt{\xi_x^2 + \xi_y^2 + \xi_z^2}}\right).$$

For all the next cases, we will use the Local One-Dimensional Inviscid (LODI) relations:

$$\begin{aligned}
&\frac{\partial \rho}{\partial \tau} + \frac{1}{c^2}\left(\frac{L_1 + L_5}{2} + L_2\right) = 0\\
&\frac{\partial u}{\partial \tau} + \frac{\xi_x}{2\rho c\sqrt{\xi_x^2 + \xi_y^2 + \xi_z^2}}\left(L_5 - L_1\right) - \frac{1}{\xi_x^2 + \xi_y^2 + \xi_z^2}\left(\xi_y L_3 + \xi_z L_4\right) = 0\\
&\frac{\partial v}{\partial \tau} + \frac{\xi_y}{2\rho c\sqrt{\xi_x^2 + \xi_y^2 + \xi_z^2}}\left(L_5 - L_1\right) + \frac{1}{\xi_x\left(\xi_x^2 + \xi_y^2 + \xi_z^2\right)}\left(\left(\xi_x^2 + \xi_z^2\right)L_3 - \xi_y \xi_z L_4\right) = 0\\
&\frac{\partial w}{\partial \tau} + \frac{\xi_z}{2\rho c\sqrt{\xi_x^2 + \xi_y^2 + \xi_z^2}}\left(L_5 - L_1\right) - \frac{1}{\xi_x\left(\xi_x^2 + \xi_y^2 + \xi_z^2\right)}\left(\xi_y \xi_z L_3 - \left(\xi_x^2 + \xi_y^2\right)L_4\right) = 0\\
&\frac{\partial p}{\partial \tau} + \frac{1}{2}\left(L_1 + L_5\right) = 0
\end{aligned} \qquad \textbf{(2.21)}$$

where the right-hand side of the equations are neglected, that is, there are no viscous nor transverse terms.

2.3.2.2. Subsonic Inflow

At $\xi = \xi_{\min}$, $\lambda_1 = U - c\sqrt{\xi_x^2 + \xi_y^2 + \xi_z^2} < 0$, $\lambda_2 = \lambda_3 = \lambda_4 = U > 0$, $\lambda_5 = U + c\sqrt{\xi_x^2 + \xi_y^2 + \xi_z^2} > 0$. Therefore, all values of L_i but L_1 must be specified from boundary conditions. All initial data must specify subsonic inflow conditions.

Then, for nonreflective boundary condition in the inflow, we need

$$L_5 = 0$$

Also, assuming that the inflow entropy is constant in the normal direction, and to set the transverse direction velocities as constants, then we take

$$L_2 = 0$$
$$L_3 = 0$$
$$L_4 = 0$$

and

$$JL_1 = \left(U - c\sqrt{\xi_x^2 + \xi_y^2 + \xi_z^2}\right)\left(-\frac{\rho c}{\sqrt{\xi_x^2 + \xi_y^2 + \xi_z^2}}\left(\xi_x \frac{\partial u}{\partial \xi} + \xi_y \frac{\partial v}{\partial \xi} + \xi_z \frac{\partial w}{\partial \xi}\right) + \frac{\partial p}{\partial \xi}\right).$$

Similarly, at $\xi = \xi_{\max}$, $\lambda_1 = U - c\sqrt{\xi_x^2 + \xi_y^2 + \xi_z^2} < 0$, $\lambda_2 = \lambda_3 = \lambda_4 = U < 0$, $\lambda_5 = U + c\sqrt{\xi_x^2 + \xi_y^2 + \xi_z^2} > 0$. Therefore, all values of L_i but L_5 must be specified from boundary conditions. All initial data must specify subsonic inflow conditions.

Then, for non-reflective boundary condition in the inflow, we need

$$L_1 = 0$$

Also, for similar reasons as above,

$$L_2 = 0$$
$$L_3 = 0$$
$$L_4 = 0$$

and

$$JL_5 = \left(U + c\sqrt{\xi_x^2 + \xi_y^2 + \xi_z^2}\right)\left(\frac{\rho c}{\sqrt{\xi_x^2 + \xi_y^2 + \xi_z^2}}\left(\xi_x \frac{\partial u}{\partial \xi} + \xi_y \frac{\partial v}{\partial \xi} + \xi_z \frac{\partial w}{\partial \xi}\right) + \frac{\partial p}{\partial \xi}\right).$$

2.3.2.3. Supersonic Inflow

At $\xi = \xi_{\min}$, $\lambda_1 = U - c\sqrt{\xi_x^2 + \xi_y^2 + \xi_z^2} < 0$, $\lambda_2 = \lambda_3 = \lambda_4 = U < 0$ $\lambda_5 = U + c\sqrt{\xi_x^2 + \xi_y^2 + \xi_z^2} > 0$. Therefore, all values of L_i must be specified from boundary conditions. All initial data must specify supersonic inflow conditions.

Also, at $\xi = \xi_{\max}$, $\lambda_1 = U - c\sqrt{\xi_x^2 + \xi_y^2 + \xi_z^2} < 0$, $\lambda_2 = \lambda_3 = \lambda_4 = U < 0$ $\lambda_5 = U + c\sqrt{\xi_x^2 + \xi_y^2 + \xi_z^2} < 0$.

Therefore, all values of L_i for both positions must be specified from boundary conditions, with

$$\begin{aligned} L_1 &= 0 \\ L_2 &= 0 \\ L_3 &= 0 \\ L_4 &= 0 \\ L_5 &= 0 \end{aligned}$$

2.3.2.4. Subsonic Outflow

In this case, at $\xi = \xi_{\min}$, $\lambda_1 = U - c\sqrt{\xi_x^2 + \xi_y^2 + \xi_z^2} < 0$, $\lambda_2 = \lambda_3 = \lambda_4 = U < 0$ $\lambda_5 = U + c\sqrt{\xi_x^2 + \xi_y^2 + \xi_z^2} > 0$. Therefore, all values of L_i but L_5must be specified by the given equations from the interior, to use one-sided derivatives. L_5 must be specified by a boundary condition. All initial data must specify subsonic outflow conditions.

Then,

$$JL_1 = \left(U - c\sqrt{\xi_x^2 + \xi_y^2 + \xi_z^2}\right)\left(-\frac{\rho c}{\sqrt{\xi_x^2 + \xi_y^2 + \xi_z^2}}\left(\xi_x \frac{\partial u}{\partial \xi} + \xi_y \frac{\partial v}{\partial \xi} + \xi_z \frac{\partial w}{\partial \xi}\right) + \frac{\partial p}{\partial \xi}\right)$$

$$JL_2 = U\left(c^2 \frac{\partial \rho}{\partial \xi} - \frac{\partial p}{\partial \xi}\right)$$

$$JL_3 = U\left(-\xi_y \frac{\partial u}{\partial \xi} + \xi_x \frac{\partial v}{\partial \xi}\right)$$

$$JL_4 = U\left(-\xi_z \frac{\partial u}{\partial \xi} + \xi_x \frac{\partial w}{\partial \xi}\right)$$

$$L_5 = 0.$$

Similarly, at $\xi = \xi_{\max}, \lambda_1 = U - c\sqrt{\xi_x^2 + \xi_y^2 + \xi_z^2} < 0$, $\lambda_2 = \lambda_3 = \lambda_4 = U > 0$, $\lambda_5 = U + c\sqrt{\xi_x^2 + \xi_y^2 + \xi_z^2} > 0$. Then, all values of L_i but L_1 are specified from the interior equations. L_1 is specified by a boundary condition.

Therefore,

$$L_1 = 0$$

$$JL_2 = U\left(c^2 \frac{\partial \rho}{\partial \xi} - \frac{\partial p}{\partial \xi}\right)$$

$$JL_3 = U\left(-\xi_y \frac{\partial u}{\partial \xi} + \xi_x \frac{\partial v}{\partial \xi}\right)$$

$$JL_4 = U\left(-\xi_z \frac{\partial u}{\partial \xi} + \xi_x \frac{\partial w}{\partial \xi}\right)$$

$$JL_5 = \left(U + c\sqrt{\xi_x^2 + \xi_y^2 + \xi_z^2}\right)\left(\frac{\rho c}{\sqrt{\xi_x^2 + \xi_y^2 + \xi_z^2}}\left(\xi_x \frac{\partial u}{\partial \xi} + \xi_y \frac{\partial v}{\partial \xi} + \xi_z \frac{\partial w}{\partial \xi}\right) + \frac{\partial p}{\partial \xi}\right).$$

2.3.2.5. Supersonic Outflow

For supersonic outflows, both at $\xi = \xi_{\min}$ and $\xi = \xi_{\max}$, all L_i values are calculated by the equations from the interior, since, at $\xi = \xi_{\min}$, $\lambda_1 = U - c\sqrt{\xi_x^2 + \xi_y^2 + \xi_z^2} < 0$, $\lambda_2 = \lambda_3 = \lambda_4 = U < 0$, $\lambda_5 = U + c\sqrt{\xi_x^2 + \xi_y^2 + \xi_z^2} < 0$, and at $\xi = \xi_{\max}$, $\lambda_1 = U - c\sqrt{\xi_x^2 + \xi_y^2 + \xi_z^2} > 0$, $\lambda_2 = \lambda_3 = \lambda_4 = U > 0$, $\lambda_5 = U + c\sqrt{\xi_x^2 + \xi_y^2 + \xi_z^2} > 0$.

Therefore,

$$JL_1 = \left(U - c\sqrt{\xi_x^2 + \xi_y^2 + \xi_z^2}\right)\left(-\frac{\rho c}{\sqrt{\xi_x^2 + \xi_y^2 + \xi_z^2}}\left(\xi_x \frac{\partial u}{\partial \xi} + \xi_y \frac{\partial v}{\partial \xi} + \xi_z \frac{\partial w}{\partial \xi}\right) + \frac{\partial p}{\partial \xi}\right)$$

$$JL_2 = U\left(c^2 \frac{\partial \rho}{\partial \xi} - \frac{\partial p}{\partial \xi}\right)$$

$$JL_3 = U\left(-\xi_y \frac{\partial u}{\partial \xi} + \xi_x \frac{\partial v}{\partial \xi}\right)$$

$$JL_4 = U\left(-\xi_z \frac{\partial u}{\partial \xi} + \xi_x \frac{\partial w}{\partial \xi}\right)$$

$$JL_5 = \left(U + c\sqrt{\xi_x^2 + \xi_y^2 + \xi_z^2}\right)\left(\frac{\rho c}{\sqrt{\xi_x^2 + \xi_y^2 + \xi_z^2}}\left(\xi_x \frac{\partial u}{\partial \xi} + \xi_y \frac{\partial v}{\partial \xi} + \xi_z \frac{\partial w}{\partial \xi}\right) + \frac{\partial p}{\partial \xi}\right).$$

SUMMARY

The non-dimensional time-dependent Navier-Stokes equations in curvilinear coordinates are given in details. The flux splitting scheme and non-reflecting boundary conditions are discussed and provided.

CHAPTER 3

Orthogonal Grid Generation

Abstract: In this chapter, a brief introduction of algebraic grid generation, transfinite interpolation and elliptic grid generation is given. In particular, a two-step elliptic grid generation method developed by Spekreijse [121] is introduced, which can generate high quality and orthogonal grids near the boundary. The smooth and orthogonal grid for the LES case of MVG and Ramp is generated by the above method.

Keywords: Algebraic grid generation, Elliptic grid generation, Orthogonal grid generation, Transfinite interpolation.

Grid generation in general means to give grid point position in a computational domain, for example $x(i, j)$, $y(i, j)$ for a 2-D computational domain. This is usually performed by giving boundary grid points such as $x(0, j)$, $y(0, j)$ $x(ni, j)$, $y(ni, j)$ $x(i, 0)$, $y(i, 0)$ $x(i, nj)$, $y(i, nj)$) to get all interior grid points such as $x(i, j)$, $y(i, j)$, $i = 1, ni - 1, j = 1, nj - 1$. This can be done by algebraic grid generation and elliptic grid generation. Here, we only limit our focus on structured body-fitted elliptic grid (shown in Fig. **3.1**) generation while we skip all other grid generation methods.

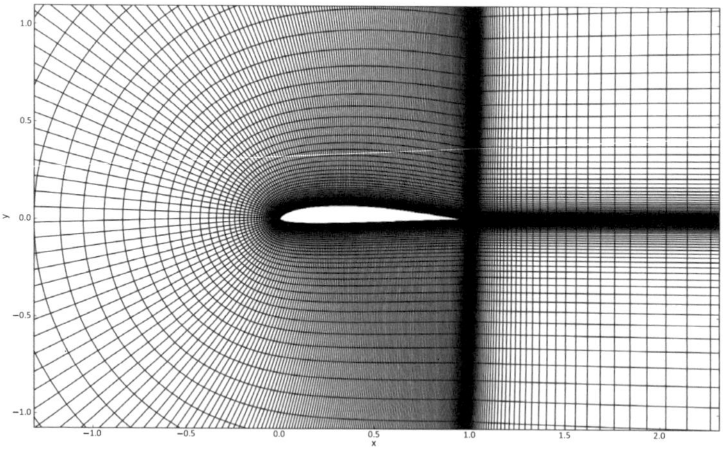

Fig. (3.1). Body-fitted grid generation.

Chaoqun Liu, Qin Li, Yonghua Yan, Yong Yang, Guang Yang, Xiangrui Dong

3.1. ALGEBRAIC GRID GENERATION

3.1.1. Algebraic Formula

The grid can be generated by a given function as an algebraic formula for 1D, 2D or 3D domain. The physical domain is in general not rectangular, but the computational domain can be rectangular through the transformation by the algebraic formula. Let us take a 2-D example (Fig. **3.2**). Assume the bottom line of a channel is a straight line, but the top shape is specified as $y = g(x)$. We can use an algebraic formula to produce a 2-D grids:

$$\begin{aligned}&\xi = x\\&\eta = \frac{y}{g(x)} \quad \text{where } \xi_i = 1 + i\frac{2-1}{ni},\ \eta_j = \frac{j}{nj}\end{aligned} \tag{3.1}$$

Although the physical domain is curved, the computational domain (ξ, η) is rectangular and we can get $x(i, j)$, $y(i, j)$ by a simple algebraic formula. This is one-to-one mapping from physical domain to computational domain (see Figs. **3.2b** and **3.2c**).

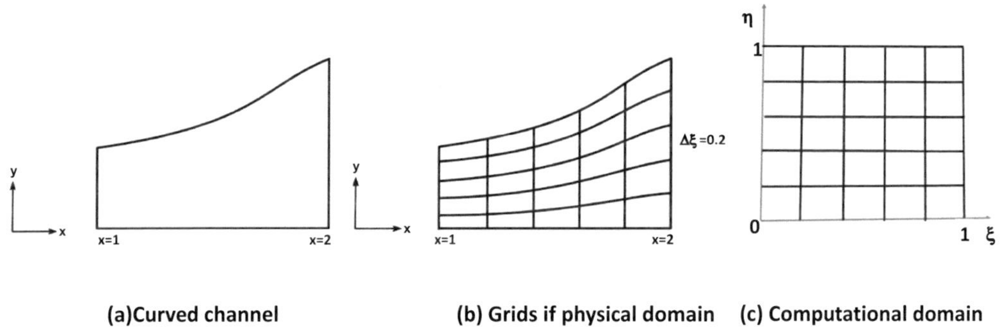

Fig. (3.2). Algebraic grid generation for a channel.

3.1.2. Transfinite Interpolation

For elliptic grid generation, in general boundary grid points must be given by the CFD users. The remained task is find the interior grids. This is also can be done by a called transfinite interpolation which was introduced by Thompson [122, 123]. Fig. (**3.3**) shows an irregular domain with curved boundaries $E1$, $E2$, $E3$, $E4$ in the right side and the left side represents a rectangular computational domain. We must give the grid point position in boundaries.

Assume:

$$
\begin{aligned}
x(0, j) = f_1(j), y(0, j) = g_1(j) \quad on\ E1 \\
x(ni, j) = f_2(j), y(ni, j) = g_2(j) \quad on\ E2 \\
x(i, 0) = f_3(i), y(i,0) = g_3(i) \quad on\ E3 \\
x(i, nj) = f_4(i), y(i,nj) = g_4(nj) \quad on\ E4
\end{aligned} \tag{3.2}
$$

Here, *f* and *g* are given functions for the boundary points.

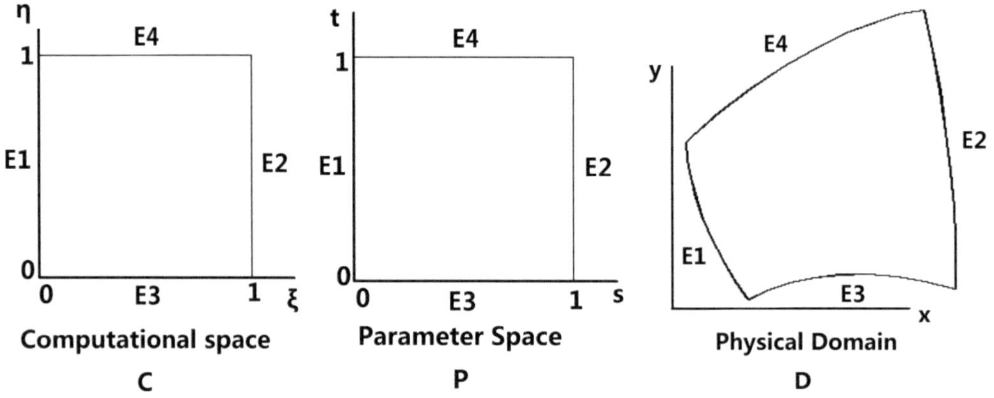

Fig. (3.3). Computational, parameter and physical domains.

All the grid points inside the computational domain can be given by the following transfinite interpolation:

$$
\begin{aligned}
x(i,j) = {} & \frac{i-1}{ni-1}x(ni,j) + \frac{ni-i}{ni-1}x(0,j) + \frac{j-1}{nj-1}x(i,nj) + \frac{nj-j}{nj-1}x(i,0) \\
& - \frac{i-1}{ni-1}\frac{j-1}{nj-1}x(ni,nj) - \frac{i-1}{ni-1}\frac{nj-j}{nj-1}x(ni,0) - \frac{ni-i}{ni-1}\frac{j-1}{nj-1}x(0,nj) - \frac{ni-i}{ni-1}\frac{nj-j}{nj-1}x(0,0) \\
y(i,j) = {} & \frac{i-1}{ni-1}y(ni,j) + \frac{ni-i}{ni-1}y(0,j) + \frac{j-1}{nj-1}y(i,nj) + \frac{nj-j}{nj-1}y(i,0) \\
& - \frac{i-1}{ni-1}\frac{j-1}{nj-1}y(ni,nj) - \frac{i-1}{ni-1}\frac{nj-j}{nj-1}y(ni,0) - \frac{ni-i}{ni-1}\frac{j-1}{nj-1}y(0,nj) - \frac{ni-i}{ni-1}\frac{nj-j}{nj-1}y(0,0)
\end{aligned} \tag{3.3}
$$

where $i = 0, ni$ and $j = 0, nj$

Note that all boundary points are given by f and g. We can rewrite the interpolation as follows:

$$\begin{aligned}
x(i,j) &= \frac{i-1}{ni-1} f_2(j) + \frac{ni-i}{ni-1} f_1(j) + \frac{j-1}{nj-1} f_4(i) + \frac{nj-j}{nj-1} f_3(i) \\
&- \frac{i-1}{ni-1}\frac{j-1}{nj-1} f_4(ni) - \frac{i-1}{ni-1}\frac{nj-j}{nj-1} f_3(ni) - \frac{ni-i}{ni-1}\frac{j-1}{nj-1} f_4(0) - \frac{ni-i}{ni-1}\frac{nj-j}{nj-1} f_3(0) \\
y(i,j) &= \frac{i-1}{ni-1} g_2(j) + \frac{ni-i}{ni-1} g_1(j) + \frac{j-1}{nj-1} g_4(i) + \frac{nj-j}{nj-1} g_3(i) \\
&- \frac{i-1}{ni-1}\frac{j-1}{nj-1} g_4(ni) - \frac{i-1}{ni-1}\frac{nj-j}{nj-1} g_3(ni) - \frac{ni-i}{ni-1}\frac{j-1}{nj-1} g_4(0) - \frac{ni-i}{ni-1}\frac{nj-j}{nj-1} g_3(0)
\end{aligned} \tag{3.4}$$

In this way, all interior grid points can be obtained by a very simple formula. However, the grids generated by transfinite interpolation may not be smooth and may have low quality. However, we can use it as an initial guess for an elliptic grid generation.

3.2. ELLIPTIC GRID GENERATION

3.2.1. Governing Equations

Assume the physical domain is *(x, y)* and the computational domain is (ξ, η). Points in the physical domain and computational domain should be one-to-one mapping. A Poisson's equation will give a unique solution of $\xi = \xi(x, y)$ and $\eta = \eta(x, y)$ or $x = x(\xi, \eta)$ and $y = y(\xi, \eta)$ if the boundary condition are given properly. The Poisson's equation can be written as:

$$\begin{aligned}
\nabla^2 \xi &= P(\xi,\eta) \\
\nabla^2 \eta &= Q(\xi,\eta)
\end{aligned}
\quad \text{or} \quad
\begin{aligned}
\frac{\partial^2 \xi}{\partial x^2} + \frac{\partial^2 \xi}{\partial y^2} &= P(\xi,\eta) \\
\frac{\partial^2 \eta}{\partial x^2} + \frac{\partial^2 \eta}{\partial y^2} &= Q(\xi,\eta)
\end{aligned} \tag{3.5}$$

However, we want to use the computational domain which is uniform and easy to do finite difference.

Let us take $\frac{\partial}{\partial \xi}$ *and* $\frac{\partial}{\partial \eta}$ to $\xi = \xi(x, y)$ and $\eta = \eta(x, y)$

$$\begin{aligned}1 &= \xi_x x_\xi + \xi_y y_\xi \\ 0 &= \xi_x x_\eta + \xi_y y_\eta\end{aligned} \quad \text{or} \quad \begin{bmatrix} x_\xi & y_\xi \\ x_\eta & y_\eta \end{bmatrix} \begin{bmatrix} \xi_x \\ \xi_y \end{bmatrix} = \begin{bmatrix} 1 \\ 0 \end{bmatrix}$$

Solve the system, we can get $\xi_x = \frac{1}{J} y_\eta, \xi_y = -\frac{1}{J} x_\eta$ and similarly, $\eta_x = -\frac{1}{J} y_\xi, \eta_y = \frac{1}{J} x_\xi$, where J is Jacobian and

$$J = x_\xi y_\eta - x_\eta y_\xi .$$

Continue to do the derivation in space *(x, y)*,

$$\xi_{xx} = (\frac{1}{J} y_\eta)_x = (\frac{1}{J} y_\eta)_\xi \cdot \xi_x + (\frac{1}{J} y_\eta)_\eta \cdot \eta_x$$

Because $(\frac{1}{J})_\xi = -\frac{J_\xi}{J^2} = -\frac{x_{\xi\xi} y_\eta + x_\xi y_{\xi\eta} - x_{\eta\xi} y_\xi - x_\eta y_{\xi\xi}}{J^2}$

We can get

$$\begin{aligned}\xi_{xx} = & [\frac{1}{J} y_{\xi\eta} - \frac{y_\eta}{J^2}(x_{\xi\xi} y_\eta + x_\xi y_{\varsigma\eta} - x_{\eta\xi} y_\xi - x_\eta y_{\xi\xi}] \cdot \frac{1}{J} y_\eta \\ & + [\frac{1}{J} y_{\eta\eta} - \frac{y_\eta}{J^2}(x_{\xi\eta} y_\eta + x_\xi y_{\eta\eta} - x_{\eta\eta} y_\xi - x_\eta y_{\xi\eta}] \cdot (-\frac{1}{J} y_\xi)\end{aligned}$$

$$\begin{aligned}\xi_{xx} = & \frac{1}{J^3}[-(y_\eta)^3 x_{\xi\xi} + 2 y_\xi (y_\eta)^2 x_{\xi\eta} - (y_\xi)^2 y_\eta x_{\eta\eta} + x_\eta (y_\eta)^2 y_{\varsigma\xi} - y_{\xi\eta}(x_\xi (y_\eta)^2 - x_\eta y_\xi y_\eta - x_\xi (y_\eta)^2 - x_\eta y_\xi y_\eta) \\ & + y_{\eta\eta}(-x_\xi y_\xi y_\eta + x_\eta (y_\xi)^2 + y_\eta x_\xi y_\xi]\end{aligned}$$

$$\xi_{xx} = \frac{1}{J^3}[-(y_\eta)^3 x_{\xi\xi} + 2 y_\xi (y_\eta)^2 x_{\xi\eta} - (y_\xi)^2 y_\eta x_{\eta\eta} + x_\eta (y_\eta)^2 y_{\varsigma\xi} - 2 y_{\xi\eta} x_\eta y_\xi y_\eta + y_{\eta\eta} x_\eta (y_\xi)^2$$

$$\xi_{xx} = \frac{y_\eta}{J^3}[-(y_\eta)^2 x_{\xi\xi} + 2 y_\xi y_\eta x_{\xi\eta} - (y_\xi)^2 x_{\eta\eta}] + \frac{x_\eta}{J^3}[(y_\eta)^2 y_{\varsigma\xi} - 2 y_{\xi\eta} y_\xi y_\eta + y_{\eta\eta}(y_\xi)^2]$$

In a similar way, we can obtain

$$\xi_{yy} = \frac{y_\eta}{J^3}[-(x_\eta)^2 x_{\xi\xi} + 2x_\xi x_\eta x_{\xi\eta} - (x_\xi)^2 x_{\eta\eta}] + \frac{x_\eta}{J^3}[(x_\eta)^2 y_{\xi\xi} - 2x_{\xi\eta} y_\xi y_\eta + y_{\eta\eta}(x_\xi)^2]$$

Merge the two terms,

$$\xi_{xx} + \xi_{yy} = -\frac{y_\eta}{J^3}\{[(x_\eta)^2 + (y_\eta)^2]x_{\xi\xi} - 2(x_\xi x_\eta + y_\xi y_\eta)x_{\xi\eta} + [(x_\xi)^2 + (y_\xi)^2]x_{\eta\eta}\}$$
$$+ \frac{x_\eta}{J^3}\{[(x_\eta)^2 + (y_\eta)^2]y_{\xi\xi} - 2(x_\xi x_\eta + y_\xi y_\eta)y_{\xi\eta} + [(x_\xi)^2 + (y_\xi)^2]y_{\eta\eta}\} = P(\xi,\eta)$$

Let $\alpha = (x_\eta)^2 + (y_\eta)^2$, $\gamma = (x_\xi)^2 + (y_\xi)^2$, $\beta = x_\xi x_\eta + y_\xi y_\eta$, we can get

$$-\frac{y_\eta}{J^3}(\alpha x_{\xi\xi} - 2\beta x_{\xi\eta} + \gamma x_{\eta\eta}) + \frac{x_\eta}{J^3}(\alpha y_{\xi\xi} - 2\beta y_{\xi\eta} + \gamma y_{\eta\eta}) = P(\xi,\eta) \qquad \textbf{(3.6)}$$

Similarly, from $\eta_{xx} + \eta_{yy} = Q(\xi,\eta)$, we can get

$$\frac{y_\xi}{J^3}(\alpha x_{\xi\xi} - 2\beta x_{\xi\eta} + \gamma x_{\eta\eta}) - \frac{x_\xi}{J^3}(\alpha y_{\xi\xi} - 2\beta y_{\xi\eta} + \gamma y_{\eta\eta}) = Q(\xi,\eta) \qquad \textbf{(3.7)}$$

Use $x_\xi \times (3.6) + x_\eta \times (3.7)$

$$x_\xi y_\eta(\alpha x_{\xi\xi} - 2\beta x_{\xi\eta} + \gamma x_{\eta\eta}) + x_\xi x_\eta(\alpha y_{\xi\xi} - 2\beta y_{\xi\eta} + \gamma y_{\eta\eta})$$
$$- x_\eta y_\xi(\alpha x_{\xi\xi} - 2\beta x_{\xi\eta} + \gamma x_{\eta\eta}) - x_\xi x_\eta(\alpha y_{\xi\xi} - 2\beta y_{\xi\eta} + \gamma y_{\eta\eta}) = -J^3 x_\xi P(\xi,\eta) - J^3 x_\eta Q(\xi,\eta)$$

$$(x_\xi y_\eta - x_\eta y_\xi)(\alpha x_{\xi\xi} - 2\beta x_{\xi\eta} + \gamma x_{\eta\eta}) = -J^3 x_\xi P(\xi,\eta) - J^3 x_\eta Q(\xi,\eta)$$

or

$$\alpha x_{\xi\xi} - 2\beta x_{\xi\eta} + \gamma x_{\eta\eta} = -J^2[x_\xi P(\xi,\eta) - x_\eta Q(\xi,\eta)]$$

And similarly

$$\alpha y_{\xi\xi} - 2\beta y_{\xi\eta} + \gamma y_{\eta\eta} = -J^2[y_\xi P(\xi,\eta) - y_\eta Q(\xi,\eta)] \qquad \textbf{(3.8)}$$

3.2.2. Control Function

In (3.8) $P(\xi, \eta)$ and $Q(\xi, \eta)$ are control functions which can generate clustered or stretched grids. If $P=0$ and $Q=0$, the elliptic grid generation equations becomes to

$$\begin{aligned} \alpha x_{\xi\xi} - 2\beta x_{\xi\eta} + \gamma x_{\eta\eta} &= 0 \\ \alpha y_{\xi\xi} - 2\beta y_{\xi\eta} + \gamma y_{\eta\eta} &= 0 \end{aligned} \quad \textbf{(3.9)}$$

which is corresponding to

$$\begin{aligned} \xi_{xx} + \xi_{yy} &= 0 \\ \eta_{xx} + \eta_{yy} &= 0 \end{aligned} \quad \textbf{(3.10)}$$

This is a Laplace equation which will give a smooth grid distribution (harmonic function) to resist any stretching and clustering (Fig. **3.4**).

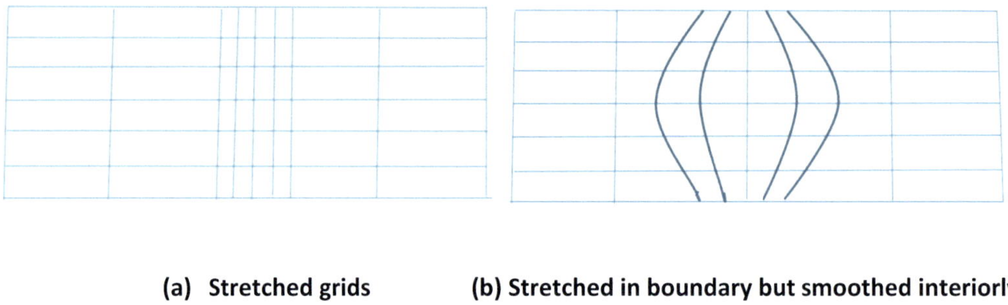

Fig. (3.4). Stretched and smooth grids.

In order to get a clustered grids, we can select proper control function P and Q. For example, we may select

$$P(\xi,\eta) = -\sum_{i=1}^{n} a_i \operatorname{sgn}(\xi-\xi_i)\exp(-c_i|\xi-\xi_i|) - \sum_{j=1}^{m} b_j \operatorname{sgn}(\xi-\xi_i)\exp(-d_j\sqrt{(\xi-\xi_i)^2+(\eta-\eta_j)^2} \quad \textbf{(3.11)}$$

It will generate grids which will be clustered to line $\xi = \xi_i$ and point (ξ_i, η_j) (see Fig. **3.5**). Here a_i, b_i represent the amplitude of attraction, c_i, d_i are decay factor with distance, $|\xi - \xi_i|$ and $|\eta - \eta_i|$ are distances, n is the number of lines for attraction and m is the number of points for attraction.

(a) Attract grid lines to $\xi = \xi_i$ **(b) Attract grid lines to point** (ξ_i, η_j)

Fig. (3.5). Use control function to make grid clustering.

Similarly, we can use

$$Q(\xi,\eta) = -\sum_{j=1}^{n} a_j \,\mathrm{sgn}(\eta-\eta_j)\exp(-c_j|\eta-\eta_j|) - \sum_{i=1}^{m} b_i \,\mathrm{sgn}(\eta-\eta_j)\exp(-d_i\sqrt{(\xi-\xi_i)^2+(\eta-\eta_j)^2} \quad \textbf{(3.12)}$$

to control the grid clustering in the η direction.

However, this kind of exponential control function is hard to make idealized grid stretching and clustering. We prefer to use a two-step elliptic grid generation developed by Spekreijse [121].

3.3. TWO-STEP ELLIPTIC GRID GENERATION (SPEKREIJSE, 1995)

There are so many progresses in grid generation recently and we have no intention to give a full introduction about these methods and we just focus on the elliptic grid generation methods which are used by UTA large eddy simulation.

3.3.1. Two-dimensional Grid Generation

An elliptic grid generation method first proposed by Spekreijse is introduced here. The elliptic grid generation method is based on a composite mapping, which consists of a non-linear transfinite algebraic transformation and an elliptic transformation. The algebraic transformation maps the computational space C onto a parameter space P, and the elliptic transformation maps the parameter space on to the physical domain D. The computational space, parameter space and the physical domain are illustrated in Fig. (**3.6**).

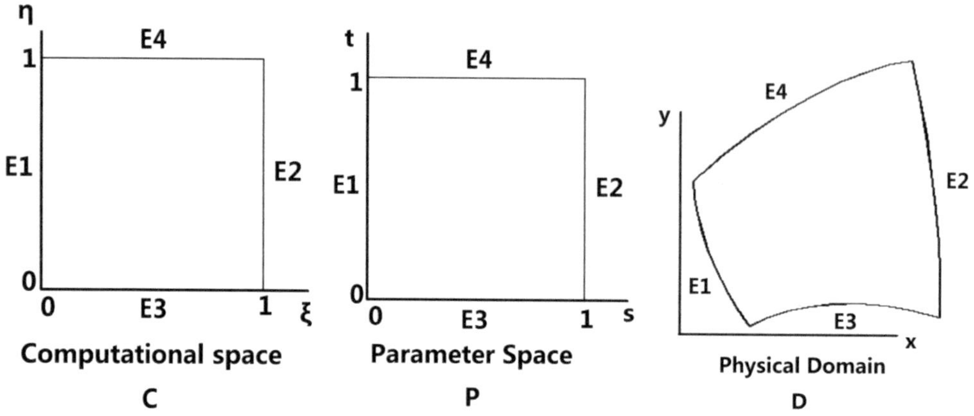

Fig. (3.6). Computational space, parameter space and the physical domain.

The computational space C is defined as the unit square in a two-dimensional space with Cartesian coordinates (ξ, η), and $\xi \in [0, 1], \eta \in [0, 1]$. The grids are uniformly distributed on the boundaries and in the interior domain of the computational space. The mesh sizes are $\frac{1}{N_\xi - 1}$ in the η direction and $\frac{1}{N_\eta - 1}$ in the η direction, where ξ and η are the grid numbers in the corresponding direction. The parameter space P is defined as a unit 2-D space with Cartesian coordinate $(s, t), and\ s \in [0, 1], t \in [0, 1]$. The boundary values of s and t are determined by the grid point distribution is the physical domain.

Consider a simply connected bounded domain D in two-dim space with Cartesian coordinates $\boldsymbol{x}=(x,y)^{\mathrm{T}}$. Suppose D is bounded with E1, E2, E3, E4 edges, which is shown in Fig. (**3.7**).

Defined computational space C as the unit square in a 2-D space with Cartesian coordinate $\boldsymbol{\xi} = (\xi, \eta)^{\mathrm{T}}$.

We wish to construct a mapping $\boldsymbol{x}$: C $|\!\rightarrow$ D such that it is one-to-one and it maps the boundary of C to the boundary of D. And it satisfies the following conditions:

1, $\xi \equiv 0\ at\ edgeE1$ and $\xi \equiv 1\ at\ edgeE2$,

2, $\eta \equiv 0\ at\ edge, E3$ and $\eta \equiv 1\ at\ edgeE4$.

This mapping will be the composition of algebraic transformation and elliptic transformation.

Parameter P space is also a unit square in a 2-D space with Cartesian coordinate $\mathbf{s} = (s, t)^T$. In the space, we have the following boundary conditions:

1, $s \equiv 0$ *at edgeE*1 and $s \equiv 1$ *at edgeE*2,

2, $t \equiv 0$ *at edge, E*3 and $t \equiv 1$ *at edgeE*4.

s is the normalized arclength along edges along *E3* and *E4*.

t is the normalized arclength along edges along *E1* and *E2*.

Of course, the value of *s, t* are determined by the grid point distribution in the physical domain D.

First we need an algebraic transformation: computational space C → parameter space P

It is differentiable, one-to-one and depends on the prescribed boundary grid points distribution at the four edges of the domain D.

The second step is an elliptic grid transformation:

parameter space P → physical domain D

It is also differentiable one-to-one and depends on the shape of domain D and is thus independent of the prescribed boundary grid points distribution. It is considered to preserve the property of the domain D.

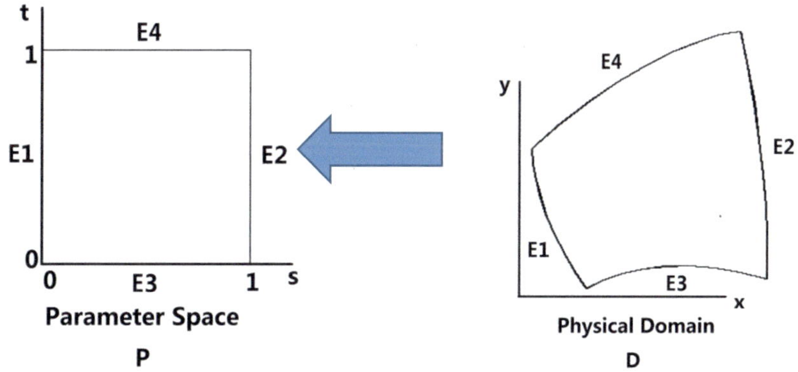

Fig. (3.7). Elliptic grid generation from parameter space P to physical domain D.

How can we get boundary values of s and t? It is determined by the grid point distribution in the physical domain D.

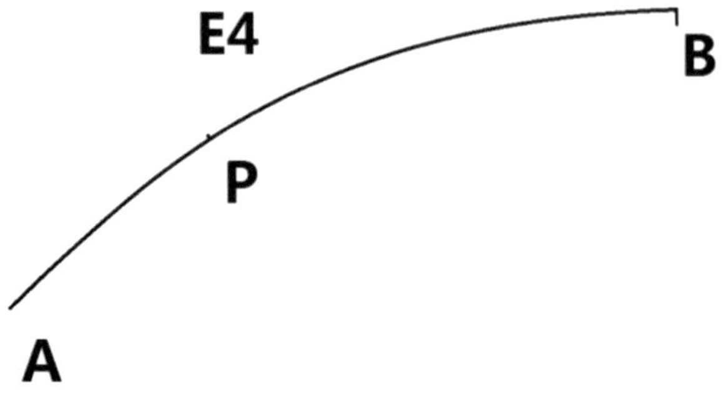

Fig. (3.8). Boundary value of s in E_4.

The boundary points can be defined as follows:

$s = 0$ *at edge E*1 and $s = 1$ *at edge E*2

$t = 0$ *at edge E*3 and $t = 1$ *at edge E*4

At a point p on E_4 $s(p) = \dfrac{\overline{AP}}{\overline{AB}}$ and similar to other boundary points (Fig. **3.8**).

In this way, we can define the algebraic transformation:

$$\begin{aligned} t &= t_{E_1}(\eta)(1-s) + t_{E_2}(\eta)s \\ s &= s_{E_3}(\xi)(1-t) + s_{E_4}(\xi)t \end{aligned} \qquad \textbf{(3.13)}$$

$s_{E_3}(\xi) = s(\xi, 0)$ is normalized arc-length along edges E3

$s_{E_4}(\xi) = s(\xi, 1)$ is normalized arc-length along edges E3

$t_{E_1}(\eta) = t(0, \eta)$ is normalized arc-length along edges E1

$t_{E_2}(\eta) = t(1, \eta)$ is normalized arc-length along edges E2

Equation (3.13) is called the algebraic straight line transformation. It defines a differentiable one-to-one mapping because of Jacobian $s_\xi t_\eta - s_\eta t_\xi > 0$.

After define the algebraic transformation, we do the elliptic transformation: $P \rightarrow D$, which is independent of the prescribed boundary grid point distribution, is defined a mapping from parameter space *P* onto the physical domain *D*. Now, the elliptic transformation is a set of Laplace equations, and unlike previous section the control function is implicitly defined by the algebraic transformation and no explicit control function *P* and *Q* are needed.

$$\begin{aligned} \Delta t &= t_{xx} + t_{yy} = 0 \\ \Delta s &= s_{xx} + s_{yy} = 0 \end{aligned} \qquad \textbf{(3.14)}$$

The elliptic transformation defined by the above Laplace equations is also differentiable and one-to-one, till now we have defined two transformations: the algebraic transformation: C → P and the elliptic transformation: P → D. Because these two transformations are differential and one to one, the composition of these two transformation is also differential and one to one.

Let us look at the composite transformation. First, in the interior of D we require that *s* and *t* are harmonic functions of *x* and *y*, thus obey the Laplace equations:

$$\Delta s = s_{xx} + s_{yy} = 0 \qquad \Delta t = t_{xx} + t_{yy} = 0$$

We then transform the Laplace equations to the computational space:

$$\Delta s = g^{11} s_{\xi\xi} + 2g^{12} s_{\xi\eta} + g^{22} s_{\eta\eta} + \Delta\xi s_\xi + \Delta\eta s_\eta$$

$$\Delta t = g^{11} t_{\xi\xi} + 2g^{12} t_{\xi\eta} + g^{22} t_{\eta\eta} + \Delta\xi t_\xi + \Delta\eta t_\eta$$

where g^{11}, g^{12}, g^{22} are the components of the contravariant metric tensor, which can be calculated from the covariant metric tensor?

$$g^{11} = \frac{1}{J^2} g_{22} = (r_\eta, r_\eta)/J^2 \qquad g^{12} = -\frac{1}{J^2} g_{12} = -(r_\xi, r_\eta)/J^2$$

$$g^{22} = \frac{1}{J^2} g_{11} = (r_\xi, r_\xi)/J^2$$

$$J = \sqrt{\det |g_{ij}|}$$

$$r = \begin{pmatrix} x \\ y \end{pmatrix}$$

The $\Delta\xi$ and $\Delta\eta$ which determine the control function can be calculated by

$$\begin{pmatrix} \Delta\xi \\ \Delta\eta \end{pmatrix} = g^{11}P_{11} + 2g^{12}P_{12} + g^{22}P_{22}$$

$$P_{11} = \begin{pmatrix} p_{11}^{(1)} \\ p_{11}^{(2)} \end{pmatrix} = -T^{-1}\begin{pmatrix} s_{\xi\xi} \\ t_{\xi\xi} \end{pmatrix} \qquad P_{12} = \begin{pmatrix} p_{12}^{(1)} \\ p_{12}^{(2)} \end{pmatrix} = -T^{-1}\begin{pmatrix} s_{\xi\eta} \\ t_{\xi\eta} \end{pmatrix}$$

$$P_{22} = \begin{pmatrix} p_{22}^{(1)} \\ p_{22}^{(2)} \end{pmatrix} = -T^{-1}\begin{pmatrix} s_{\eta\eta} \\ t_{\eta\eta} \end{pmatrix} \qquad T = \begin{pmatrix} s_{\xi} & s_{\eta} \\ t_{\xi} & t_{\eta} \end{pmatrix}$$

In physical domain, the curvilinear coordinate system satisfies a system of Laplace equations:

$$\Delta r = 0 \qquad r = (x, y)^T$$

transformed to the computational space C:

$$g^{11}r_{\xi\xi} + 2g^{12}r_{\xi\eta} + g^{22}r_{\eta\eta} + \Delta\xi r_{\xi} + \Delta\eta r_{\eta} = 0$$

$$\begin{pmatrix} \Delta\xi \\ \Delta\eta \end{pmatrix} = g^{11}P_{11} + 2g^{12}P_{12} + g^{22}P_{22}$$

This will give a controlled elliptic grid generation equation:

$$\begin{aligned} & g^{11}r_{\xi\xi} + 2g^{12}r_{\xi\eta} + g^{22}r_{\eta\eta} + (g^{11}p_{11}^{(1)} + 2g^{12}p_{12}^{(1)} + g^{22}p_{22}^{(1)})r_{\xi} \\ & + (g^{11}p_{11}^{(2)} + 2g^{12}p_{12}^{(2)} + g^{22}p_{22}^{(2)})r_{\eta} = 0 \end{aligned} \qquad \textbf{(3.15)}$$

The elliptic transformation is carried out by solving a set of Poisson's equations. The control functions are specified by the algebraic transformation and it is, therefore, not needed to compute the control functions at the boundary and then to interpolate them to the interior points as required by standard well-known elliptic grid generation systems described by Poison's equations.

In general, the grids computed by the above method are stretched and smooth but not orthogonal at the boundary.

3.3.2. Orthogonal Grid Generation Near the Boundary

The algebraic transformations can be redefined to obtain a grid which is orthogonal at the boundary. First, redefine the elliptic transformation P → D by imposing following boundary conditions for s and t:

- $s = 0$ *at edgeE*1 and $s = 1$ *at edge E*2
- $t = 0$ *at edge E*3 and $t = 1$ *at edge E*4
- $\frac{\partial s}{\partial n} = 0$ along edges E3 and E4, where n is the outward normal direction
- $\frac{\partial t}{\partial n} = 0$ along edges E1 and E2, where n is the outward normal direction

Second redefine the algebraic transformation s: C → P

$$s = s_{E3}(\xi)H_0(t) + s_{E4}(\xi)H_1(t)$$

⇨ New s and t => New $\Delta\xi$and $\Delta\eta$=> New x and y **(3.16)**

$$t = t_{E1}(\eta)H_0(s) + t_{E2}(\eta)H_1(s)$$

Where H_0 and H_1 are cubic Hermite interpolation functions defined as

$$H_0(s) = (1+2s)(1-s)^2$$
$$H_1(s) = (3-2s)s^2$$

In summary, grid orthogonality at boundaries is obtained in three steps:

1. Compute an initial grid based on the Poisson's grid generation with control functions specified according to the algebraic straight line transformation defined in (3.15)

2. Solve the two Laplace equations given by (3.15) with the specified boundary conditions
3. Re-compute the grid based on the Poisson's system with control functions specified according to the algebraic transformation defined by (3.16).

A two-dimensional grid near the leading edge of a Joukowsky airfoil is given in Fig. (**3.9**). Detailed method can be found from the paper given by Spekreijse. Finally we have the grids that are smooth, stretched and orthogonal.

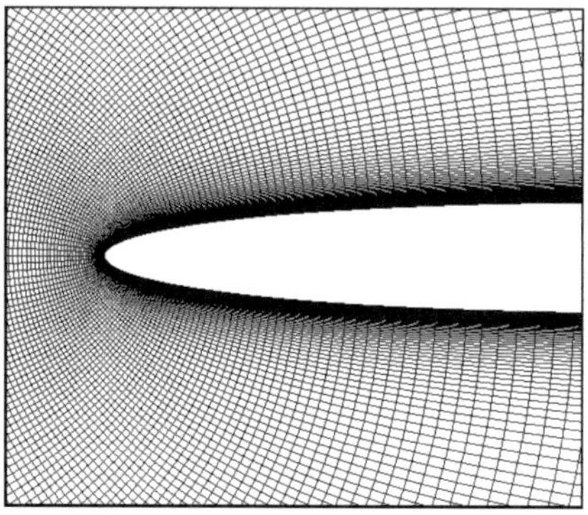

Fig. (3.9). Orthogonal grids near the leading edge of a Joukowsky airfoil.

3.3.3. Three-dimensional Grid Generation

The basic idea of three-dimensional grid generation is similar to that of the two-dimensional case.

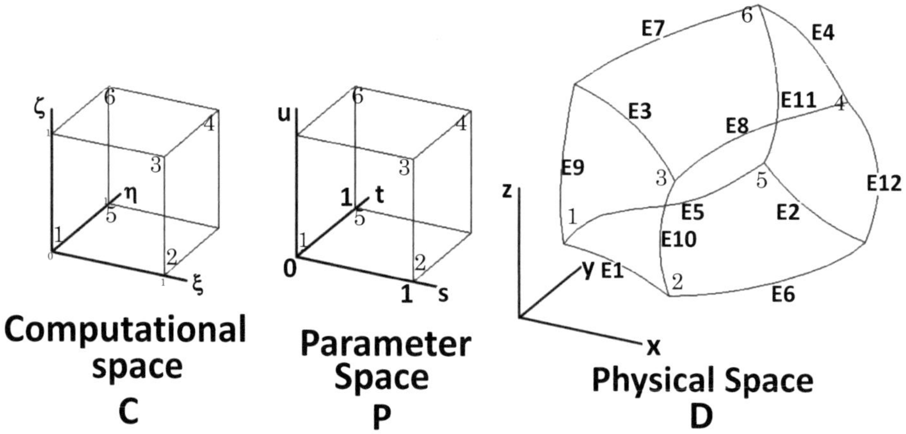

Fig. (3.10). Computational, parameter and physical domain.

The computational space is a unit cubic with $\xi \in [0, 1], \eta \in [0, 1], \varsigma \in [0, 1]$. The parameter space is a unit cubic with $s \in [0, 1], t \in [0, 1], u \in [0, 1]$ (see Fig. **3.10**).

- $s = 0$ at Face 1 and $s = 1$ at Face 2 where s is the normalized arc-length along edges E1, E2, E3, E4.
- $t = 0$ at Face 3 and $t = 1$ at Face 4 where t is the normalized arc-length along edges E5, E6, E7, E8.>
- $u = 0$ at Face 3 and $u = 1$ at Face 5 where u is the normalized arc-length along edges E9, E10, E11, E12.

Let $s_{E1}(\xi) = s(\xi,0,0), s_{E2}(\xi) = s(\xi,1,0), s_{E3}(\xi) = s(\xi,0,1), s_{E4}(\xi) = s(\xi,1,1)$ denote the normalized arc-length along edge E1, E2, E3, E4; $t_{E5}(\eta) = t(0,\eta,0), t_{E6}(\eta) = t(1,\eta,0), t_{E7}(\eta) = t(0,\eta,1), t_{E8}(\eta) = t(1,\eta,1)$ denote the normalized arc-length along edge E5, E6, E7, E8; $u_{E9}(\varsigma) = u(0,0,\varsigma), u_{E10}(\varsigma) = u(1,0,\varsigma), u_{E11}(\varsigma) = u(1,0,\varsigma), u_{E12}(\varsigma) = u(1,1,\varsigma)$ denote the normalized arc-length along edge E9, E10, E11, E12;

The new algebraic transformation from computational space C to parameter space P is defined as

$$\begin{aligned}
s &= s_{E1}(\xi)(1-t)(1-u) + s_{E2}(\xi)t(1-u) + s_{E3}(\xi)(1-t)u + s_{E4}(\xi)tu \\
t &= t_{E5}(\eta)(1-s)(1-u) + t_{E6}(\eta)s(1-u) + t_{E7}(\eta)(1-s)u + t_{E8}(\eta)su \\
u &= u_{E9}(\varsigma)(1-s)(1-t) + u_{E10}(\varsigma)s(1-t) + u_{E11}(\varsigma)(1-s)t + u_{E12}(\varsigma)st
\end{aligned} \tag{3.17}$$

Equation (3.17) is an algebraic straight line transformation from computational space to parameter space, or $C \to P$.

The elliptic transformation from parameter space to physical domain, or $P \to D$, which is independent of the prescribed boundary grid point distribution, is governed by a set of Laplace equations

$$\begin{aligned}
s_{xx} + s_{yy} + s_{zz} &= 0 \\
t_{xx} + t_{yy} + t_{zz} &= 0 \\
u_{xx} + u_{yy} + u_{zz} &= 0
\end{aligned} \tag{3.18}$$

The elliptic transformation defined above is also differentiable and one-to-one.

This will give a relationship between parameter space and physical domain, or $\Delta r = 0$ where $r = (x, y, z)^{\mathrm{T}}$

We now transform $\Delta r = 0$ from parameter space to the computational space. A Poisson's system will be obtained:

$$\begin{aligned}
&g^{11}r_{\xi\xi} + g^{22}r_{\eta\eta} + g^{33}r_{\varsigma\varsigma} + 2g^{12}r_{\xi\eta} + 2g^{13}r_{\xi\varsigma} + 2g^{23}r_{\eta\varsigma} \\
&+(g^{11}P^{1}_{11} + g^{22}P^{1}_{22} + g^{33}P^{1}_{33} + 2g^{12}P^{1}_{12} + 2g^{13}P^{1}_{13} + 2g^{23}P^{1}_{23})r_{\xi} \\
&+(g^{11}P^{21}_{11} + g^{22}P^{2}_{22} + g^{33}P^{21}_{33} + 2g^{12}P^{2}_{12} + 2g^{13}P^{2}_{13} + 2g^{23}P^{2}_{23})r_{\eta} \\
&+(g^{11}P^{3}_{11} + g^{22}P^{3}_{22} + g^{33}P^{3}_{33} + 2g^{12}P^{3}_{12} + 2g^{13}P^{3}_{13} + 2g^{23}P^{3}_{23})r_{\varsigma} = 0
\end{aligned} \qquad \textbf{(3.19)}$$

where g^{11}, g^{22}, g^{33}, g^{12}, g^{13}, g^{23} are contravariant tensor, which are calculated through the covariant metric tensor:

$$\begin{aligned}
g^{11} &= \frac{1}{J^2}(g_{22}g_{33} - g_{23}^2) \\
g^{22} &= \frac{1}{J^2}(g_{11}g_{33} - g_{13}^2) \\
g^{33} &= \frac{1}{J^2}(g_{11}g_{22} - g_{12}^2) \\
g^{12} &= \frac{1}{J^2}(g_{13}g_{23} - g_{12}g_{33}) \\
g^{13} &= \frac{1}{J^2}(g_{12}g_{23} - g_{13}g_{22}) \\
g^{23} &= \frac{1}{J^2}(g_{12}g_{13} - g_{23}g_{11})
\end{aligned}$$

The control functions are defined through the composite transformation

$$\begin{aligned}
&P_{11} = -T^{-1}\begin{pmatrix} s_{\xi\xi} \\ t_{\xi\xi} \\ u_{\xi\xi} \end{pmatrix}, \quad P_{22} = -T^{-1}\begin{pmatrix} s_{\eta\eta} \\ t_{\eta\eta} \\ u_{\eta\eta} \end{pmatrix}, \quad P_{33} = -T^{-1}\begin{pmatrix} s_{\varsigma\varsigma} \\ t_{\varsigma\varsigma} \\ u_{\varsigma\varsigma} \end{pmatrix}, \\
&P_{12} = -T^{-1}\begin{pmatrix} s_{\xi\eta} \\ t_{\xi\alpha\eta} \\ u_{\xi\eta} \end{pmatrix}, \quad P_{13} = -T^{-1}\begin{pmatrix} s_{\xi\varsigma} \\ t_{\xi\varsigma} \\ u_{\xi\varsigma} \end{pmatrix}, \quad P_{23} = -T^{-1}\begin{pmatrix} s_{\eta\varsigma} \\ t_{\eta\varsigma} \\ u_{\eta\varsigma} \end{pmatrix},
\end{aligned} \qquad \textbf{(3.20)}$$

And the matrix T is defined as

$$T = \begin{pmatrix} s_\xi & s_\eta & s_\varsigma \\ t_\xi & t_\eta & t_\varsigma \\ u_\xi & u_\eta & u_\varsigma \end{pmatrix}$$

The computed grids are in general not orthogonal at the boundary. However, the algebraic transformation can be redefined in a similar way as described for the two-dimensional case to obtain a grid which is orthogonal at the boundary. A three-dimensional grid around a delta wing is shown in Fig. (**3.11**).

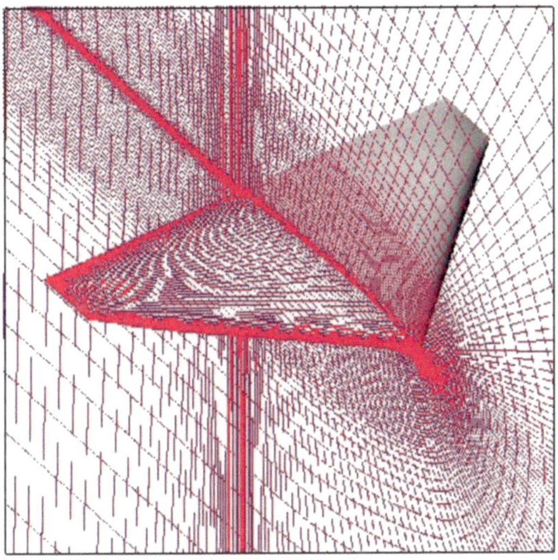

Fig. (3.11). Orthognal grids around a Delta wing.

3.4. HIGH QUALITY GRID GENERATION FOR MVG

In order to make comparison between the supersonic ramp flow with and without MVG control, two cases are designed. The configuration of the first case is a MVG mounted ahead of the ramp (MVG-ramp). The configuration of the second case is the ramp only. Two cases have the same streamwise and spanwise length. In order to preserve the accuracy of the geometry and reduce numerical errors as much as possible while using 5th order *WENO* scheme, the strategy of body-fitted grids is adopted. Results in later parts of the paper testify that such grid frame is very helpful to obtain the high resolution of the flow structure.

3.4.1. The Grid Generation for Case 1

The geometry of MVG is shown in Fig. (**3.12**). In order to alleviate the difficulty to grid generation caused by original vertical trailing-edge, a modification is made by declining the edge to 70°. The other geometric parameters in the figure are the same as those given by Babinsky [51], *i.e.*, $c = 7.2h$, $\alpha = 24°$ and $s = 7.5h$, where h is the height of MVG and s is the distance between the center lines of two adjacent MVGs. So the distance from the center line to the spanwsie boundary of the computation domain is 3.75h.

According to experiments by Babinsky [51], the ratio $h/\delta_0$of the models has the range from 0.3~1. The appropriate distance from the trailing-edge to the control area is around 19~56h or 8~19δ_0. So in this study, the height of MVG h is assumed to be $\delta_0/2$ and the horizontal distance from the apex of MVG to the ramp corner is set to be 19.5h or 9.75δ_0. The distance from the end of the ramp to the apex is 32.2896h. The distance from the starting point of the domain to the apex of MVG is 17.7775h. The height of the domain is from 10h to 15h and the width of the half domain is 3.75h. The geometric relation of the half of the domain can be seen in Fig. (**3.13**), where the symmetric plane is the central plane.

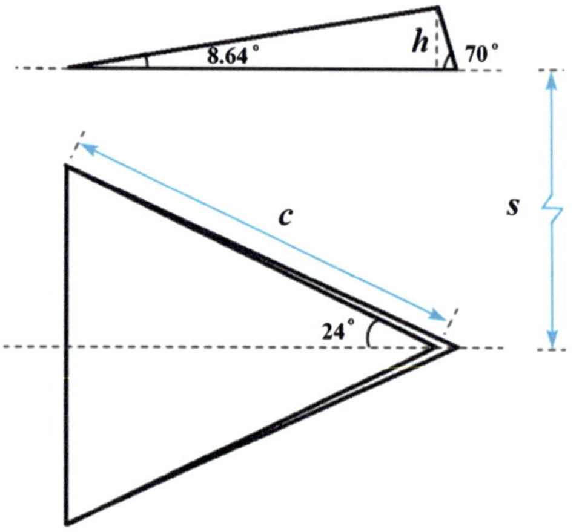

Fig. (3.12). The geometry of MVG.

Because the singularity of the geometry, it is difficult to use one technique to generate the whole grid system. A general grid partition technique is used in this grid generation. As shown in Fig. (**3.13**), three regions are divided as: the ramp region, MVG region and fore-region. Between each two regions, there is a grid transition buffer. Because of the symmetry of the grid distribution, only half of the grids need to be generated. The grid number for the whole system is: $n_{spanwise} \times n_{normal} \times n_{streamwise} = 128 \times 192 \times 1600$.

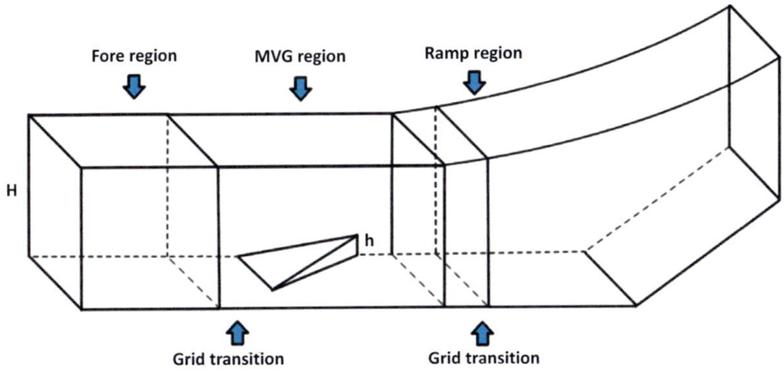

Fig. (3.13). The schematic of the half grid system of case 1.

3.4.1.1. The Grid Generation for the Ramp Region

The grid generation includes two steps: first using analytical methods to generate the algebraic grids; next using elliptic grid generation equation to improve the orthogonal and smooth property of the grid.

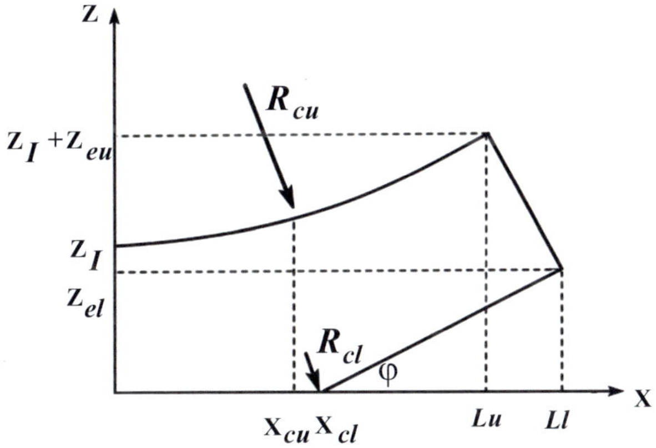

Fig. (3.14). The geometry sketch of the ramp [73]

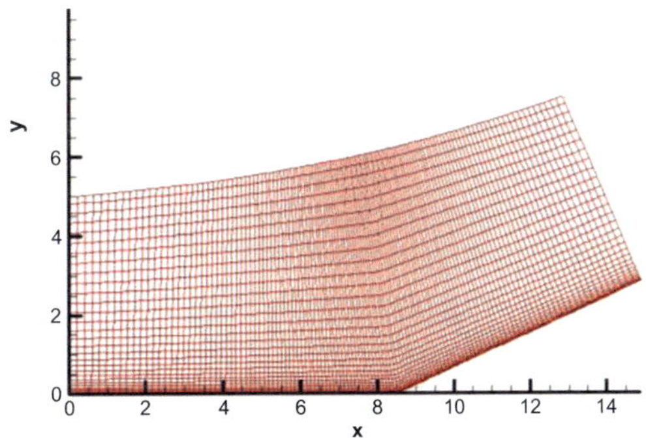

Fig. (3.15). The grid system of case 1.

a. The algebraic grid generation

The schematic figure can be shown in Fig. (**3.14**) and the definition of the variables in the figure can be found in Li and Liu [73]. The procedures for grid generation are: first generating the boundary grids of the lower and upper boundaries; next generating the inner grids by interpolation. The grid number is 993×192. Other parameters in the grid generation [73] have the values: z_l=5h, z_{el}=2.5h, φ=24°, x_{cl}=8.5h, x_{cu}=7.65h, R_{cl}=0.085h, R_{cu}=10h, and the length of the slope is 7h. The unit length is defined as the nominal boundary layer thickness or two times of the height of MVG. Grid concentration is made to ensure half of the grids are located under the height 1.2δ_0. Fig. (**3.15**) shows the generated mesh with a grid interval of 8 in streamwise and of 6 in normal directions.

b. Optimization of the algebraic grids

Because of the specification of the boundary conditions on the body surface, an orthogonal grid is very important to ensure the high accuracy of the computation. This is particularly important to the case with complex geometry, where the zero normal gradient condition is usually realized by using the derivative along the normal grid line. To make grids orthogonal and smooth, a grid solver was developed by the group of the second author based on Laplace equations and algebraic transformations (see Li and Liu [73]).

3.4.1.2. The Surface Grid Generation for the MVG Region

In order to generate the complete grids of MVG, it is essential to generate the surface grids. According to our experience, the quality of the surface grids will directly influence the quality of the 3-D volume grids, and the accurate description of the geometry by the surface grids can enhance the accuracy of the computation. Because the surface of MVG is of high singularity, it failed to use the automatic grid generation technique like projection by some commercial software. In this study, some manual work had to be done by the following steps: first a modification is made by smoothing the trailing-edge using a very small arc (see the yellow part in Fig. **3.16**); next the surface is divided into many small patches so that the singularity of the shape is reduced in each patch; thirdly, the skeleton grid lines are constructed manually in the patch using some grid generation software in an interactive manner. Afterwards the lines are discretized into grid points. Careful adjustments are made to make the distribution of grids as smooth as possible; and finally, optimizations are made to let grid lines transit fluently between patches. The final surface grids can be seen in Fig. (**3.17**).

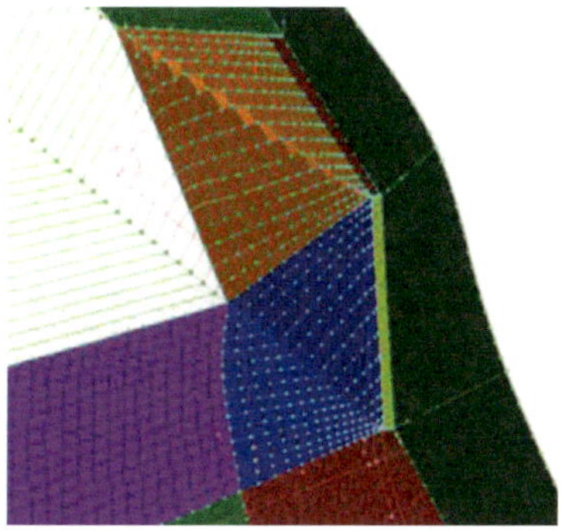

Fig. (3.16). The local view of grids around the trailing-edge.

Fig. (3.17). The surface grids of MVG in patches.

3.4.1.3. The Connection of the MVG and Ramp Region

Using the beginning section of the ramp grid as a reference, the frame grid lines of the MVG region can be determined. Using some grid generation software, it is easy to generate the other surface grids. Special orthogonal procedures are taken on the central symmetric grid plane. After all surface grids are generated the volume grids are constructed using transfinite interpolation.

Further optimization on grids is conducted to make the volume grids more orthogonal near the wall surface and smoothly change in space. This is done by running an elliptic solver several times, while making sure that there is no additional distortion introduced in each run. Also a grid transition zone is divided from the ramp region and the grids in the zone are re-distributed to make the mesh transition smoothly from the MVG region to the ramp region.

3.4.1.4. The Grid Generation for Fore-region

Using the beginning section of the MVG grid, the frame grid lines of the fore-region can be determined. A transfinite interpolation method is used to generate the volume grids. After that, a grid transition zone is divided from the fore-region and the grids in it are re-distributed to make the mesh transition smoothly from the MVG region to the ramp region.

The final grids can be seen in Figs. (**3.18** & **3.19**).

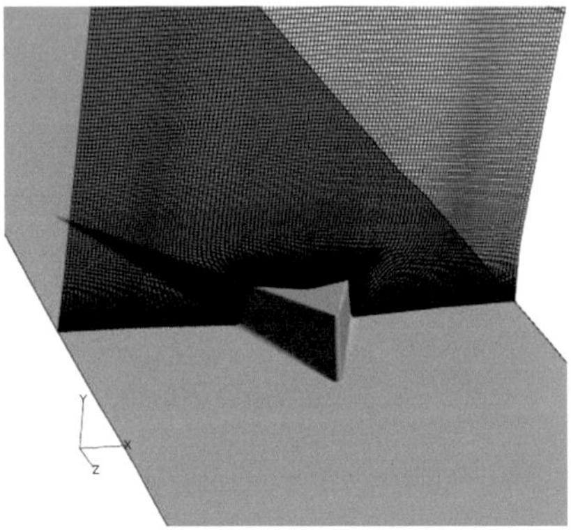

Fig. (3.18). The grids in certain cross-section.

Fig. (3.19). The grids at the foot of the trailing-edge.

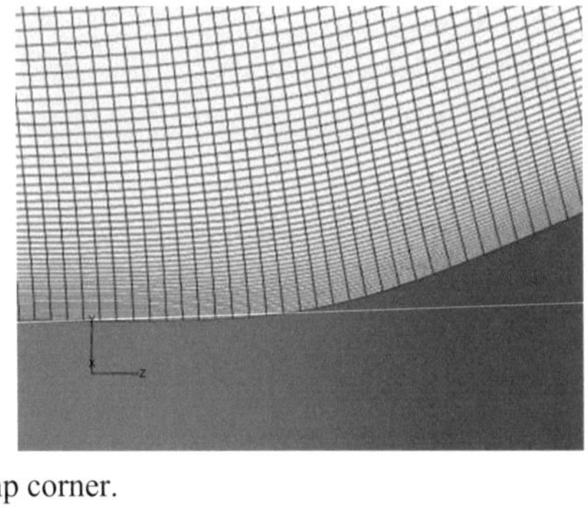

Fig. (3.20). The grids at the ramp corner.

3.4.2. The Grid Generation for Case 2

In order to make comparison with Case 1, the grids of Case 2 is generated by doing some modifications to the grids of Case1. The detailed procedures include:

a. Remove the volume grids and body surface grids of the MVG region.
b. Re-generate the frame grid line on original MVG geometry by straight line; redistribute the grid points on the new frame grid lines to diminish the influence of MVG.
c. Re-generate the volume grids using the transfinite interpolation. Optimization is done again to improve the quality of grids. Fig. (**3.20**) shows a local view of the grids at the ramp corner.

Using the inflow flow profile described in the next section, a summary is given in Table **3.1** about the geometric parameters of the grid system. Because the existence of MVG geometry, there are inevitable grid deformation and local clustering at the location where the shape changes discontinuously. To void such extreme cases, the grid intervals in wall unit are gauged at the entrance of the domain of Case 1, and the results of Case 2 are the same as that of Case 1.

Table 3.1. The geometric parameters for the computation.

L_x	L_y	L_z	Δx^+	Δy^+	Δz^+
$3.75\delta_0$	$5\text{-}7.5\delta_0$	$25.03355\delta_0$	26.224	1.357-38.376	12.788

CONCLUSION

High quality grid generation which is smooth and orthogonal near the wall boundary is critical to success to high order LES. A two-step orthogonal grid generation method proposed by Spekreijse [121] is adopted for MVG cases.

CHAPTER 4

High Order Schemes for Shock Capturing

Abstract: In this chapter, a series of high order shock capturing schemes including WENO scheme, weighted compact scheme, modified upwinding compact scheme are introduced, which can get sharp shock capturing but keep high order accuracy in the smooth area. These schemes are particularly useful for LES of shock-turbulence interaction where both high resolution and sharp shock capturing are important. All these schemes are somehow successful in high order LES for SBLI. An efficient shock detector is introduced as a part of weighted compact and modified upwinding schemes. In addition, the accuracy, truncation errors, dissipation and dispersion of these schemes are analyzed.

Keywords: Compact scheme, Dispersion, Dissipation, High order scheme, Shock capturing, Shock detector, Truncation error, Upwinding scheme, WENO.

4.1. A SHORT REVIEW ON SHOCK CAPTURING RELATED SCHEMES

The flow filed is a system of time dependent partial differential equations and in general governed by another system called Navier-Stokes. However, for external flow, only the viscosity of boundary layers is significantly large while the main flow can still be considered as inviscid. And the governing system can be dominated by the time-dependent hyperbolic Euler equations. Shock capturing is a numerical solution to solve this difficult problem and it can be regarded as a discontinuity or mathematical singularity, which is no classical unique solution and no bounded derivatives. In the shock area, only the weak solution can be obtained from an integration form and the governing Euler equations will lose continuity and differentiability. But in some cases, the shock can be improved because the Euler equation is non-linear and hyperbolic. To get high order accuracy and resolution, it is expected that the equation should be solved by high order compact scheme. High order accuracy is appropriate for resolving small length scale in the process of transition and turbulence flow. However, for the hyperbolic system, the studies already show the existences of characteristic lines and Riemann invariants. Obviously, the upwind finite difference scheme is corresponding to the physics for a hyperbolic system. Not only the eigenvalues and eigenvectors of the Jacobian system should be considered, but also non-linearity the Rankine-Hugoniot shock relations.

Chaoqun Liu, Qin Li, Yonghua Yan, Yong Yang, Guang Yang, Xiangrui Dong

From the point of above-stated view, upwind scheme is applicable for the hyperbolic system. Many upwind or bias upwind schemes, such as Godnov [124], Roe [125], MUSCL by Van [126], TVD by Harten [127], ENO by Harten *et al.* [128], Shu *et al.* [129, 130] and WENO [131, 132], have achieved great success in capturing the shocks sharply. These schemes for capturing the shock are all based on upwind or bias upwind technology. They are nice for hyperbolic system, however, are not favorable to the N-S system. The small length scale is really sensitive to any artificial numerical dissipation. Therefore, high order compact scheme [133, 134] is more applicable for flow transition and turbulence simulation since it is central and non-dissipative with high order of accuracy and resolution.

The shock-boundary layer interaction, which is common in high-speed flow, is a complicated problem which exit shock (discontinuity), boundary layer (viscosity), separation, transition, expansion fans, fully developed turbulence, and reattachment. For shock-boundary layer interaction, there are elliptic areas (separation, transition, and turbulence) and hyperbolic areas (main flow, shocks and expansion fans); all these items could make the accurate numerical simulation extremely complex and difficult. The computational domain can be divided into several parts: the elliptic, hyperbolic, and mixed. The division or detection can be automatically created by the switch function like shock detectors, which gives 1 at shock area and 0 when meeting others. Although the switch function could give a good result for shock-boundary layer interaction, however, it still has too many logical statements in the code, due to its increase of the computation time as well as the time of convergence for steady problems.

It can be desirable that if the compact scheme and WENO scheme are combined. There are some achievements for the combination of WENO and standard central [135, 136] as well as WENO and upwinding compact (UCS) scheme [137, 138]. However, these mixing functions are some kind complex and may be case related on adjustable coefficients.

To conquer the side effect of the CS scheme, local dependency in shock regions and the global dependency in smooth regions should be achieved. According to this basic idea, a combination of local dependent scheme, Modified Compact Scheme (MCS) and Modified Upwinding Compact Scheme (MUCS) come out.

This above-introduced method adopt WENO to improve 7th order upwinding compact scheme, as we called "modified upwinding compact scheme (MUCS)". While the improved 6th order weighted compact scheme is called "modified compact scheme". This method is to use a new shock detector for capturing the shock as well as a control function for mixing compact schemes (UCS or WCS) and WENO. The mixing function is proposed as following ways: the new scheme with high order of accuracy and resolution becomes bias automatically once meeting the shock, while rapidly recovers to be upwinding compact. The hybrid scheme by merging weighted compact scheme with WENO (Modified weighted compact scheme or MWCS) is also developed by the UTA team.

Some black box type subroutines are carried out and users can adopt compact scheme in the cases with shock and shock boundary layer interaction through calling these subroutines (The reader should contact cliu@uta.edu to get these black-box type subroutines).

4.2. HIGH ORDER SCHEMES FOR SHOCK CAPTURING

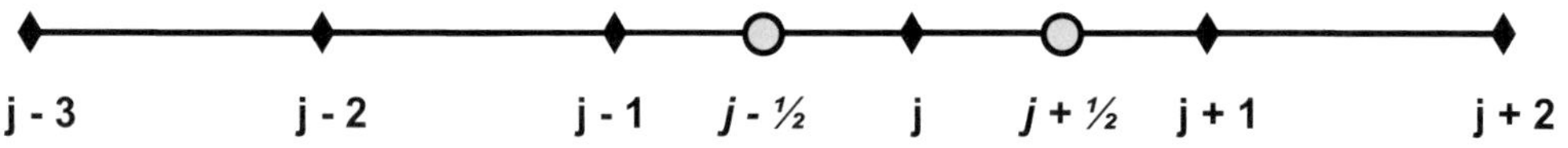

Fig. (4.1). Grid points for original and primitive functions.

Compact scheme is well-known for high order accuracy and high resolution. However, due to its global dependency, the compact scheme cannot directly be used for shock capturing. On the other hand, WENO scheme can be used for high order shock capturing. However, WENO scheme is still considered to have too much dissipation which will smear the high frequency waves and small length scales. Let us first briefly introduce compact scheme and WENO scheme.

4.2.1. Primitive Function for Conservation

To capture the shock, primitive function for conservation should be used here.

For 1-D conservation laws:

$$u_t(x,t) + f_x(u(x,t)) = 0 \tag{4.1}$$

A semi-discrete conservative form of the equation (4.1) is written in the following:

$$\frac{du_j}{dt} = -\frac{1}{\Delta x}(\hat{f}_{j+(1/2)} - \hat{f}_{j-(1/2)}) \tag{4.2}$$

where $f_j = \frac{1}{\Delta x}\int_{x_j-\Delta x/2}^{x_j+\Delta x/2} \hat{f}(\xi)d\xi$ and then $(f_x)_j = -\frac{1}{\Delta x}(\hat{f}_{j+(1/2)} - \hat{f}_{j-(1/2)})$. Note that f is the original function, but $\hat{f}$ is the flux defined by the above integration which is an exact expression of the flux but is different from f (see Fig. **4.1**). Let H be the primitive function of $\hat{f}$ defined below:

$$H(x_{j+(1/2)}) = \int_{-\infty}^{x_j+\Delta x/2} \hat{f}(\xi)d\xi = \sum_{i=-\infty}^{i=j} \int_{x_i-\Delta x/2}^{x_i+\Delta x/2} \hat{f}(\xi)d\xi = \Delta x \sum_{i=-\infty}^{j} f_i \quad \textbf{(4.3)}$$

H is easy to be calculated, but is a discrete data set. The numerical flux $\hat{f}$ at the cell interfaces is the derivative of its primitive function H. *i.e.*:

$$\hat{f}_{j+(1/2)} = H'_{j+(1/2)} \quad \textbf{(4.4)}$$

All above-given formulae are exact without approximations. Whereas, the primitive function H is a discrete data set or discrete function and numerical method is used to get the derivatives, which can introduce numerical errors.

This procedure of $f \rightarrow H \rightarrow \hat{f} \rightarrow f_x$, was introduced by Shu & Osher [129, 130]. There is one problem left for numerical methods, which is how to solve (4.4) or how to get accurate derivatives for a discrete data set.

4.2.2. High-order Compact Schemes

The idea of Lagrange interpolation are used for traditional finite difference schemes. A stencil covering $n + 1$ grid points is needed to obtain nth order of accuracy. In other words, the derivative at a certain grid point depends upon the function values at these $n + 1$ grid points and only these grid points. On the contrary, standard compact schemes [133, 134] adopt the idea of Hermitian interpolation. Through using both derivatives and function values, a compact scheme can obtain high order accuracy without increasing the stencils width. From Lele's paper, a compact scheme has both high order accuracy and high resolution. By Fourier analysis, with the same order of accuracy, a compact scheme has better spectral resolution than the traditional explicit finite difference scheme. For this reason, compact schemes have advantage in the simulation of turbulent flows where small-length-scale structures are dominant, compact schemes usually give us a tri-diagonal or penta-diagonal system due to the usage of derivatives. Although the tri-diagonal matrix is sparse, the inverse of a tri-diagonal matrix is dense, which means the derivative at a certain grid point depends on all the grid points along a single line of grid. The suggestion of compact schemes indicates that the global dependency is very important for high resolution, while the global dependency is good for resolution, however, not so applicable for capturing the shock.

A pade-type compact scheme can be constructed based on the Hermite interpolation where both function and derivatives at grid points are used, for example, an 8th order finite difference scheme can be constructed if both the function and first order derivative are involved at 5 grid points. For a function *f*, a compact scheme with five grid points [10] are written as follows:

$$\beta_- f'_{j-2} + \alpha_- f'_{j-1} + f'_j + \alpha_+ f'_{j+1} + \beta_+ f'_{j+2} = (b_- f_{j-2} + a_- f_{j-1} + c f_j + a_+ f_{j+1} + b_+ f_{j+2}) / \Delta\xi \quad \textbf{(4.5)}$$

We can get eighth order accuracy by using the above-mentioned formula based on Taylor series.

If a symmetric and tri-diagonal system is used by setting $\beta_- = \beta_+ = 0$, we can get a one parameter family of compact scheme [133]:

$$\alpha f'_{i-1} + f'_i + \alpha f'_{i+1} = \left[-\frac{1}{12}(4\alpha - 1) f_{i-2} - \frac{1}{3}(\alpha + 2) f_{j-1} + \frac{1}{3}(\alpha + 2) f_{j+1} + \frac{1}{12}(4\alpha - 1) f_{i+2} \right] / h \quad \textbf{(4.6)}$$

If $\alpha = \frac{1}{3}$, we will get a standard 6th order compact scheme

$$\frac{1}{3} f'_{i-1} + f'_i + \frac{1}{3} f'_{i+1} = \left[-\frac{1}{36} f_{i-2} - \frac{7}{9} f_{j-1} + \frac{7}{9} f_{j+1} + \frac{1}{36} f_{i+2} \right] / h \quad \textbf{(4.7)}$$

It should be note that only CS for primitive function is used to get flux:

$$\hat{f}_{j+(1/2)} = H'_{j+(1/2)}$$

4.2.3. Upwinding Compact Scheme

The standard compact scheme which is a type of non-dissipative scheme does not have dissipation and needs filter even at the smooth area. The upwinding compact scheme can keep high order without the filter. By using this approach of conservative primitive function, the 7th order up-winding scheme shows in the following:

For the primitive function of positive flux H^{+}:

1) Interior points

$$\frac{1}{2}H'_{j-\frac{3}{2}}+H'_{j-\frac{1}{2}}+\frac{1}{4}H'_{j+\frac{1}{2}}=\frac{\left(\frac{1}{240}H_{j-\frac{7}{2}}-\frac{1}{12}H_{j-\frac{5}{2}}-\frac{11}{12}H_{j-\frac{3}{2}}+\frac{1}{3}H_{j-\frac{1}{2}}+\frac{31}{48}H_{j+\frac{1}{2}}+\frac{1}{60}H_{j+\frac{3}{2}}\right)}{h} \quad \textbf{(4.8)}$$

2) For boundary points:

Point 0

$$H'_{j-\frac{1}{2}}+6H'_{j+\frac{1}{2}}=\frac{\left(-\frac{69}{20}H_{j-\frac{1}{2}}-\frac{17}{10}H_{j+\frac{1}{2}}+\frac{15}{2}H_{j+\frac{3}{2}}-\frac{10}{3}H_{j+\frac{5}{2}}+\frac{5}{4}H_{j+\frac{7}{2}}-\frac{3}{10}H_{j+\frac{9}{2}}+\frac{1}{30}H_{j+\frac{11}{2}}\right)}{h} \quad \textbf{(4.9)}$$

Point 1

$$\frac{1}{10}H'_{j-\frac{3}{2}}+H'_{j-\frac{1}{2}}+H'_{j+\frac{1}{2}}=\frac{\left(-\frac{227}{600}H_{j-\frac{3}{2}}-\frac{13}{12}H_{j-\frac{1}{2}}+\frac{7}{6}H_{j+\frac{1}{2}}+\frac{1}{3}H_{j+\frac{3}{2}}-\frac{1}{24}H_{j+\frac{5}{2}}+\frac{1}{300}H_{j+\frac{7}{2}}\right)}{h} \quad \textbf{(4.10)}$$

Point 2

$$\frac{1}{4}H'_{j-\frac{3}{2}}+H'_{j-\frac{1}{2}}+\frac{1}{2}H'_{j+\frac{1}{2}}=\frac{\left(-\frac{1}{60}H_{j-\frac{5}{2}}-\frac{31}{48}H_{j-\frac{3}{2}}-\frac{1}{3}H_{j-\frac{1}{2}}+\frac{11}{12}H_{j+\frac{1}{2}}+\frac{1}{12}H_{j+\frac{3}{2}}-\frac{1}{240}H_{j+\frac{5}{2}}\right)}{h} \quad \textbf{(4.11)}$$

Point N-1

$$H'_{j-\frac{3}{2}}+H'_{j-\frac{1}{2}}+\frac{1}{10}H'_{j+\frac{1}{2}}=\frac{\left(-\frac{1}{300}H_{j-\frac{9}{2}}+\frac{1}{24}H_{j-\frac{7}{2}}-\frac{1}{3}H_{j-\frac{5}{2}}-\frac{7}{6}H_{j-\frac{3}{2}}+\frac{13}{12}H_{j-\frac{1}{2}}+\frac{227}{600}H_{j+\frac{1}{2}}\right)}{h} \quad \textbf{(4.12)}$$

Point N

$$6H'_{j-\frac{3}{2}}+H'_{j-\frac{1}{2}}=\frac{\left(-\frac{1}{30}H_{j-\frac{13}{2}}+\frac{3}{10}H_{j-\frac{11}{2}}-\frac{5}{4}H_{j-\frac{9}{2}}+\frac{10}{3}H_{j-\frac{7}{2}}-\frac{15}{2}H_{j-\frac{5}{2}}+\frac{17}{10}H_{j-\frac{3}{2}}+\frac{69}{20}H_{j-\frac{1}{2}}\right)}{h} \quad \textbf{(4.13)}$$

For the negative primitive function H^-:

For Interior points

$$\frac{1}{4}H'_{j-\frac{3}{2}}+H'_{j-\frac{1}{2}}+\frac{1}{2}H'_{j+\frac{1}{2}}=\frac{\left(-\frac{1}{60}H_{j-\frac{5}{2}}-\frac{31}{48}H_{j-\frac{3}{2}}-\frac{1}{3}H_{j-\frac{1}{2}}+\frac{11}{12}H_{j+\frac{1}{2}}+\frac{1}{12}H_{j+\frac{3}{2}}-\frac{1}{240}H_{j+\frac{5}{2}}\right)}{h} \quad \textbf{(4.14)}$$

For Boundary points:

Point 0

$$H'_{j-\frac{1}{2}}+6H'_{j+\frac{1}{2}}=\frac{\left(-\frac{69}{20}H_{j-\frac{1}{2}}-\frac{17}{10}H_{j+\frac{1}{2}}+\frac{15}{2}H_{j+\frac{3}{2}}-\frac{10}{3}H_{j+\frac{5}{2}}+\frac{5}{4}H_{j+\frac{7}{2}}-\frac{3}{10}H_{j+\frac{9}{2}}+\frac{1}{30}H_{j+\frac{11}{2}}\right)}{h} \quad \textbf{(4.15)}$$

Point 1

$$\frac{1}{10}H'_{j-\frac{3}{2}}+H'_{j-\frac{1}{2}}+H'_{j+\frac{1}{2}}=\frac{\left(-\frac{227}{600}H_{j-\frac{3}{2}}-\frac{13}{12}H_{j-\frac{1}{2}}+\frac{7}{6}H_{j+\frac{1}{2}}+\frac{1}{3}H_{j+\frac{3}{2}}-\frac{1}{24}H_{j+\frac{5}{2}}+\frac{1}{300}H_{j+\frac{7}{2}}\right)}{h} \quad \textbf{(4.16)}$$

Point N-2

$$\frac{1}{2}H'_{j-\frac{3}{2}}+H'_{j-\frac{1}{2}}+\frac{1}{4}H'_{j+\frac{1}{2}}=\frac{\left(\frac{1}{240}H_{j-\frac{7}{2}}-\frac{1}{12}H_{j-\frac{5}{2}}-\frac{11}{12}H_{j-\frac{3}{2}}+\frac{1}{3}H_{j-\frac{1}{2}}+\frac{31}{48}H_{j+\frac{1}{2}}+\frac{1}{60}H_{j+\frac{3}{2}}\right)}{h} \quad \textbf{(4.17)}$$

Point N-1

$$H'_{j-\frac{3}{2}}+H'_{j-\frac{1}{2}}+\frac{1}{10}H'_{j+\frac{1}{2}}=\frac{\left(-\frac{1}{300}H_{j-\frac{9}{2}}+\frac{1}{24}H_{j-\frac{7}{2}}-\frac{1}{3}H_{j-\frac{5}{2}}-\frac{7}{6}H_{j-\frac{3}{2}}+\frac{13}{12}H_{j-\frac{1}{2}}+\frac{227}{600}H_{j+\frac{1}{2}}\right)}{h} \quad \textbf{(4.18)}$$

Point N

$$6H'_{j-\frac{3}{2}}+H'_{j-\frac{1}{2}}=\frac{\left(-\frac{1}{30}H_{j-\frac{13}{2}}+\frac{3}{10}H_{j-\frac{11}{2}}-\frac{5}{4}H_{j-\frac{9}{2}}+\frac{10}{3}H_{j-\frac{7}{2}}-\frac{15}{2}H_{j-\frac{5}{2}}+\frac{17}{10}H_{j-\frac{3}{2}}+\frac{69}{20}H_{j-\frac{1}{2}}\right)}{h} \quad \textbf{(4.19)}$$

Here, H^+ and H^- respectively denote primitive function of positive flux and negative flux.

4.2.4. WENO Scheme (Jiang & Su [132])

The basic ideas of the weighted schemes like WENO [132] are as follows,

1. Apply basic grid stencils and difference schemes on them;
2. Combine these schemes on different stencils and get linear weights to obtain higher order;
3. Obtain nonlinear weights to make the scheme adaptive to discontinuity like shock waves.

4.2.4.1. Conservation Form of Derivative

For integrity, the 5th order WENO will be described as following. Considering the one dimensional hyperbolic equation:

$$\frac{\partial U}{\partial t}+\frac{\partial F}{\partial x}=0 \tag{4.20}$$

The ENO [128] reconstruction can provide a semi-discretization for the derivative:

$$\frac{\partial F}{\partial x}=\frac{\hat{F}_{i+\frac{1}{2}}-\hat{F}_{i-\frac{1}{2}}}{\Delta x}$$

where $\hat{F}$ is the flux which must be accurately obtained.

The 5th Order WENO (bias upwind)

1) Flux approximation

To get high order approximation for $\hat{F}_{j-\frac{1}{2}}=H'_{j-\frac{1}{2}}$, we can use three different candidates (Fig. **4.2**) which are all third order polynomials: $E_0: H_{j-\frac{7}{2}}, H_{j-\frac{5}{2}}, H_{j-\frac{3}{2}}, H_{j-\frac{1}{2}}$; $E_1: H_{j-\frac{5}{2}}, H_{j-\frac{3}{2}}, H_{j-\frac{1}{2}}, H_{j+\frac{1}{2}}$; $E_2: H_{j-\frac{3}{2}}, H_{j-\frac{1}{2}}, H_{j+\frac{1}{2}}, H_{j+\frac{3}{2}}$.

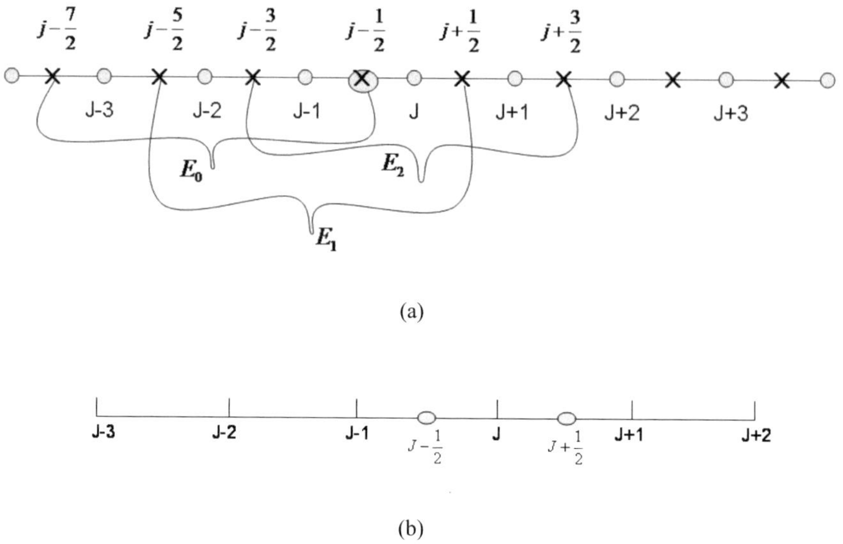

Fig. (4.2). (**a**) WENO candidates (**b**) 5th order WENO Scheme.

Let us look at candidate E_0 first. Assume H is a third order polynomial:

$H = a_0 + a_1(x - x_{j-1/2}) + a_2(x - x_{j-1/2})^2 + a_3(x - x_{j-3/2})^3$, we have

$$
\begin{aligned}
H_{j-1/2} &= a_0 \\
H_{j-3/2} &= a_0 - a_1 h + a_2 h^2 - a_3 h^3 \\
H_{j-5/2} &= a_0 - 2a_1 h + 4a_2 h^2 - 8a_3 h^3 \\
H_{j-7/2} &= a_0 - 3a_1 h + 9a_2 h^2 - 27a_3 h^3
\end{aligned}
\tag{4.21}
$$

Further by subtraction, we can get

$$
\begin{aligned}
H_{j-1/2} - H_{j-3/2} &= hF_{j-1} = a_1 h - a_2 h^2 + a_3 h^3 \\
H_{j-1/2} - H_{j-5/2} &= h(F_{j-1} + F_{j-2}) = 2a_1 h - 4a_2 h^2 + 8a_3 h^3 \\
H_{j-1/2} - H_{j-7/2} &= h(F_{j-1} + F_{j-2} + F_{j-3}) = 3a_1 h - 9a_2 h^2 + 27a_3 h^3
\end{aligned}
\tag{4.22}
$$

Deleting a_3, we can get

$$\begin{aligned} &8F_{j-1}-(F_{j-1}+F_{j-2})=6a_1-4a_2h \\ &27F_{j-1}-(F_{j-1}+F_{j-2}+F_{j-3})=24a_1-18a_2h \end{aligned} \tag{4.23}$$

or

$$\begin{aligned} &63F_{j-1}-9F_{j-2}=54a_1 \\ &52F_{j-1}-2F_{j-2}-2F_{j-3}=48a_1 \end{aligned} \tag{4.24}$$

Then,

$$6a_1=2F_{j-3}-7F_{j-2}+11F_{j-1}$$

Or

$$E_0:\hat{F}_{j-\frac{1}{2}}=H'_{j-1/2}=a_1=\frac{1}{3}F_{j-3}-\frac{7}{6}F_{j-2}+\frac{11}{6}F_{j-1} \tag{4.25}$$

Finally, we have

$$\begin{aligned} &E_0:\hat{F}_{j-\frac{1}{2}}=\frac{1}{3}F_{j-3}-\frac{7}{6}F_{j-2}+\frac{11}{6}F_{j-1} \\ &E_1:\hat{F}_{j-\frac{1}{2}}=-\frac{1}{6}F_{j-2}+\frac{5}{6}F_{j-1}+\frac{1}{3}F_{j} \\ &E_2:\hat{F}_{j-\frac{1}{2}}=\frac{1}{3}F_{j-1}+\frac{5}{6}F_{j}-\frac{1}{6}F_{j+1} \end{aligned} \tag{4.26}$$

2) Optimal weights for high order of accuracy

Three candidates combine this final scheme: $E = C_0E_0 + C_1E_1 + C_2E_2$

If we set $C_0 = \frac{1}{10}, C_1 = \frac{6}{10}, C_2 = \frac{3}{10}$, we will have

$$\hat{F}_{j-\frac{1}{2}} = \frac{1}{30}F_{j-3} - \frac{13}{60}F_{j-2} + \frac{47}{60}F_{j-1} + \frac{27}{60}F_j - \frac{1}{20}F_{j+1}$$

$$\hat{F}_{j+\frac{1}{2}} = \frac{1}{30}F_{j-2} - \frac{13}{60}F_{j-1} + \frac{47}{60}F_j + \frac{27}{60}F_{j+1} - \frac{1}{20}F_{j+2} \quad \textbf{(4.27)}$$

$$\frac{\partial F}{\partial x} = \frac{\hat{F}_{j+\frac{1}{2}} - \hat{F}_{j=\frac{1}{2}}}{\Delta x} = (-\frac{1}{30}F_{j-3} + \frac{1}{4}F_{j-2} - F_{j-1} + \frac{1}{3}F_j + \frac{1}{2}F_{j+1} - \frac{1}{20}F_{j+2})/\Delta x + O(\Delta x^5)$$

Using Taylor expansion for F_{j-k}, we find

$$\frac{\partial F}{\partial x} = \frac{\hat{F}_{j+\frac{1}{2}} - \hat{F}_{j-\frac{1}{2}}}{\Delta x} = F'_j - \frac{1}{60}(\Delta x)^5 F_j^{(6)} + \frac{1}{140}(\Delta x)^6 F_j^{(7)} + \ldots, \quad \textbf{(4.28)}$$

It is easy to verify by a Taylor series expansion that equation (4.27) is a 5th order approximation to the discrete derivative F'_j.

3) Bias up-wind weights:

According to WENO, a bias weight for each candidate can be defined as:

$$\omega_k = \frac{\gamma_k}{\sum_{i=0}^{2}\gamma_i},\ \gamma_k = \frac{C_k}{(\varepsilon + IS_k)^p} \quad \textbf{(4.29)}$$

where

$$IS_i = \int_{x_{j-1/2}}^{x_{j+1/2}} \sum_{k=1}^{\infty} [p_2(x)^{(k)}]^2 h^{2k-1} dx$$

$$IS_0 = \frac{13}{12}(F_{j-2} - 2F_{j-1} + F_j)^2 + \frac{1}{4}(F_{j-2} - 4F_{j-1} + 3F_j)^2$$

$$IS_1 = \frac{13}{12}(F_{j-1} - 2F_j + F_{j+1})^2 + \frac{1}{4}(F_{j-1} - F_{j+1})^2$$

$$IS_2 = \frac{13}{12}(F_j - 2F_{j+1} + F_{j+2})^2 + \frac{1}{4}(F_{j+2} - 4F_{j+1} + 3F_j)^2$$

The 5th order WENO can be obtained

$$\hat{F}_{j-1/2} = \omega_0 E_0 + \omega_1 E_1 + \omega_2 E_2 \quad \textbf{(4.30)}$$

$$\hat{F}_{j-1/2} = \omega_{0,j-1/2}(\frac{1}{3}F_{j-3} - \frac{7}{6}F_{j-2} + \frac{11}{6}F_{j-1}) + \omega_{1,j-1/2}(-\frac{1}{6}F_{j-2} + \frac{5}{6}F_{j-1} + \frac{1}{3}F_j) + \omega_{2,j-1/2}(\frac{1}{3}F_{j-1} + \frac{5}{6}F_j - \frac{1}{6}F_{j+1}) \quad \textbf{(4.31)}$$

For many users, WENO is a great scheme with great successes. However, this scheme has 5th order dissipation everywhere and 3rd order dissipation near the shock, and DNS/LES researchers complain it will be too dissipative working on the transition and turbulence. It is turned into central and compact schemes.

4.3. MODIFIED WEIGHTED COMPACT SCHEME

To resolve small length scales, compact scheme is involved. However, if the simulation is approaching the shock, a new modified compact scheme proposed to remove the weakness. For shocks or large gradient, it is not feasible; this comes Modified Weighted Compact Scheme.

4.3.1. Effective New Shock Detector

An effective shock detector, proposed by C. Liu [139], has two steps. The first step is to check the ratio of the truncation errors on the coarse and fine grids. The second step is to check the local ratio of the left and right hand slopes. There are some popular shock/discontinuity detectors from Harten, Jameson and WENO which can detect shock, however, they may mistake high frequency waves and critical points as shock and then damp the physically important high frequency waves. Current results show that the new shock detector is very delicate and can capture both strong and weak shocks and 1st, 2nd, and 3rd order derivatives without artificial case related constants, but never mistake high frequency waves, critical points and expansion waves as shock. The bottle neck problem for the shock and boundary layer interaction, shock-acoustic interaction, image process, porous media flow, multiple phase flow, detonation wave can be overcome. To introduce this shock/discontinuity detector, some popular shock detectors need to be introduced firstly.

1) Harten's Switch Function and Jameson's Shock Detector

An automatic switch function is defined by Harten [140] for detecting large changes in the variation of the function values f_i. It is the value between 0 and 1, where 0 is considered as smooth and 1 is considered as non-smooth.

The switch is defined as

$$\theta_{j+1/2} = \max\left(\hat{\theta}_j, \hat{\theta}_{j+1}\right) \tag{4.32}$$

with

$$\hat{\theta}_i = \begin{cases} \left| \dfrac{|\alpha_{i+1/2}| - |\alpha_{i-1/2}|}{|\alpha_{i+1/2}| + |\alpha_{i-1/2}|} \right|^p, & \text{if } |\alpha_{i+1/2}| + |\alpha_{i-1/2}| > \varepsilon \\ 0, & \text{otherwise} \end{cases}$$

where $\alpha_{i+1/2} = f_{i+1} - f_i$ and ε is a suitably chosen measure of insignificant variation in f.

Jameson's shock detector [141] is similar, which can be described as:

$$v_i = \frac{|p_{i-1} - 2p_i + p_{i+1}|}{|p_{i-1}| + 2|p_i| + |p_{i+1}|} \tag{4.33}$$

It is related to the 2nd order derivative of the pressure.

2) WENO

Smoothness measurements are adopted by the WENO weights to evaluate the variation of the function values f_i. We assume the three weights have equal contribution, it can be determined that a function is smooth if all values are approximately 1/3.

$$\omega_i = \frac{\alpha_i}{\sum_j \alpha_j}; \quad \alpha_i = \frac{1}{(IS_i + \varepsilon)^2}, \quad i = 0, 1, 2 \tag{4.34}$$

where

$$\begin{aligned} IS_0 &= \frac{13}{12}(f_{i-2} - 2f_{i-1} + f_i)^2 + \frac{1}{4}(f_{i-2} - 4f_{i-1} + 3f_i)^2; \\ IS_1 &= \frac{13}{12}(f_{i-1} - 2f_i + f_{i+1})^2 + \frac{1}{4}(f_{i-1} - f_{i+1})^2; \\ IS_2 &= \frac{13}{12}(f_{i+2} - 2f_{i+1} + f_i)^2 + \frac{1}{4}(f_{i+2} - 4f_{i+1} + 3f_i)^2. \end{aligned} \tag{4.35}$$

3) New Two Step Shock/Discontinuity Locator

Step one: Determine the multigrid ratio approximation of the sum (4th, 5th and 6th) truncation errors by [$F = f +$ smooth sine wave of small amplitude] and pick the points where its ratio is smaller than 4. Theoretically, on coarse or fine grids, the ratio of 4th order truncation error should be 16, however, any function with a ratio of 4 will be treated smooth and considered passing the test. The points which have a ratio smaller than 4 will be selected out for the second left and right hand slope ratio check.

The multigrid truncation error ratio check is:

$$MR(i,h) = \frac{T_C(i,h)}{T_F(i,h) + \varepsilon}, \text{ where}$$

$$T_C(i,h) = T_4(i,2h) + T_5(i,2h) + T_6(i,2h) = \frac{|f_i^{(4)}|(2h)^4}{4!} + \frac{|f_i^{(5)}|(2h)^5}{5!} + \frac{|f_i^{(6)}|(2h)^6}{6!} \text{ and}$$

$$T_F(i,h) = T_4(i,h) + T_5(i,h) + T_6(i,h) = \frac{|f_i^{(4)}|(h)^4}{4!} + \frac{|f_i^{(5)}|(h)^5}{5!} + \frac{|f_i^{(6)}|(h)^6}{6!} \tag{4.36}$$

where $T_F(i, h)$ is the sum of these 4th, 5th and 6th truncation errors obtained from the fine grid which has n points, $T_C(i, h)$ is the sum of these truncation errors obtained from the coarse grid which has n/2 points. $T_F(i, h)$ and $T_C(i, h)$ have 4th, 5th and 6th order derivatives which are all obtained by using 6th order compact scheme.

Step two: Calculate both the local left and right slope ratio check only at the points which have first ratio smaller than 4.

The new local left and right slope ratio check is:

$$LR(i) = \frac{\left\| \frac{f'_R}{f'_L} \right| - \left| \frac{f'_L}{f'_R} \right\|}{\left| \frac{f'_R}{f'_L} \right| + \left| \frac{f'_L}{f'_R} \right| + \varepsilon} = \left| \frac{(f'_R)^2 - (f'_L)^2}{(f'_R)^2 + (f'_L)^2 + \varepsilon} \right|$$

$$LR(i) = \frac{|f'_L / f'_R| - |f'_R / f'_l|}{|f'_L / f'_R| + |f'_R / f'_L|} = \left| \frac{\alpha_R^2 - \alpha_L^2}{\alpha_R^2 + \alpha_L^2 + \varepsilon} \right| \quad \textbf{(4.37)}$$

where $f'_R = 3f_i - 4f_{i+1} + f_{i+2}$, $f'_L = 3f_i - 4f_{i-1} + f_{i-2}$ and ε is a far small number for avoiding division by zero.

Step three (optional): Set 0.8 as a cutoff value to create a 0/1 switch function for the result obtained from Step two. If the value is 0, f is considered locally smooth, while if the value is 1, f is shock/discontinuity at this point.

It is noted that the first step of Liu always checks $f + \sigma \sin(k\pi x + \phi)$ but not f, here σ is a small number. For two step checks, due to all derivatives are obtained by using a subroutine with standard compact scheme, the cost is not expensive relatively.

For finding universal formula, the data set should be normalized, u(i), i = 1, …, n:

$$u_{diff} = | u_{\max} - u_{\min} | \quad \textbf{(4.38)}$$

$$\bar{u} = (u - u_{\min}) / u_{diff} \quad \textbf{(4.39)}$$

where u_{max} and u_{min} respectively are the maximum and minimum of the value u, and $\bar{u}$ is normalized value. To be simplicity, the hat of u can be neglected and $u(i)$ is used as the normalized data set.

4.3.2. Numerical Tests of the New Shock Detector

Since it is claimed that the new shock detector can detect discontinuity for the function, 1st, 2nd, and 3rd derivatives for any function, a wide range of functions should be chosen for testing this detector. Here eight cases (Figs. **4.3-4.10**) are selected to do the comparison between our new shock detector with detector from Harten and WENO detector.

1) Jump function $f(x)=\begin{cases}0, & -1\le x\le 0\\ 1, & 0<x\le 1\end{cases},\quad n=81$ **(4.40)**

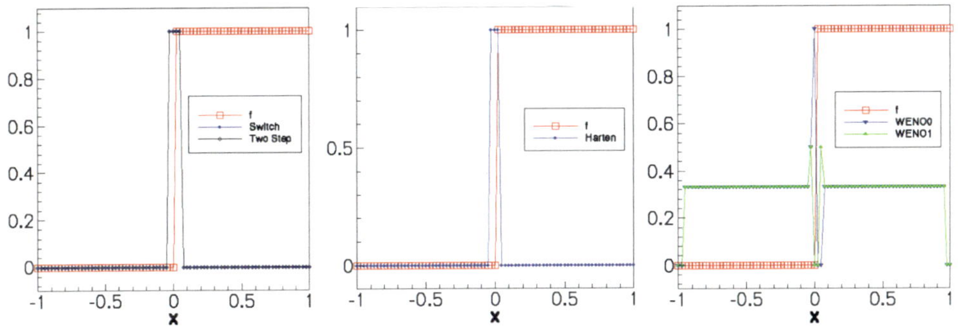

(a) Liu's **(b) Harten's** **(c) WENO**

Fig. (4.3). Shock detector for jump function.

For jump function, all three shock detectors work (Fig. **4.3**).

2) Jump slope: $f(x)=\begin{cases}1+x, & -1\le x\le 0\\ 1, & 0<x\le 1\end{cases},\quad n=81$ **(4.41)**

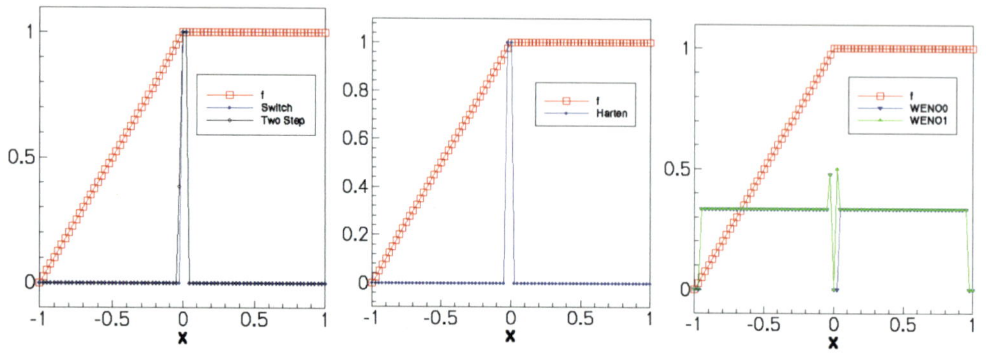

(a) Liu's **(b) Harten's** **(c) WENO**

Fig. (4.4). Shock detector for jump slope.

For jump slope, all three shock detectors work (Fig. **4.4**).

3) High frequency sound waves (eight points per wave):

$$f(x) = \sin\left(\frac{(n-1)\pi x}{8}\right), -1 \le x \le 1, n = 81 \qquad \textbf{(4.42)}$$

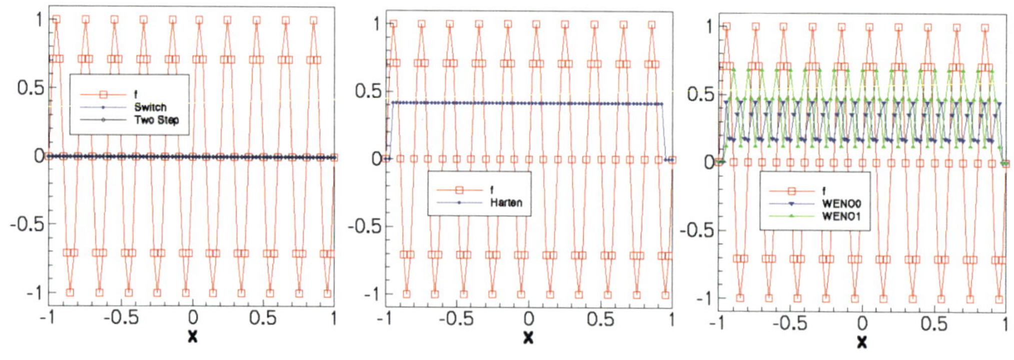

(a) Liu's **(b) Harten's** **(c) WENO**

Fig. (4.5). Shock detector for high frequncy sound waves.

For sound waves with high frequency (eight grid points for each wave), both Harten's and WENO have difficulty in solving serious obstacles and treating the sound waves as shock. However, it is found that the new detector of Liu are all smooth (Fig. **4.5**).

4) Smooth function with large slope:

$$f(x) = \exp(-300x^2), \; -1 \le x \le 0, n = 81 \tag{4.43}$$

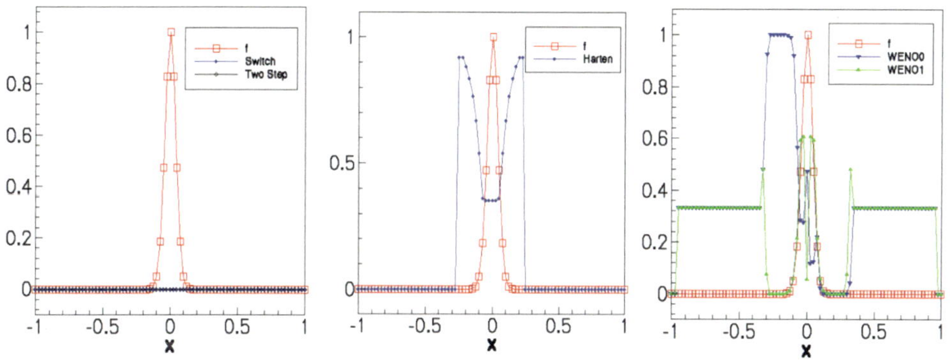

(a) Liu's **(b) Harten's** **(c) WENO**

Fig. (4.6). Shock detector for smooth function with large slope.

For the smooth exponetial function with large slope, both Harten and WENO have serious troubles on determining a shock correctly, but for Liu's shock detector, it is found to be a smooth function. However, both Harten's and WENO will seriously smear the solution at the bottom. For this case, WENO is even worse than Harten's (Fig. **4.6**).

5) Smooth function with two jumps:

$$f(x) = \begin{cases} \sqrt{1-\left(\frac{10}{3}x\right)^2}, & -\frac{3}{10} \le x \le \frac{3}{10}, n = 81 \\ 0, & \text{otherwise} \end{cases} \tag{4.44}$$

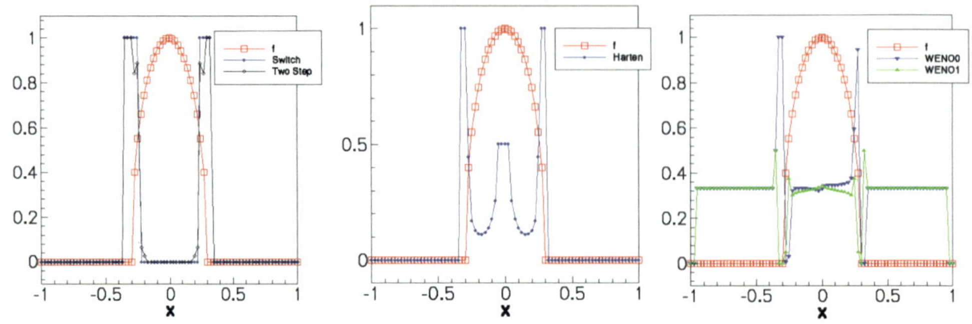

(a) Liu's **(b) Harten's** **(c) WENO**

Fig. (4.7). Shock detector for smooth function with two jumps.

For this mixed function with two jump points at each conner, Harten's mistreats the critical point as shock, WENO even captures more points, while Liu's detector just finds the exact jump points (Fig. **4.7**).

6) Medium frequency with two jump points:

$$f(x)=\begin{cases}\sin\left(\dfrac{(n-1)\pi x}{16}\right), & -1\le x\le 0\\ 1+\sin\left(\dfrac{(n-1)\pi x}{16}\right), & 0<x\le 1\end{cases},\ n=81 \qquad \textbf{(4.45)}$$

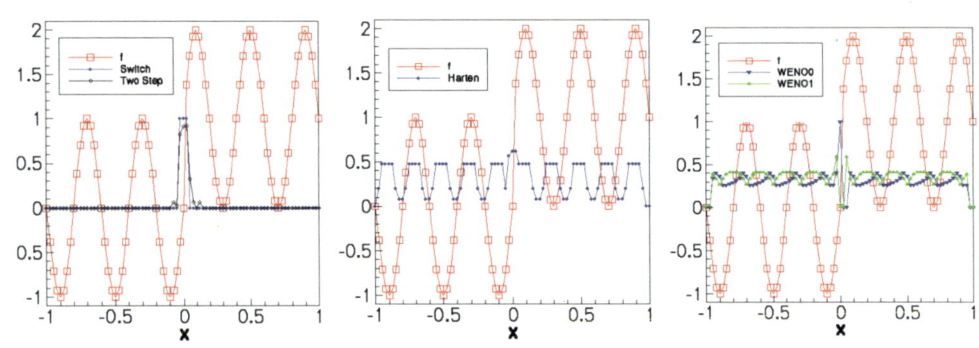

(a) Liu's **(b) Harten's** **(c) WENO**

Fig. (4.8). Shock detector for smooth function with two jump points.

For a medium frequency sound wave with two jump points, Harton's and WENO didn't do very well in separating the shock from high frequecy waves, however, by using Liu's finds, three points are picked out to be shock area and the rest are smooth (Fig. **4.8**).

7) 1-D Shock Tube Problem (T = 2.0, n = 101)

For testing our new detector, a case of the 1-D shock tube is chosen.

The governing equations are 1D Euler equations:

$$\frac{\partial U}{\partial t}+\frac{\partial F}{\partial x}=0 \text{ where}$$

$$U=(\rho, \rho u, E)^T; \quad F=(\rho, \rho u+p, u(E+p))^T \tag{4.46}$$

The initial conditions are given as follows:

$$(\rho, u, p) = \begin{cases}(1, 0, 1), & x<0; \\ (0.125, 0, 0.1) & x\geq 0.\end{cases} \tag{4.47}$$

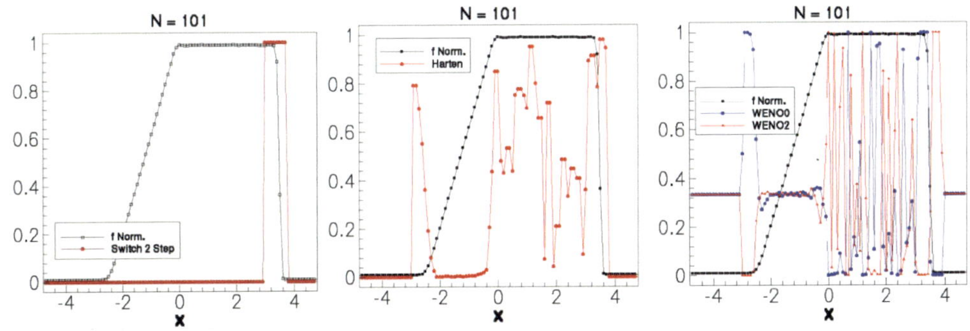

(a) Liu's **(b) Harten's** **(c) WENO**

Fig. (4.9). Shock detector for 1-D shock tube.

Although all these three detectors can capture shocks, the expansion waves are mistreated as shocks by Harten's and WENO (Fig. **4.9**). That is why WENO smears the sound wave in the case of the grids are not fine enough, see Fig. (**4.10**).

8) 1-D Shock/Entropy Wave Interaction (T = 1.8, N = 201):

Here I the 1-D problem of shock/entropy wave interaction is selected as a case for testing the capability of our new shock detector. The following are the initial conditions for solving the 1-D Euler equations:

$$(\rho, u, p)_0 = \begin{cases} (3.857143, 2.629369, 10.33333), & x < -4; \\ (1 + 0.2\sin(5x), 0, 1) & x \geq -4. \end{cases} \tag{4.48}$$

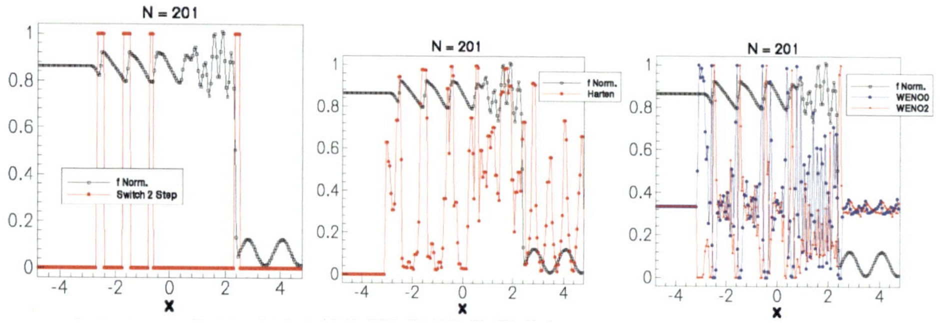

a) Liu's (b) Harten's (c) WENO

Fig. (4.10). Shock detector for 1-D shock entropy interaction.

4.3.3. Control Function for Using WENO to Improve Compact Scheme (CS)

4.3.3.1. Basic Idea of the Control Function

Although the accurate location of weak and strong shock, oblique shock as well as discontinuity in function, 1st, 2nd and 3rd order derivatives, can obtained by using the new shock detector, it is a switch function which gives 1 at shock area and 0 when meeting others. Due to the above-mentioned switch function cannot be directly used to mix CS (4.49) and WENO, a rather smooth function should be developed:

$$\frac{1}{3}H'_{j-\frac{3}{2}}+H'_{j-\frac{1}{2}}+\frac{1}{3}H'_{j+\frac{1}{2}}=\left[-\frac{1}{36}H_{j-\frac{5}{2}}-\frac{7}{9}H_{j-\frac{3}{2}}+\frac{7}{9}H_{j+\frac{1}{2}}+\frac{1}{36}H_{j+\frac{3}{2}}\right]/h \quad \textbf{(4.49)}$$

$$\begin{aligned}H'_{j-1/2}=&\omega_{0,j-1/2}(\frac{1}{3}F_{j-3}-\frac{7}{6}F_{j-2}+\frac{11}{6}F_{j-1})+\omega_{1,j-1/2}(-\frac{1}{6}F_{j-2}+\frac{5}{6}F_{j-1}+\frac{1}{3}F_j)\\&+\omega_{2,j-1/2}(\frac{1}{3}F_{j-1}+\frac{5}{6}F_j-\frac{1}{6}F_{j+1})\end{aligned} \quad \textbf{(4.50)}$$

Where F is the original function and H is a primitive function of F and H' is the flux $\hat{F}=H'$

We defined a new control function Ω:

$$\Omega*CS+(1-\Omega)*WENO \quad \textbf{(4.51)}$$

which will lead a tri-diagonal matrix system that is the core of our new scheme:

$$\begin{aligned}&\frac{1}{3}\Omega H'_{j-\frac{3}{2}}+H'_{j-\frac{1}{2}}\frac{1}{3}\Omega H'_{j+\frac{1}{2}}=\Omega*[\frac{1}{36}(H_{j+\frac{3}{2}}-H_{j-\frac{5}{2}})+\frac{7}{9}(H_{j+\frac{1}{2}}-H_{j-\frac{3}{2}})]\\&+(1-\Omega)*[\omega_{0,j-1/2}(\frac{1}{3}F_{j-3}-\frac{7}{6}F_{j-2}+\frac{11}{6}F_{j-1})+\omega_{1,j-1/2}(-\frac{1}{6}F_{j-2}+\frac{5}{6}F_{j-1}+\frac{1}{3}F_j)\\&\quad+\omega_{2,j-1/2}(\frac{1}{3}F_{j-1}+\frac{5}{6}F_j-\frac{1}{6}F_{j+1})]\end{aligned} \quad \textbf{(4.52)}$$

When Ω = 1.0, the equation become a standard sixth order compact scheme, but when Ω = 0.0 the scheme is a standard WENO scheme.

For the modified upwinding compact scheme (MUCS), the final matrix becomes:

$$\begin{aligned}&\frac{1}{4}\Omega H'_{j-\frac{3}{2}}+H'_{j-\frac{1}{2}}\frac{1}{2}\Omega H'_{j+\frac{1}{2}}=\Omega*[-\frac{1}{60}H_{j-\frac{5}{2}}-\frac{31}{48}H_{j-\frac{3}{2}}-\frac{1}{3}H_{j-\frac{1}{2}}+\frac{11}{12}H_{j+\frac{1}{2}}+\frac{1}{12}H_{j+\frac{3}{2}}-\frac{1}{240}H_{j+\frac{5}{2}}]\\&+(1-\Omega)*[\omega_{0,j-1/2}(\frac{1}{3}F_{j-3}-\frac{7}{6}F_{j-2}+\frac{11}{6}F_{j-1})+\omega_{1,j-1/2}(-\frac{1}{6}F_{j-2}+\frac{5}{6}F_{j-1}+\frac{1}{3}F_j)\\&\quad+\omega_{2,j-1/2}(\frac{1}{3}F_{j-1}+\frac{5}{6}F_j-\frac{1}{6}F_{j+1})]\end{aligned} \quad \textbf{(4.53)}$$

4.3.3.2. Construction of the Control Function

For the new shock detector, $MR(i,h)=\dfrac{T_C(i,h)}{T_F(i,h)+\varepsilon}$ can be defined as a ratio of coarse grid truncation over the find grid truncation. If the function has at least 6th order continuous derivatives, MR should be about16.0. Ω is defined as:

$$\Omega=1.0-\min[1.0,\quad 8.0/MR(i,h)]*LR \tag{4.54}$$

Here *MR* is the multigrid global truncation error ratio while *LR* is local ratio of left and right side angle ratio. If the shock exist in the case, MR is small, LR is 1 and Ω = 0.0. The WENO will be used and the CS is fully blocked. If the area is smooth, MR should be around 16.0 and LR is close to 0 (left and right angle are same). Additional requirement is set that any point must compare with left and right neighboring points and the largest Ω is picked among these neighboring points.

The 4th order continuous function can be regard as smooth function and only half of LR for Ω is needed, so we pick 8.0.This shock detector is really reliable and the new scheme is very robust which can be found in our computation.

4.3.4. The Modified Weighted Compact Scheme (MWCS)

4.3.4.1. The Weighted Compact Scheme (WCS)

The WCS idea [142] is to use the weighted average of 3rd order and one 4th order approximations for the numerical flux, where each approximation involves the primitive function H and its derivative at different points H', constructed by Hermite polynomials. To obtain the approximations for $\hat{F}_{j+1/2}$, three candidate stencils are used:

$$E_0 : H_{j-3/2}, H_{j-1/2}, H_{j+1/2}$$
$$E_1 : H_{j-1/2}, H_{j+1/2}, H_{j+3/2}$$
$$E_2 : H_{j+1/2}, H_{j+3/2}, H_{j+5/2}$$

The approximations for the numerical flux $H'_{j+1/2} = \hat{F}_{j+1/2}$ are obtained making of compact schemes [133] for the three stencils E_0 and E_1 and E_2 respectively, as

$$E_0:\ 2H'_{j-1/2} + H'_{j+1/2} \approx \left(-\frac{1}{2}H_{j-3/2} - 2H_{j-1/2} + \frac{5}{2}H_{j+1/2}\right)\frac{1}{h_j},$$

$$E_1:\ \frac{1}{4}H'_{j-1/2} + H'_{j+1/2} + \frac{1}{4}H'_{j+3/2} \approx \frac{3(H_{j+3/2} - H_{j-1/2})}{4h_j},$$

$$E_2:.\ H'_{j+1/2} + 2H'_{j+3/2} \approx \left(-\frac{5}{2}H_{j+1/2} + 2H_{j+3/2} + \frac{1}{2}H_{j+5/2}\right)\frac{1}{h_j}.$$

Through Taylor expansion, if a weighted average of the three stencils is considered, with different constant weights

$$C_0 = \frac{1}{18}, \qquad C_1 = \frac{8}{9}, \qquad C_2 = \frac{1}{18}, \tag{4.55}$$

by E_0 and E_1 and E_2, we have for $j + 1/2$:

$$\frac{1}{3}H'_{j-1/2} + H'_{j+1/2} + \frac{1}{3}H'_{j+3/2} \approx \left(-\frac{1}{36}H_{j-3/2} - \frac{7}{9}H_{j-1/2} + \frac{7}{9}H_{j+3/2} + \frac{1}{36}H_{j+5/2}\right)\frac{1}{h_j}. \tag{4.56}$$

If WENO weights are used instead of the current C_0, C_1, and C_2, the non-linear weights with parameter $p = 1$ is used to obtain the WCS scheme expression,

$$\begin{aligned}
&-\left[+3\left(\omega_{0,j-1/2} - \omega_{0,j+1/2}\right) + \frac{3}{2}\left(\omega_{1,j-1/2} - \omega_{1,j+1/2}\right) + 3\left(\omega_{2,j-1/2} - \omega_{2,j+1/2}\right)\right]\frac{H'_{j+1/2}}{h_j} + \\
&\left(2\omega_{0,j-1/2} + \frac{1}{4}\omega_{1,j-1/2}\right)F'_{j-1} + \left(3\omega_{0,j-1/2} + \frac{5}{4}\omega_{1,j-1/2} + \omega_{2,j-1/2} - 2\omega_{0,j+1/2} - \frac{1}{4}\omega_{1,j+1/2}\right)F'_j + \\
&\left(\frac{1}{4}\omega_{1,j+1/2} + 2\omega_{2,j+1/2}\right)F'_{j+1} \approx \\
&\left[-\frac{1}{2}\omega_{0,j-1/2}F_{j-2} + \left(-\frac{5}{2}\omega_{0,j-1/2} - \frac{3}{4}\omega_{1,j-1/2} + \frac{1}{2}\omega_{0,j+1/2}\right)F_{j-1} + \right. \\
&\left(-\frac{3}{4}\omega_{1,j-1/2} - \frac{5}{2}\omega_{2,j-1/2} + \frac{5}{2}\omega_{0,j+1/2} + \frac{3}{4}\omega_{1,j+1/2}\right)F_j + \\
&\left.\left(-\frac{1}{2}\omega_{2,j-1/2} + \frac{3}{4}\omega_{1,j+1/2} + \frac{5}{2}\omega_{2,j+1/2}\right)F_{j+1} + \frac{1}{2}\omega_{2,j+1/2}F_{j+2}\right]\frac{1}{h_j}.
\end{aligned} \tag{4.57}$$

For the WCS (4.57), we should solve the tri-diagonal matrix of size N + 1. The j-th row contains the j − 1/2-th numerical flux $H'_{j+1/2} = \hat{F}_{j+1/2}$. The WCS also involves the derivatives at different points, F'_{j-1} and F'_{j+1}, therefore, the results in WCS have global dependency.

4.3.4.2. Modified Weighted Compact Scheme

The MWCS, which is the combination of WENO and WCS, includes the following aspects; first, we move the weights of the WENO and WCS outside the derivative subroutine, for example, using global weights. Second, we use the mean value of the WENO and WCS weights which are calculated from both density and pressure to be the weights of the whole scheme for all variables. Here, we only need to calculate the weights once before carrying out the Runge-Kutta method.

Finally, in order to combine WENO and WCS, a combined function is adopted here for the mean value of the weights of those four fluxes.

On the one hand, this mixing function aims to linearly combine these two schemes for ensuring stability of the simulation, on the other hand, capturing sharp shock and obtaining a good resolution on small length scales. The final formulation of the MWCS numerical flux:

$$\hat{F}_{j-1/2}^{(MWCS)} = (1\text{-}\alpha_j)\hat{F}_{j-1/2}^{(WCS)} + \alpha_j \hat{F}_{j-1/2}^{(WENO)}, \tag{4.58}$$

For consistency of the scheme, $0 \leq \alpha \leq 1$ must be satisfied. Virtually, with $\alpha = 1$ the WENO scheme is recovered. Here we choose α as following [144],

$$\alpha = 1 - 0.5 * \left(1 - \frac{(IS0 - IS1)^2 + (IS1 - IS2)^2 + (IS2 - IS0)^2}{2 * (IS0^2 + IS1^2 + IS2^2)}\right), \tag{4.59}$$

where $IS0$, $IS1$ and $IS2$ are the smoothness indicators obtained from WENO scheme.

4.4. DISPERSION AND DISSIPATION ANALYSIS

A continuous and periodic function can be expanded by Fourier series:

$$f(x) = \sum_{k=-\pi/2}^{k=\pi/2} \hat{f}_k \exp\left(\frac{2\pi i k x}{L}\right) \tag{4.60}$$

The wave number $w = 2\pi kh / L = 2\pi k / N$.

The exact first derivative generates a function with Fourier coefficients $\hat{f}_k' = iw\hat{f}_k$ but a numerical scheme will give a different coefficient $(\hat{f}_k')_{fd} = iw'\hat{f}_k$ w' is called modified wave number which is not same as the exact wave number w.

Let $w' = \hat{k}_e$ be the effective wavenumber.

Fourier analysis of dispersion error and dissipation error in the two publications by Vichnevetsky [143] and Anderson [145] provides us an effective insight into resolution and diffusion properties of numerical schemes, hence for MWCS, and possibly can be used to do further research in improving the mixing function proposed in equation (4.59).

Following Vichnevetsky and Bowles [143], the effective wavenumber $i\hat{k}_e$ for WENO, WCS and MWCS need to be calculated, where i is the imaginary unit $i = \sqrt{-1}$ Practically, we suppose $\omega_{i,j-1/2} = \omega_{i,j+1/2} = \omega_i$ in WENO, WCS and MWCS. For MWCS, the linear combination weight function α in equation (4.59) is not constant, which is not suitable for performing dispersion and dissipation analysis, hence it is assumed to be constant for references.

The effective wave number $i\hat{k}_e$ of WENO scheme, following Vichnevetsky and Bowles [143], can be calculated as

$$
\begin{aligned}
i\hat{k}_e^{(WENO)} &= \frac{11}{6}\omega_0 + \frac{1}{2}\omega_1 - \frac{1}{2}\omega_2 + (-3\omega_0 - \frac{2}{3}\omega_1 + \frac{2}{3}\omega_2)\cos\hat{k} + \\
&(\frac{3}{2}\omega_0 - \frac{1}{6}\omega_1 - \frac{1}{6}\omega_2)\cos 2\hat{k} - \frac{1}{3}\omega_0 \cos 3\hat{k} + \\
&i[(3\omega_0 + \frac{4}{3}\omega_1 + \frac{4}{3}\omega_2)\sin\hat{k} + (-\frac{3}{2}\omega_0 - \frac{1}{6}\omega_1 - \frac{1}{6}\omega_2)\sin 2\hat{k} + \\
&\frac{1}{3}\omega_0 \sin 3\hat{k}]
\end{aligned}
\qquad \textbf{(4.61)}
$$

Similarly, the effective wavenumber of WCS is

$$\begin{aligned}
i\hat{k}_e^{(WCS)} &= [(\omega_0-\omega_2)(5-4\cos\hat{k}-\cos 2\hat{k})+ \\
&i((4\omega_0+3\omega_1+4\omega_2)\sin\hat{k}+(\omega_0+\omega_2)\sin 2\hat{k})]/ \\
&[(4\omega_0+\omega_1+4\omega_2)\cos\hat{k}+2(\omega_0+\omega_1+\omega_2)-4i(\omega_0-\omega_2)\sin\hat{k}]
\end{aligned} \tag{4.62}$$

The dispersion error, or the scheme resolution, could be quantified by the imaginary part of the effective wave number Im($i\hat{k}_e$), however, the real part Re($i\hat{k}_e$) is related to the dissipation. Dispersion errors are waves which travel at different velocity, corresponding to different wavenumbers. The imaginary part of the effective wavenumber Im($i\hat{k}_e$) represents dispersion. Dissipation or diffusion errors related to the negative real part of the effective wavenumber Re($i\hat{k}_e$), constitute the amplification error introduced by the numerical scheme.

The following subsections discuss the Fourier analysis of the dispersion and dissipation in smooth region and shock region when using WENO, WCS and MWCS, the linear combination function is assumed to be 0.1, 0.4 and 0.8 for references.

4.4.1. Smooth Regions

For smooth region, we only need to consider optimal weights in WENO, WCS and MWCS.

For the dispersion error, the imaginary parts of the resulting effective wavenumbers have to be considered, *i.e.*,

$$\begin{aligned}
\mathrm{Im}(i\hat{k}_e^{(WENO)}) &= \frac{3}{2}\sin\hat{k}-\frac{3}{10}\sin 2\hat{k}+\frac{1}{30}\sin 3\hat{k} \\
\mathrm{Im}(i\hat{k}_e^{(WCS)}) &= \frac{(14+\cos\hat{k})\sin\hat{k}}{9+6\cos\hat{k}}
\end{aligned} \tag{4.63}$$

Fig. (**4.11a**) shows that

- MWCS is of higher resolution than WENO on smooth region;
- WCS achieves the highest resolution of the three.

Similarly, dissipation error can be derived by considering real parts of (4.61) and (4.62), which read as

$$\begin{aligned} \operatorname{Re}(i\hat{k}_e^{(WENO)}) &= \frac{1}{3} - \frac{1}{2}\cos\hat{k} + \frac{1}{5}\cos 2\hat{k} - \frac{1}{30}\cos 3\hat{k} \\ \operatorname{Re}(i\hat{k}_e^{(WCS)}) &= 0 \end{aligned} \tag{4.64}$$

Fig. (**4.11b**) presents the comparison of dissipation error, and it can be concluded that

- WCS is characterized by absence of dissipation error;
- WENO scheme if of highest dissipation error of high range of wavenumber $\hat{k}$;
- MWCS is of low dissipation error over the middle and high wavenumber range.

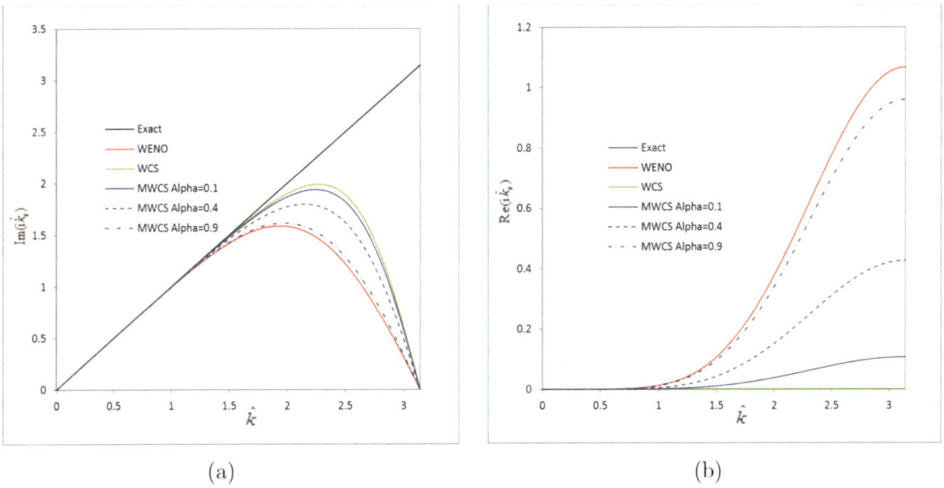

Fig. (4.11). Error analysis in smooth region (**a**) dispersion; (**b**) dissipation.

4.4.2. Shock Region: Using Only Stencil E_0

We assume that the stencils E_1 and E_2 defined in section 3.8 contain shocks, therefore we substitute the following values of the linear weights

$$\omega_0 = 1, \qquad \omega_1 = \omega_2 = 0 \tag{4.65}$$

into the expressions of effective wavenumbers (3.19) and (3.19) for WENO and WCS, respectively.

The dispersion errors are

$$\mathrm{Im}(i\hat{k}_e^{(WENO)}) = 3\sin\hat{k} - \frac{3}{2}\sin 2\hat{k} + \frac{1}{3}\sin 3\hat{k}$$

$$\mathrm{Im}(i\hat{k}_e^{(WCS)}) = \frac{(8+\cos\hat{k})\sin\hat{k}}{5+4\cos\hat{k}} \tag{4.66}$$

And the dissipation errors become

$$\mathrm{Re}(i\hat{k}_e^{(WENO)}) = \frac{11}{6} - 3\cos\hat{k} + \frac{3}{2}\cos 2\hat{k} - \frac{1}{3}\cos 3\hat{k}$$

$$\mathrm{Re}(i\hat{k}_e^{(WCS)}) = -\frac{4\sin(\hat{k}/2)^4}{5+4\cos\hat{k}} \tag{4.67}$$

The following two Figs. (**4.12a** and **4.12b**) show us that

- WENO is of lowest resolution, while MWCS is better and better as α decreasing, and WCS is of highest resolution;
- WENO scheme is dissipative predominantly over high wavenumber range, but WCS is negatively dissipative at high wavenumber range, MWCS, on other hand, tends to have lower dissipative error than WENO scheme.

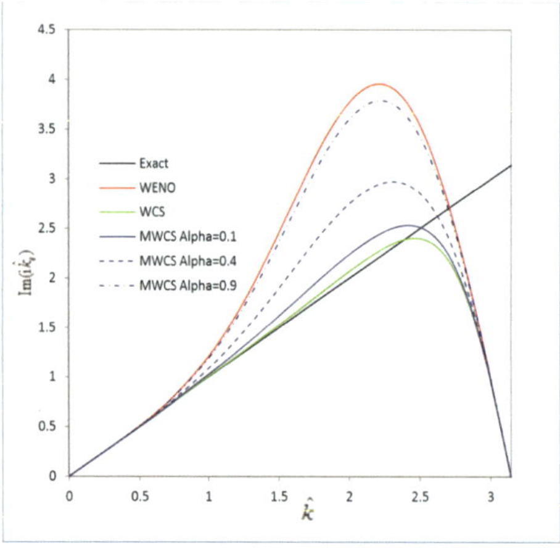

(a)

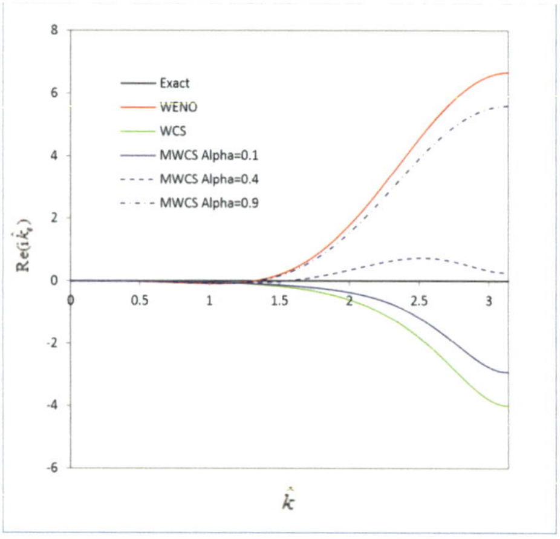

(b)

Fig. (4.12). Error analysis in in stencil E_0 (**a**) dispersion; (**b**) dissipation.

4.4.3. Shock Region: Using Only Stencil E_1

In this case, we assume that the stencils E_0 and E_2 contain shocks, therefore assume the weights,

$$\omega_0 = 0, \qquad \omega_1 = 1, \qquad \omega_2 = 0 \tag{4.68}$$

which induce the following dispersion errors

$$\begin{aligned} \mathrm{Im}(i\hat{k}_e^{(WENO)}) &= \frac{4}{3}\sin\hat{k} - \frac{1}{6}\sin 2\hat{k} \\ \mathrm{Im}(i\hat{k}_e^{(WCS)}) &= \frac{3\sin\hat{k}}{2+\cos\hat{k}} \end{aligned} \tag{4.69}$$

and the following dissipation errors

$$\begin{aligned} \mathrm{Re}(i\hat{k}_e^{(WENO)}) &= \frac{1}{2} - \frac{2}{3}\cos\hat{k} + \frac{1}{6}\cos 2\hat{k} \\ \mathrm{Re}(i\hat{k}_e^{(WCS)}) &= 0 \end{aligned} \tag{4.70}$$

The next two Figs. (**4.13a** and **4.13b**) demonstrate the conclusions that

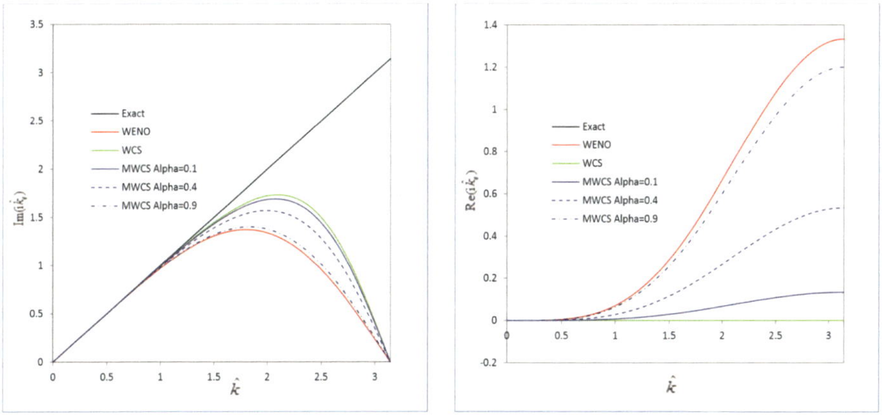

(a) (b)

Fig. (4.13). Error analysis in in stencil E_1 (**a**) dispersion; (**b**) dissipation.

- On stencil E_1, the resolution of WENO scheme is confined to the low wavenumber range;
- MWCS is better than WENO in resolution, while WCS is the best;
- WCS assumes no dissipative error which is better than MWCS, while WENO scheme is of the most dissipative one.

4.4.4. Shock Region: Using Only Stencil E_2

This last case deals with the stencil E_2, *i.e.*, considering the case that E_0 and E_1 contain shocks. Hence the following weights should be assumed,

$$\omega_0 = \omega_1 = 0, \qquad \omega_2 = 1. \tag{4.71}$$

The corresponding dispersion errors become

$$\mathrm{Im}(i\hat{k}_e^{(WENO)}) = \frac{4}{3}\sin\hat{k} - \frac{1}{6}\sin 2\hat{k}$$

$$\mathrm{Im}(i\hat{k}_e^{(WCS)}) = \frac{(8+\cos\hat{k})\sin\hat{k}}{5+4\cos\hat{k}} \tag{4.72}$$

and the dissipative errors are

$$\mathrm{Re}(i\hat{k}_e^{(WENO)}) = -\frac{1}{2} + \frac{2}{3}\cos\hat{k} - \frac{1}{6}\cos 2\hat{k}$$

$$\mathrm{Re}(i\hat{k}_e^{(WCS)}) = \frac{4\sin(\hat{k}/2)^4}{5+4\cos\hat{k}} \tag{4.73}$$

The following two Figs. (**4.14a** and **4.14b**) show us that

- MWCS has improved the resolution with respect to WENO scheme, as α decreases, while WCS achieves the highest resolution;
- On stencils E_1 and E_2, WENO scheme has the same resolution. Meanwhile, on stencils E_0 and E_2, MWCS has the same resolution;
- WENO scheme is negatively dissipated at middle and high wavenumber range, while WCS is positively dissipated. The combination of them, MWCS, achieves smaller dissipation error with respect to both WENO and WCS;
- On stencils E_1 and E_2, WENO scheme has the equal dissipation error, but with opposite signs. Meanwhile, on stencils E_0 and E_2, WCS achieves the same

property.

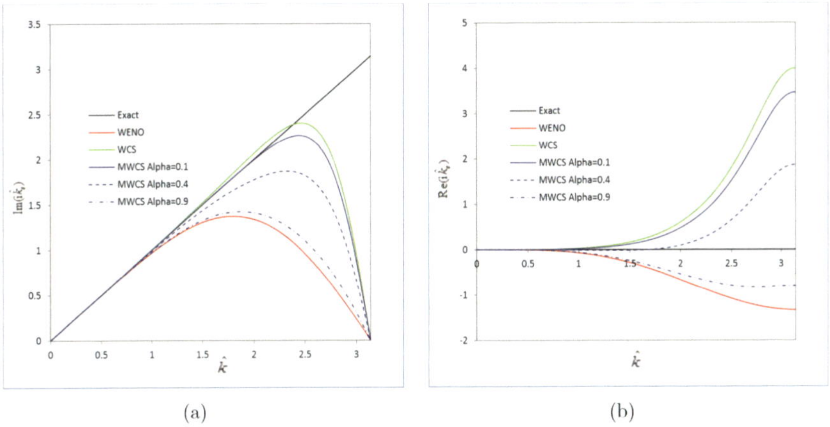

Fig. (4.14). Error analysis in in stencil E_2 (**a**) dispersion; (**b**) dissipation.

4.4.5. Numerical Results

The Euler equations which are selected one and two dimensional test cases are solved by the proposed MWCS and compared the usage of WENO.

4.4.5.1. One-dimensional Case

The one-dimensional Euler equations in vector and conservative form show as follows:

$$\frac{\partial \mathbf{U}}{\partial t}+\frac{\partial \mathbf{F}}{\partial x}=0$$
$$\mathbf{U}=\left(\rho,\rho u,E_t\right)^T \tag{4.74}$$
$$\mathbf{F}=\left(\rho u,\rho u^2+p,u\left(E_t+p\right)\right)^T.$$

where $x\in(-5,5)$, and the grid is uniform with size $h = 0.05$ (201 grid points). Steger-Warming [120] flux-splitting is used, the third-order Runge-Kutta scheme is used for the time marching.

4.4.5.2. Sod Shock-tube Problem

The capability of the MWCS on shock capturing is tested by the Sod shock-tube problem. The following is the initial conditions,

$$(\rho,u,p)=\begin{cases}(1,0,1), & t=0, x\le 0;\\ (0.125,0,0.1), & t=0, x>0,\end{cases} \quad \textbf{(4.75)}$$

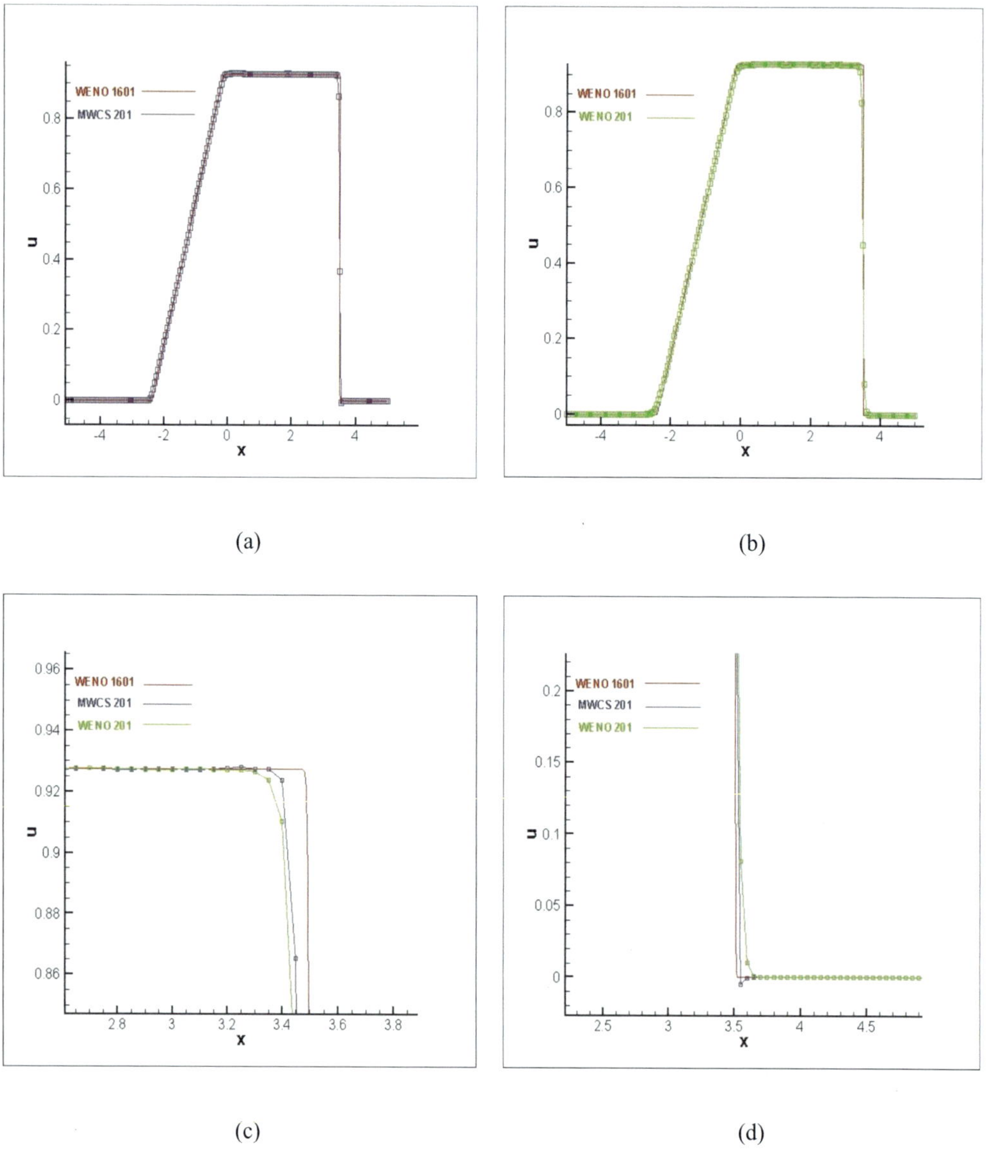

Fig. (4.15). Solution of shock tube problem (**a**) MWCS; (**b**) WENO; (**c**) enlarged comparison; (**d**) enlarged comparison.

Fig. (**4.15a-d**) display the solved velocity u at time $t = 2$. Figs. (**4.15a** and **4.15b**) respectively show the solutions of MWCS and WENO schemes in whole domain. The reference solution is treated as the one carried out by the 5th order WENO scheme based on a grid with 1601 points, labeled as WENO 1601. All the others are obtained on a coarser grid with 201 points. The solutions of MWCS (labeled as MWCS 201) and WENO scheme (labeled as WENO 201) which are different from that of WCS, are free from visible oscillations. Figs. (**4.15c** and **4.15d**) show the enlargements of the shock areas by using the three different schemes. Compared with the usage of 5th order WENO scheme, MWCS scheme could capture the discontinuity more sharply, and the solution does not present visible oscillations.

4.4.5.3. Shu-Osher Problem

Let us take a shock-entropy interaction case by Shu and Osher [130], which is solved in testing the proposed method's shock capturing capability to compare WENO with MWCS. The entropy waves are so sensitive to numerical dissipation introduced by a numerical scheme that it can be excessively damped. Equations (4.74) are solved. The initial conditions are given as follows:

$$(\rho, u, p)_0 = \begin{cases} (3.857143, 2.629369, 10.33333), & x < -4; \\ (1 + 0.2\sin(5x), 0, 1) & x \geq -4. \end{cases} \tag{4.76}$$

Fig. (**4.16**) shows the result for the solved pressure distributionat time $t = 18$ by using MWCS and WENO respectively. Figs. (**4.17-4.19**) are the comparisons of the results for the solved density distribution ρ. The reference solution is regarded as the one obtained by the fifth-order WENO scheme using a mesh of 1601 points, labeled as WENO 1601. All the other calculations are made on a coarser mesh of 201 points. Comparing with WENO (labeled WENO 201), higher resolution and sharper shock capturing are shown in the figure of the MWCS scheme (labeled MWCS 201). Figs. (**4.18** and **4.19**) shows enlargements of discontinuity areas in the shock region in detail, comparing the two different schemes. The results which are obtained by using WCS has numerical oscillations. However, it can be found that the MWCS solution is free from numerical oscillations and can capture the shock sharply and has better resolution properties than WENO.

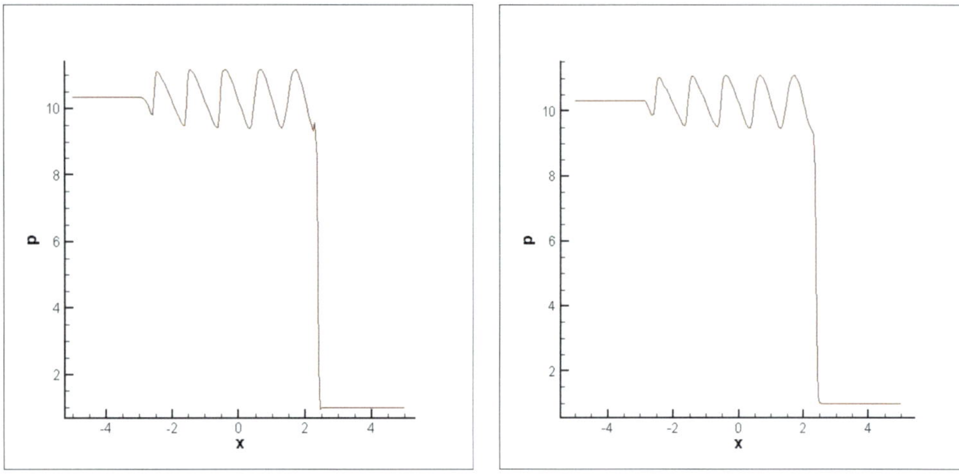

(a) (b)

Fig. (4.16). Solution of pressure to Shu-Osher problem (**a**) MWCS; (**b**) WENO.

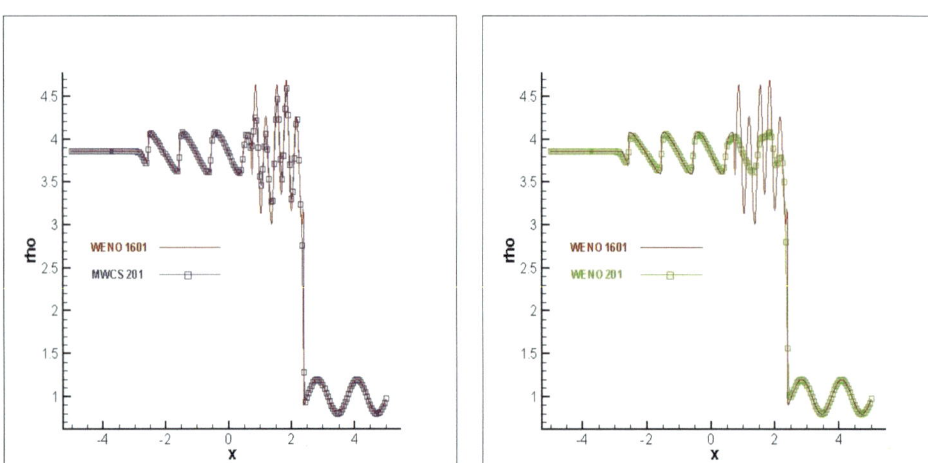

(a) (b)

Fig. (4.17). Solution of density to Shu-Osher problem (**a**) MWCS; (**b**) WENO.

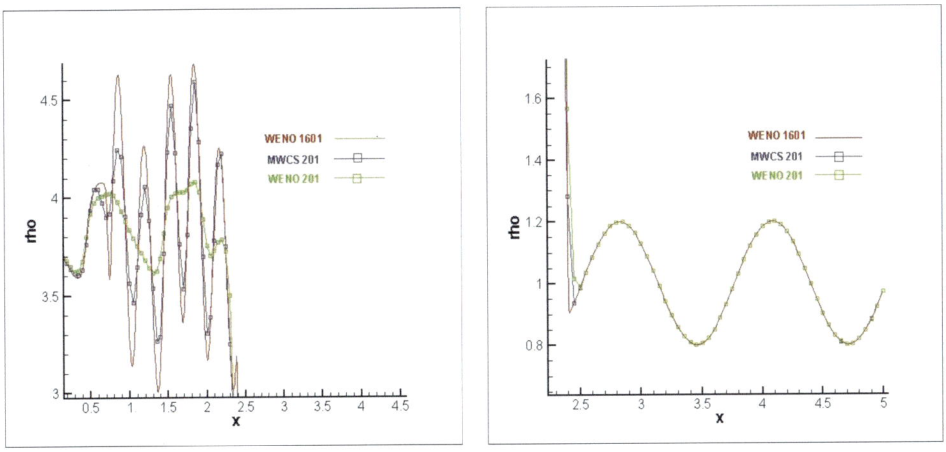

(a) (b)

Fig. (4.18). Enlarged solution of pressure to Shu-Osher problem (**a**) MWCS; (**b**) WENO.

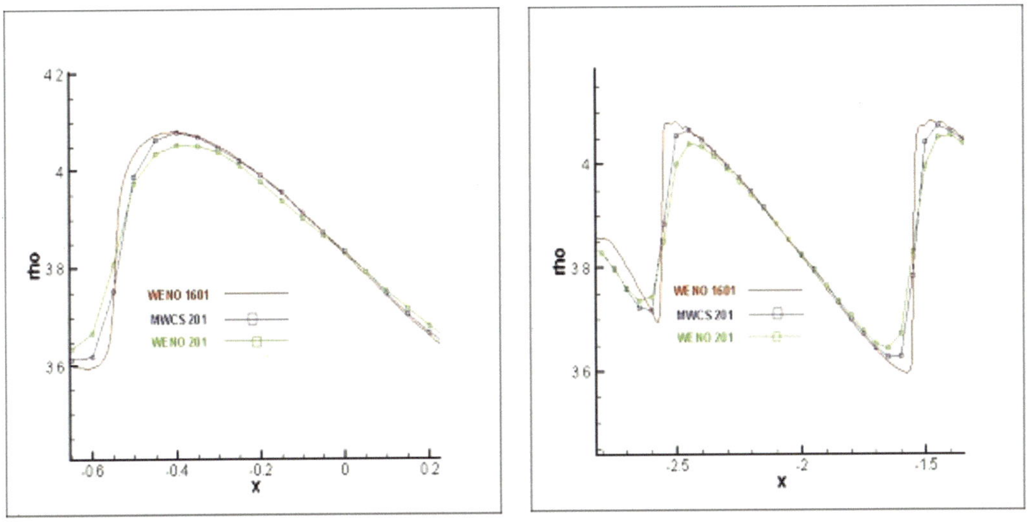

(a) (b)

Fig. (4.19). Enlarged solution of pressure to Shu-Osher problem (**a**) MWCS; (**b**) WENO.

4.4.5.4. Two-dimensional Case

The two-dimensional Euler equations in vector and conservative form is shown as,

$$\begin{aligned} &\frac{\partial \mathbf{U}}{\partial t}+\frac{\partial \mathbf{F}}{\partial x}+\frac{\partial \mathbf{G}}{\partial y}=0 \\ &\mathbf{U}=\left(\rho,\rho u,\rho v,E_t\right)^T \\ &\mathbf{F}=\left(\rho u,\rho u^2+p,\rho uv,u\left(E_t+p\right)\right)^T \\ &\mathbf{G}=\left(\rho v,\rho uv,\rho v^2+p,v\left(E_t+p\right)\right)^T. \end{aligned} \qquad \mathbf{(4.77)}$$

An incident shock case with an inflow Mach number of 2 and attach angle of 35.24° was chosen as a sample problem to compare the WENO and MWCS results with the exact solution. Since the incident shock has exact solution, it is a good prototype problem for scheme validation. It is also a very difficult problem to get sharp shock without visible oscillation for any high order scheme. Three sizes of grids are selected, a coarse grid 33×33, a middle grid 65×65, and a fine grid 129×129. The computational domain is $x \in (0,2)$, $y \in (0,1.1)$ and a uniform grids was used. We find that modified weighted compact scheme or MWCS does not have serious oscillation around the shock, better than WENO. On the other hand MWCS captured shock sharper than WENO for all grids. All of the comparisons are made by using same code and same boundary treatment but different subroutines (WENO or MWCS) for derivatives only. Figs. **(4.20-4.21)** give the pressure distribution. Comparing the two schemes, it can be found that the MWCS can capture the shock more sharply than WENO scheme; however, there are few visible numerical oscillations.

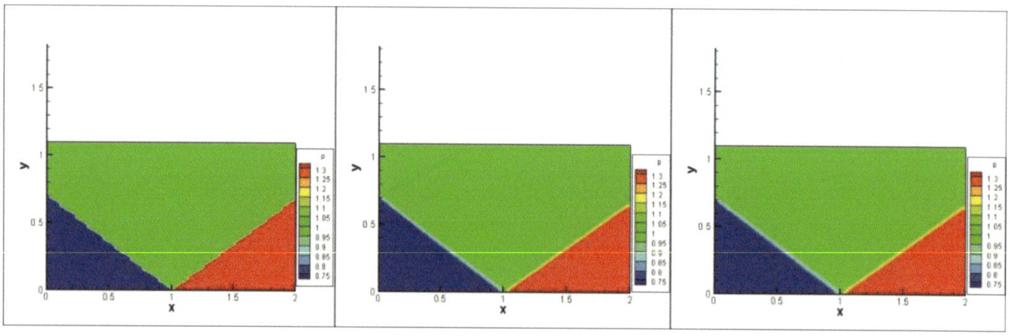

Fig. (4.20). Solution contour pressure of oblique shock reflection (**a**) analytic solution; (**b**) by MWCS; (**c**) by WENO.

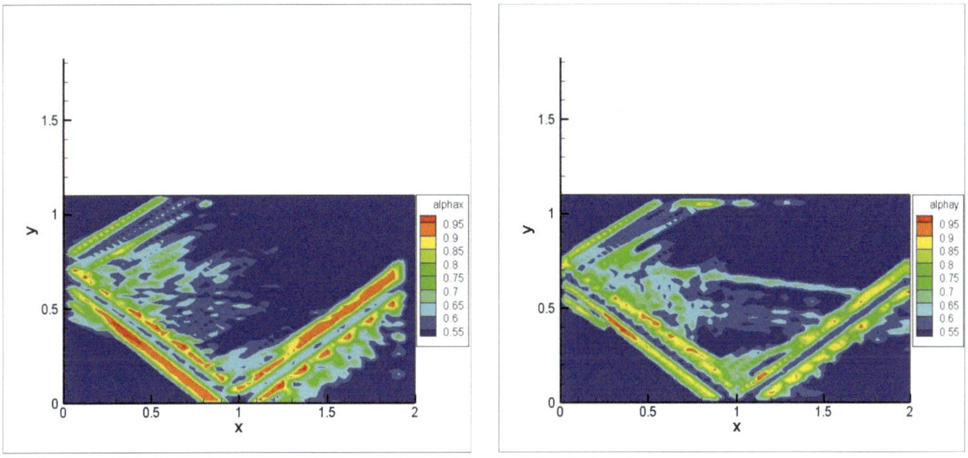

(a) **(b)**

Fig. (4.21). α distribution on 2D grids (**a**) x direction; (**b**) y direction.

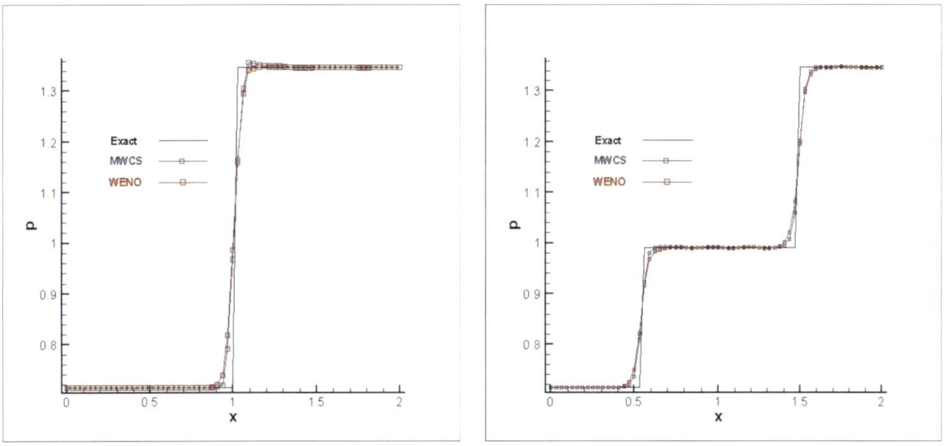

(a) **(b)**

Fig. (4.22). Pressure distribution (**a**) $y = 0$; (**b**) $y = 0.34$.

Figs. (**4.22a** and **4.22b**) respectively shows the pressure distribution for the Mach 2 oblique shock reflection at $y = 0$ and $y = 0.34$. by using MWCS and WENO. It is can be found that the MWCS can capture the sharper shock than WENO, and without generating visible simulation oscillations. Enlargements of shock regions for the pressure distribution at $y = 0.34$ are reported in Figs. (**4.23a-b**), confirming that MWCS smears the shock less than WENO, without generating visible numerical oscillations.

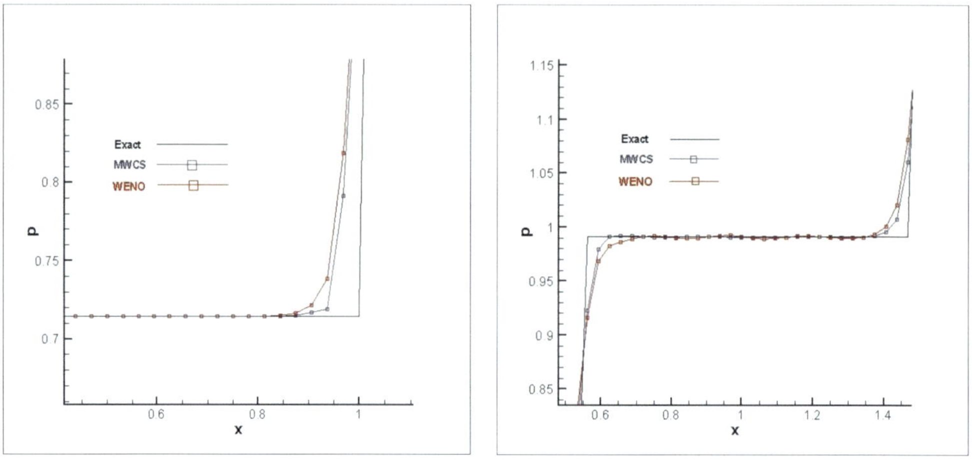

(a) (b)

Fig. (4.23). Enlarged pressure distribution (**a**) $y = 0$; (**b**) $y = 0.34$.

4.4.5.5. Some Concluding Remarks

Compared with the standard WENO, the MWCS, which is a linear combination of the WCS and WENO schemes, is found capable to capture shock and small length scales. A new mixing function formulation according to the smoothness shows that compared with the well-established WENO scheme, the proposed MWCS has higher resolution as well as lower dissipation. Numerical results carried out for one and two dimensional inviscid flow problems indicate that this method has the capability to capture the sharper shock without the generation of the visible numerical oscillation.

4.5. ANALYSIS OF LOCAL TRUNCATION ERROR

A detailed analysis of the local truncation errors, the dissipation and dispersion terms is done for the 5th-order WENO scheme and the Weighted Compact Scheme. A Fourier analysis and an investigation near shocks are performed on both schemes.

4.5.1. Local Truncation Error, Dissipation and Dispersion Terms

The WENO scheme gives following equation for derivatives:

$$F'_j \approx \frac{\hat{F}_{j+\frac{1}{2}} - \hat{F}_{j-\frac{1}{2}}}{h} = \left[-\frac{1}{3}\omega_{0,j-1/2}F_{j-3} + \left(\frac{7}{6}\omega_{0,j-1/2} + \frac{1}{3}\omega_{0,j+1/2} + \frac{1}{6}\omega_{1,j-1/2} \right) F_{j-2} \right.$$
$$+ \left(-\frac{11}{6}\omega_{0,j-1/2} - \frac{7}{6}\omega_{0,j+1/2} - \frac{5}{6}\omega_{1,j-1/2} - \frac{1}{6}\omega_{1,j+1/2} - \frac{1}{3}\omega_{2,j-1/2} \right) F_{j-1}$$
$$+ \left(\frac{11}{6}\omega_{0,j+1/2} - \frac{1}{3}\omega_{1,j-1/2} + \frac{5}{6}\omega_{1,j+1/2} - \frac{5}{6}\omega_{2,j-1/2} + \frac{1}{3}\omega_{2,j+1/2} \right) F_j$$
$$\left. + \left(\frac{1}{3}\omega_{1,j+1/2} + \frac{1}{6}\omega_{2,j-1/2} + \frac{5}{6}\omega_{2,j+1/2} \right) F_{j+1} - \frac{1}{6}\omega_{2,j+1/2}F_{j+2} \right] \Big/ h$$

By using the Taylor series expansion around j, the truncation error τ_{WENO} of the above equation is

$$\tau_{WENO} = \left(\frac{-\omega_{0,j-1/2} + \omega_{0,j+1/2} + \omega_{1,j-1/2} - \omega_{1,j+1/2}}{6} \right) h^2 \frac{\partial^3 F_j}{\partial x^3} +$$
$$+ \left(\frac{61\omega_{0,j-1/2} - 25\omega_{0,j+1/2} - 11\omega_{1,j-1/2} - \omega_{1,j+1/2} + \omega_{2,j-1/2} + 11\omega_{2,j+1/2}}{144} \right) h^3 \frac{\partial^4 F_j}{\partial x^4} +$$
$$+ \left(\frac{-91\omega_{0,j-1/2} + 19\omega_{0,j+1/2} + 9\omega_{1,j-1/2} - \omega_{1,j+1/2} - \omega_{2,j-1/2} + 9\omega_{2,j+1/2}}{240} \right) h^4 \frac{\partial^5 F_j}{\partial x^5} +$$
$$+ \left(\frac{1021\omega_{0,j-1/2} - 121\omega_{0,j+1/2} - 59\omega_{1,j-1/2} - \omega_{1,j+1/2} + \omega_{2,j-1/2} + 59\omega_{2,j+1/2}}{4320} \right) h^5 \frac{\partial^6 F_j}{\partial x^6} +$$
$$+ \left(\frac{-1163\omega_{0,j-1/2} + 83\omega_{0,j+1/2} + 41\omega_{1,j-1/2} - \omega_{1,j+1/2} - \omega_{2,j-1/2} + 41\omega_{2,j+1/2}}{10080} \right) h^6 \frac{\partial^7 F_j}{\partial x^7} + \quad \textbf{(4.78)}$$
$$+ \left(\frac{11341\omega_{0,j-1/2} - 505\omega_{0,j+1/2} - 251\omega_{1,j-1/2} - \omega_{1,j+1/2} + \omega_{2,j-1/2} + 251\omega_{2,j+1/2}}{241920} \right) h^7 \frac{\partial^8 F_j}{\partial x^8} +$$
$$+ \left(\frac{-11931\omega_{0,j-1/2} + 339\omega_{0,j+1/2} + 169\omega_{1,j-1/2} - \omega_{1,j+1/2} - \omega_{2,j-1/2} + 169\omega_{2,j+1/2}}{725760} \right) h^8 \frac{\partial^9 F_j}{\partial x^9} +$$
$$+ \left(\frac{110941\omega_{0,j-1/2} - 2041\omega_{0,j+1/2} - 1019\omega_{1,j-1/2} - \omega_{1,j+1/2} + \omega_{2,j-1/2} + 227\omega_{2,j+1/2}}{21772800} \right) h^9 \frac{\partial^{10} F_j}{\partial x^{10}} +$$
$$+ \dots$$

From (4.78), we can determine the dissipation error and the dispersion error, which are respectively the even derivative terms and the odd derivative terms of τ_{WENO}:

Dissipation error:

$$\begin{aligned}E_{WENO,dissip} = &\left(\frac{61\omega_{0,j-1/2} - 25\omega_{0,j+1/2} - 11\omega_{1,j-1/2} - \omega_{1,j+1/2} + \omega_{2,j-1/2} + 11\omega_{2,j+1/2}}{144}\right) h^3 \frac{\partial^4 F_j}{\partial x^4} + \\
&+ \left(\frac{1021\omega_{0,j-1/2} - 121\omega_{0,j+1/2} - 59\omega_{1,j-1/2} - \omega_{1,j+1/2} + \omega_{2,j-1/2} + 59\omega_{2,j+1/2}}{4320}\right) h^5 \frac{\partial^6 F_j}{\partial x^6} + \\
&+ \left(\frac{11341\omega_{0,j-1/2} - 505\omega_{0,j+1/2} - 251\omega_{1,j-1/2} - \omega_{1,j+1/2} + \omega_{2,j-1/2} + 251\omega_{2,j+1/2}}{241920}\right) h^7 \frac{\partial^8 F_j}{\partial x^8} + \\
&+ \left(\frac{110941\omega_{0,j-1/2} - 2041\omega_{0,j+1/2} - 1019\omega_{1,j-1/2} - \omega_{1,j+1/2} + \omega_{2,j-1/2} + 227\omega_{2,j+1/2}}{21772800}\right) h^9 \frac{\partial^{10} F_j}{\partial x^{10}} + \\
&+ \ldots\end{aligned} \quad \textbf{(4.79)}$$

Dispersion error:

$$\begin{aligned}E_{WENO,disp} = &\left(\frac{-\omega_{0,j-1/2} + \omega_{0,j+1/2} + \omega_{1,j-1/2} - \omega_{1,j+1/2}}{6}\right) h^2 \frac{\partial^3 F_j}{\partial x^3} + \\
&+ \left(\frac{-91\omega_{0,j-1/2} + 19\omega_{0,j+1/2} + 9\omega_{1,j-1/2} - \omega_{1,j+1/2} - \omega_{2,j-1/2} + 9\omega_{2,j+1/2}}{240}\right) h^4 \frac{\partial^5 F_j}{\partial x^5} + \\
&+ \left(\frac{-1163\omega_{0,j-1/2} + 83\omega_{0,j+1/2} + 41\omega_{1,j-1/2} - \omega_{1,j+1/2} - \omega_{2,j-1/2} + 41\omega_{2,j+1/2}}{10080}\right) h^6 \frac{\partial^7 F_j}{\partial x^7} + \\
&+ \left(\frac{-11931\omega_{0,j-1/2} + 339\omega_{0,j+1/2} + 169\omega_{1,j-1/2} - \omega_{1,j+1/2} - \omega_{2,j-1/2} + 169\omega_{2,j+1/2}}{725760}\right) h^8 \frac{\partial^9 F_j}{\partial x^9} + \\
&+ \ldots\end{aligned} \quad \textbf{(4.80)}$$

On a similar analysis for the WCS, from equation (4.57),

$$
\begin{aligned}
&\left(-3\omega_{0,j-1/2}-\frac{3}{2}\omega_{1,j-1/2}-3\omega_{2,j-1/2}+3\omega_{0,j+1/2}+\frac{3}{2}\omega_{1,j+1/2}+3\omega_{2,j+1/2}\right)\frac{H'_{j+1/2}}{h}+\\
&+\left(2\omega_{0,j-1/2}+\frac{1}{4}\omega_{1,j-1/2}\right)F'_{j-1}+\\
&+\left(3\omega_{0,j-1/2}+\frac{5}{4}\omega_{1,j-1/2}+\omega_{2,j-1/2}-2\omega_{0,j+1/2}-\frac{1}{4}\omega_{1,j+1/2}\right)F'_{j}+\\
&+\left(\frac{1}{4}\omega_{1,j+1/2}+2\omega_{2,j+1/2}\right)F'_{j+1}\approx\\
&\approx\left[-\frac{1}{2}\omega_{0,j-1/2}F_{j-2}+\left(-\frac{5}{2}\omega_{0,j-1/2}-\frac{3}{4}\omega_{1,j-1/2}+\frac{1}{2}\omega_{0,j+1/2}\right)F_{j-1}+\right.\\
&+\left(-\frac{3}{4}\omega_{1,j-1/2}-\frac{5}{2}\omega_{2,j-1/2}+\frac{5}{2}\omega_{0,j+1/2}+\frac{3}{4}\omega_{1,j+1/2}\right)F_{j}+\\
&+\left(-\frac{1}{2}\omega_{2,j-1/2}+\frac{3}{4}\omega_{1,j+1/2}+\frac{5}{2}\omega_{2,j+1/2}\right)F_{j+1}+\\
&\left.+\frac{1}{2}\omega_{2,j+1/2}F_{j+2}\right]\Big/h
\end{aligned}
\tag{4.81}
$$

by using the Taylor series expansion around j, the truncation error τ_{WCS} of the above equation is

$$
\begin{aligned}
\tau_{WCS}=&\left(\left(\omega_{0,j+1/2}-\omega_{0,j-1/2}\right)+\frac{1}{2}\left(\omega_{1,j+1/2}-\omega_{1,j-1/2}\right)+\left(\omega_{2,j+1/2}-\omega_{2,j-1/2}\right)\right)\frac{3}{h}\left(H'_{j+1/2}-F_j\right)+\\
&-\left(\left(\omega_{0,j+1/2}-\omega_{0,j-1/2}\right)+\frac{1}{2}\left(\omega_{1,j+1/2}-\omega_{1,j-1/2}\right)+\left(\omega_{2,j+1/2}-\omega_{2,j-1/2}\right)\right)\frac{3}{2}\frac{\partial F_j}{\partial x}+\\
&-\left(\left(\omega_{0,j+1/2}-\omega_{0,j-1/2}\right)+\frac{1}{2}\left(\omega_{1,j+1/2}-\omega_{1,j-1/2}\right)+\left(\omega_{2,j+1/2}-\omega_{2,j-1/2}\right)\right)\frac{h}{4}\frac{\partial^2 F_j}{\partial x^2}+\\
&+\left(\left(\omega_{0,j+1/2}-\omega_{0,j-1/2}\right)-\left(\omega_{2,j+1/2}-\omega_{2,j-1/2}\right)\right)\frac{h^2}{12}\frac{\partial^3 F_j}{\partial x^3}+
\end{aligned}
$$

$$
\begin{aligned}
&+\left(5\omega_{0,j-1/2}-\omega_{0,j+1/2}-\frac{1}{2}\omega_{1,j-1/2}+\frac{1}{2}\omega_{1,j+1/2}+\omega_{2,j-1/2}-5\omega_{2,j+1/2}\right)\frac{h^3}{48}\frac{\partial^4 F_j}{\partial x^4}+\\
&+\left(-17\omega_{0,j-1/2}+\omega_{0,j+1/2}+\omega_{1,j-1/2}+\omega_{1,j+1/2}+\omega_{2,j-1/2}-17\omega_{2,j+1/2}\right)\frac{h^4}{240}\frac{\partial^5 F_j}{\partial x^5}+\\
&+\left(45\omega_{0,j-1/2}-\omega_{0,j+1/2}-\frac{3}{2}\omega_{1,j-1/2}+\frac{3}{2}\omega_{1,j+1/2}+\omega_{2,j-1/2}-45\omega_{2,j+1/2}\right)\frac{h^5}{1440}\frac{\partial^6 F_j}{\partial x^6}+\\
&+\left(-105\omega_{0,j-1/2}+\omega_{0,j+1/2}+2\omega_{1,j-1/2}+2\omega_{1,j+1/2}+\omega_{2,j-1/2}-105\omega_{2,j+1/2}\right)\frac{h^6}{10080}\frac{\partial^7 F_j}{\partial x^7}+\\
&+\left(229\omega_{0,j-1/2}-\omega_{0,j+1/2}-\frac{5}{2}\omega_{1,j-1/2}+\frac{5}{2}\omega_{1,j+1/2}+\omega_{2,j-1/2}-229\omega_{2,j+1/2}\right)\frac{h^7}{80640}\frac{\partial^8 F_j}{\partial x^8}+\\
&+\left(-481\omega_{0,j-1/2}+\omega_{0,j+1/2}+3\omega_{1,j-1/2}+3\omega_{1,j+1/2}+\omega_{2,j-1/2}-481\omega_{2,j+1/2}\right)\frac{h^8}{725760}\frac{\partial^9 F_j}{\partial x^9}+\\
&+\left(989\omega_{0,j-1/2}-\omega_{0,j+1/2}-\frac{7}{2}\omega_{1,j-1/2}+\frac{7}{2}\omega_{1,j+1/2}+\omega_{2,j-1/2}-989\omega_{2,j+1/2}\right)\frac{h^9}{7257600}\frac{\partial^{10} F_j}{\partial x^{10}}+\\
&+\left(-2009\omega_{0,j-1/2}+\omega_{0,j+1/2}+4\omega_{1,j-1/2}+4\omega_{1,j+1/2}+\omega_{2,j-1/2}-2009\omega_{2,j+1/2}\right)\frac{h^{10}}{79833600}\frac{\partial^{11} F_j}{\partial x^{11}}+\\
&+\ldots
\end{aligned}
\quad (4.82)
$$

Equation (4.82) is not purely dependent on the derivatives of F at j; it contains the derivative $H'_{j+1/2}$. To remove this dependence on H, an expansion on F around $j \pm 1/2$ needs to be considered.

To satisfy $F'_j = \dfrac{H'_{j+1/2}-H'_{j-1/2}}{h} = \dfrac{\hat{F}_{j+1/2}-\hat{F}_{j-1/2}}{h}$, we must have

$$\hat{F}_{j\pm 1/2} = F_{j\pm 1/2} + \sum_{k=1}^{m-1} c_k h^{2k} \frac{\partial^{2k}}{\partial x^{2k}} F_{j\pm 1/2} + O\left(h^{2m+1}\right),$$

where c_k are constants.

For $m = 7$, for example,

$$c_1 = -\frac{1}{24};\quad c_2 = \frac{7}{5760};\quad c_3 = -\frac{31}{967680};$$
$$c_4 = \frac{127}{154828800};\quad c_5 = -\frac{73}{3503554560};\quad c_6 = \frac{1414477}{2678117105664000}.$$

By Taylor expansion around j,

$$\frac{H'_{j+1/2}}{h}=\frac{\hat{F}_{j+1/2}}{h}=\frac{F_j}{h}+\frac{1}{2}F'_j+\frac{h}{12}F''_j-\frac{h^3}{720}F_j^{(4)}+\frac{h^5}{30240}F_j^{(6)}-\frac{h^7}{1209600}F_j^{(8)}+$$
$$+\frac{31h^9}{743178240}F_j^{(10)}+O\left(h^{11}\right)$$

Since $\omega_{0,j\pm1/2}+\omega_{1,j\pm1/2}+\omega_{2,j\pm1/2}=1$, the truncation error becomes

$$\begin{aligned}
\tau_{WCS}&=\left(-\omega_{0,j-1/2}+\omega_{2,j-1/2}+\omega_{0,j+1/2}-\omega_{2,j+1/2}\right)\frac{h^2}{12}\frac{\partial^3F_j}{\partial x^3}+\\
&+\left(5\omega_{0,j-1/2}-\frac{3}{5}\omega_{1,j-1/2}+\omega_{2,j-1/2}-\omega_{0,j+1/2}+\frac{3}{5}\omega_{1,j+1/2}-5\omega_{2,j+1/2}\right)\frac{h^3}{48}\frac{\partial^4F_j}{\partial x^4}+\\
&+\left(-17\omega_{0,j-1/2}+\omega_{1,j-1/2}+\omega_{2,j-1/2}+\omega_{0,j+1/2}+\omega_{1,j+1/2}-17\omega_{2,j+1/2}\right)\frac{h^4}{240}\frac{\partial^5F_j}{\partial x^5}+\\
&+\left(45\omega_{0,j-1/2}-\frac{10}{7}\omega_{1,j-1/2}+\omega_{2,j-1/2}-\omega_{0,j+1/2}+\frac{10}{7}\omega_{1,j+1/2}-45\omega_{2,j+1/2}\right)\frac{h^5}{1440}\frac{\partial^6F_j}{\partial x^6}+\\
&+\left(-105\omega_{0,j-1/2}+2\omega_{1,j-1/2}+\omega_{2,j-1/2}+\omega_{0,j+1/2}+2\omega_{1,j+1/2}-105\omega_{2,j+1/2}\right)\frac{h^6}{10080}\frac{\partial^7F_j}{\partial x^7}+\\
&+\left(229\omega_{0,j-1/2}-\frac{13}{5}\omega_{1,j-1/2}+\omega_{2,j-1/2}-\omega_{0,j+1/2}+\frac{13}{5}\omega_{1,j+1/2}-229\omega_{2,j+1/2}\right)\frac{h^7}{80640}\frac{\partial^8F_j}{\partial x^8}+\\
&+\left(-481\omega_{0,j-1/2}+3\omega_{1,j-1/2}+\omega_{2,j-1/2}+\omega_{0,j+1/2}+3\omega_{1,j+1/2}-481\omega_{2,j+1/2}\right)\frac{h^8}{725760}\frac{\partial^9F_j}{\partial x^9}+\\
&+\left(989\omega_{0,j-1/2}-\frac{36}{11}\omega_{1,j-1/2}+\omega_{2,j-1/2}-\omega_{0,j+1/2}+\frac{36}{11}\omega_{1,j+1/2}-989\omega_{2,j+1/2}\right)\frac{h^9}{7257600}\frac{\partial^{10}F_j}{\partial x^{10}}+\\
&+\left(-2009\omega_{0,j-1/2}+4\omega_{1,j-1/2}+\omega_{2,j-1/2}+\omega_{0,j+1/2}+4\omega_{1,j+1/2}-2009\omega_{2,j+1/2}\right)\frac{h^{10}}{79833600}\frac{\partial^{11}F_j}{\partial x^{11}}+\\
&+\ldots
\end{aligned} \qquad \textbf{(4.83)}$$

From (4.83), we can determine the dissipation error and the dispersion error, which are respectively the even derivative terms and the odd derivative terms of τ_{WCS}:

Dissipation error:

$$\begin{aligned}E_{WCS,dissip} &= \left(5\omega_{0,j-1/2} - \frac{3}{5}\omega_{1,j-1/2} + \omega_{2,j-1/2} - \omega_{0,j+1/2} + \frac{3}{5}\omega_{1,j+1/2} - 5\omega_{2,j+1/2}\right)\frac{h^3}{48}\frac{\partial^4 F_j}{\partial x^4} + \\ &+\left(45\omega_{0,j-1/2} - \frac{10}{7}\omega_{1,j-1/2} + \omega_{2,j-1/2} - \omega_{0,j+1/2} + \frac{10}{7}\omega_{1,j+1/2} - 45\omega_{2,j+1/2}\right)\frac{h^5}{1440}\frac{\partial^6 F_j}{\partial x^6} + \\ &+\left(229\omega_{0,j-1/2} - \frac{13}{5}\omega_{1,j-1/2} + \omega_{2,j-1/2} - \omega_{0,j+1/2} + \frac{13}{5}\omega_{1,j+1/2} - 229\omega_{2,j+1/2}\right)\frac{h^7}{80640}\frac{\partial^8 F_j}{\partial x^8} + \\ &+\left(989\omega_{0,j-1/2} - \frac{36}{11}\omega_{1,j-1/2} + \omega_{2,j-1/2} - \omega_{0,j+1/2} + \frac{36}{11}\omega_{1,j+1/2} - 989\omega_{2,j+1/2}\right)\frac{h^9}{7257600}\frac{\partial^{10} F_j}{\partial x^{10}} + \\ &+\ldots\end{aligned} \tag{4.84}$$

Dispersion error:

$$\begin{aligned}E_{WCS,disp} &= \left(-\omega_{0,j-1/2} + \omega_{2,j-1/2} + \omega_{0,j+1/2} - \omega_{2,j+1/2}\right)\frac{h^2}{12}\frac{\partial^3 F_j}{\partial x^3} + \\ &+\left(-17\omega_{0,j-1/2} + \omega_{1,j-1/2} + \omega_{2,j-1/2} + \omega_{0,j+1/2} + \omega_{1,j+1/2} - 17\omega_{2,j+1/2}\right)\frac{h^4}{240}\frac{\partial^5 F_j}{\partial x^5} + \\ &+ -105\omega_{0,j-1/2} + 2\omega_{1,j-1/2} + \omega_{2,j-1/2} + \omega_{0,j+1/2} + 2\omega_{1,j+1/2} - 105\omega_{2,j+1/2} \frac{h^6}{10080}\frac{\partial^7 F_j}{\partial x^7} + \\ &+\left(-481\omega_{0,j-1/2} + 3\omega_{1,j-1/2} + \omega_{2,j-1/2} + \omega_{0,j+1/2} + 3\omega_{1,j+1/2} - 481\omega_{2,j+1/2}\right)\frac{h^8}{725760}\frac{\partial^9 F_j}{\partial x^9} + \\ &+\ldots\end{aligned} \tag{4.85}$$

It is important to mention that the local truncation errors obtained in above equations are valid for any type of function. In the case of continuous smooth functions (*e.g.* sine wave), the values of the nonlinear weights $\omega_{0,j\pm1/2}$, $\omega_{1,j\pm1/2}$, and $\omega_{2,j\pm1/2}$ revert back to the optimal weights C_0, C_1 and C_2, respectively (WENO: $C_0 = 1/10$, $C_1 = 6/10$, $C_2 = 3/10$; WCS: $C_0 = 1/18$, $C_1 = 8/9$, $C_2 = 1/18$), since the values of the "smoothness" indicators $IS_{i,j\pm1/2}$, ($i = 1,2,3$) are the same. Then, the WCS and the WENO scheme achieve their highest possible order, 6th and 5th, respectively. In the case of a function with discontinuities, since the values of the "smoothness" indicators vary at regions near these discontinuities, the nonlinear weights differ from the optimal values, and so the order of the truncation errors of the schemes is reduced at these regions. The use of nonlinear weights in the calculations allows the schemes to generate results with less oscillation near discontinuities, while keeping high order accuracy everywhere else.

4.6. MODIFIED UPWIND COMPACT SCHEME (MUCS)

For the modified upwinding compact scheme (MUCS), which is a combination of upwinding compact scheme and WENO, the final matrix becomes:

$$\begin{aligned}&\frac{1}{4}\Omega H'_{j-\frac{3}{2}}+H'_{j-\frac{1}{2}}+\frac{1}{2}\Omega H'_{j+\frac{1}{2}}=\Omega*[-\frac{1}{60}H_{j-\frac{5}{2}}-\frac{31}{48}H_{j-\frac{3}{2}}-\frac{1}{3}H_{j-\frac{1}{2}}+\frac{11}{12}H_{j+\frac{1}{2}}+\frac{1}{12}H_{j+\frac{3}{2}}-\frac{1}{240}H_{j+\frac{5}{2}}]\\&+(1-\Omega)*[\omega_{0,j-1/2}(\frac{1}{3}F_{j-3}-\frac{7}{6}F_{j-2}+\frac{11}{6}F_{j-1})+\omega_{1,j-1/2}(-\frac{1}{6}F_{j-2}+\frac{5}{6}F_{j-1}+\frac{1}{3}F_{j})\\&\quad+\omega_{2,j-1/2}(\frac{1}{3}F_{j-1}+\frac{5}{6}F_{j}-\frac{1}{6}F_{j+1})]\end{aligned} \tag{4.86}$$

4.6.1. MUCS for 1-D Euler Equation

A shock-entropy interaction case is given to compare WENO with MUCS.The governing equations are 1D Euler equations:

$$\frac{\partial U}{\partial t}+\frac{\partial F}{\partial x}=0 \tag{4.87}$$

$$U=(\rho,\rho u,E)^T;\quad F=(\rho u,\rho uu+p,u(E+p))^T$$

The initial conditions are given as follows:

$$(\rho,u,p)_0=\begin{cases}(3.857143,2.629369,10.33333), & x<-4;\\(1+0.2\sin(5x),0,1) & x\geq-4.\end{cases} \tag{4.88}$$

For solving the Euler equations, we adopt a three step TVD Runge-Kutta scheme in time marching along with a Lax-Friedrich flux vector splitting. The derivatives in terms of splitting flux F^+, F^- are calculated through using WENO and MUCS.

The MUCS scheme has much better resolution for small length scales than the fifth order WENO on a coarser grid, which gives N = 200 (Figs. **4.24a-b**).

4.6.2. MUCS for 2-D Euler Equations

The WENO and MUCS results with the exact solution are compared based on an incident shock case with an inflow Mach number of 2 and attach angle of 35.24^o.

The incident shock is a typical prototype problem for scheme validation cause it has exact solution. No matter using any high order scheme, it is still a difficult problem to get sharp shocks without visible oscillation. Three sizes of grids, which respectively are a coarse grid 33×33, a middle grid 65×65, and a fine grid 129×129, are chosen here. The computational domain is defined as $x \in (0,2)$, $y \in (0,1.1)$ and a uniform grids was adopted. It is found that this MCS scheme did well on grids with coarse and middle size, while on the fine grids of 129×129, still has oscillations. However, MUCS does not present serious oscillation, which is better than WENO after the second shock. Secondly, for all grids, MUCS can capture the sharper shock than WENO. The mixing function block the UCS at red area by using WENO at blue area by UCS. All of the comparisons of different subroutines (WENO or MUCS) are made based on same code and same boundary conditions for derivatives. Figs. (**4.25** to **4.27**) give the numerical results on a coarse grid of 33×33. Figs. (**4.28** to **4.30**) show the numerical results on a middle grid of 65×65. The results on the fine grid 129×129 are shown on Figs. (**4.31** to **4.34**). From these figures we can find the seventh order MUCS results have a good agreement with the exact solution and are better than the result obtained by fifth order WENO scheme.

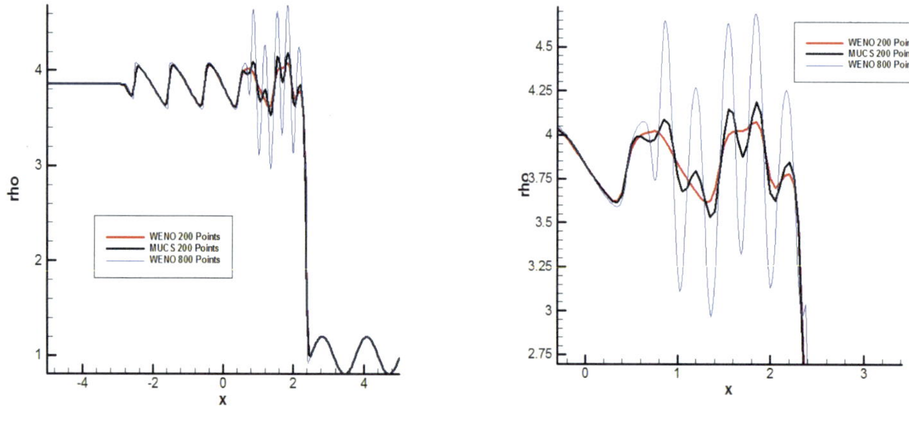

(a) (b)

Fig. (4.24). Numerical tests for 1-D shock-entropy wave interaction problem, t = 1.8, N = 200. (b) is locally enlarged.

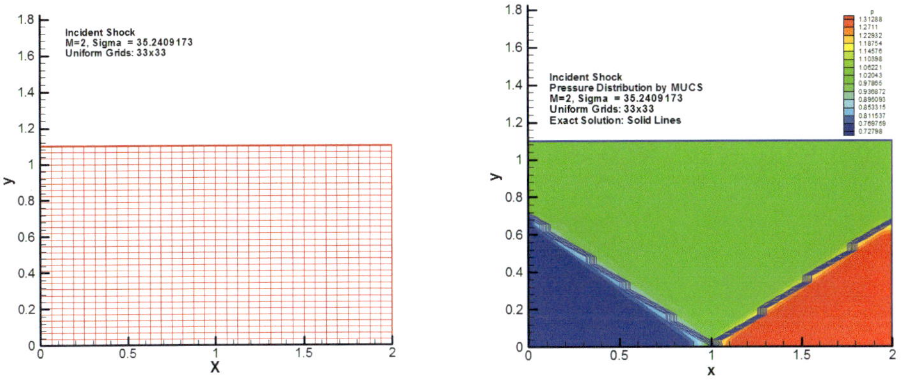

(a) **(b)**

Fig. (4.25). Numerical test for 2D incident shock on coarse grids (**a**) Grids (33×33) (**b**) Pressure Contour.

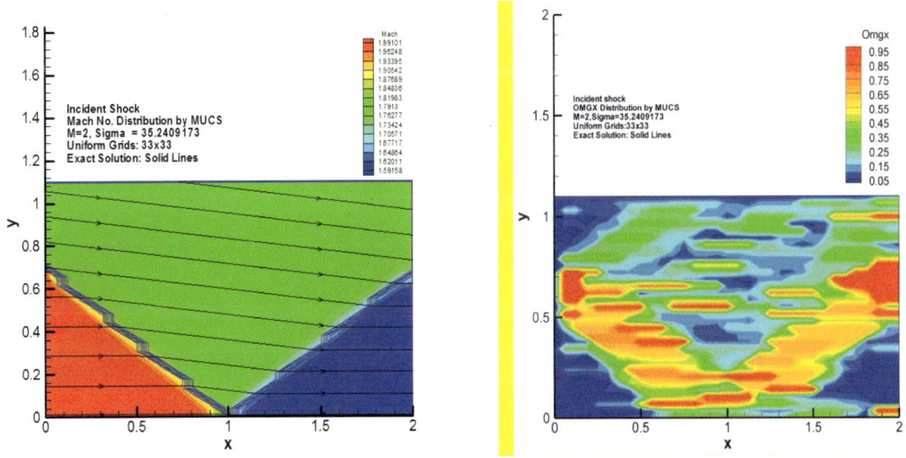

(a) **(b)**

Fig. (4.26). Numerical test for 2-D incident shock on coarse grids (**a**) Mach number (**b**) Control function (Red is WENO and blue is UCS).

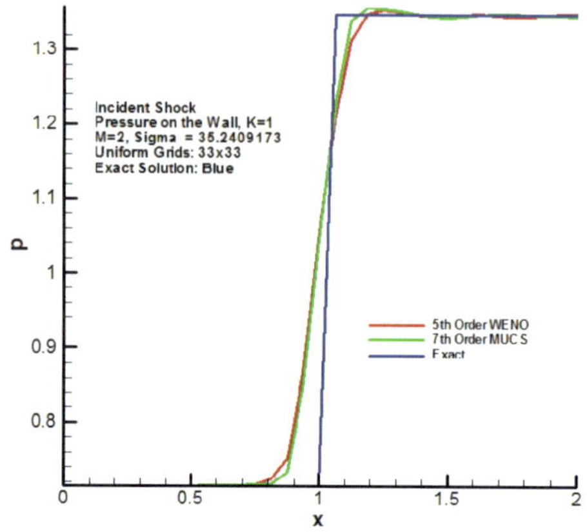

(a)

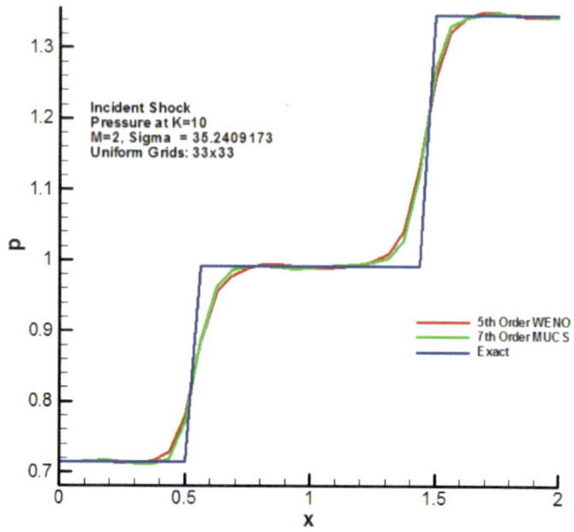

(b)

Fig. (4.27). Pressure distribution on coarse grids (33×33) (**a**) On the wall k = 1 (**b**) k = 10.

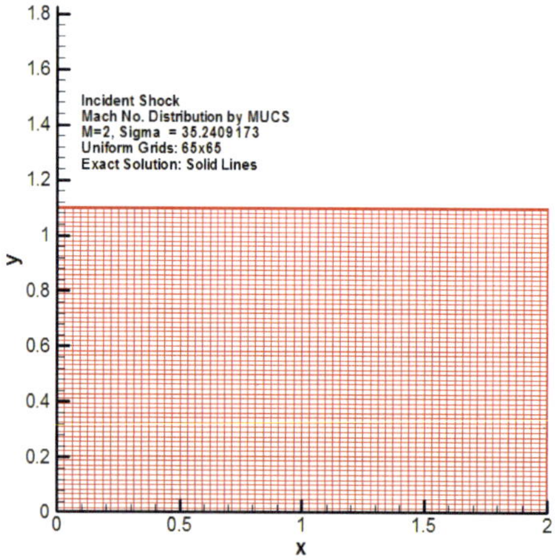

(a)

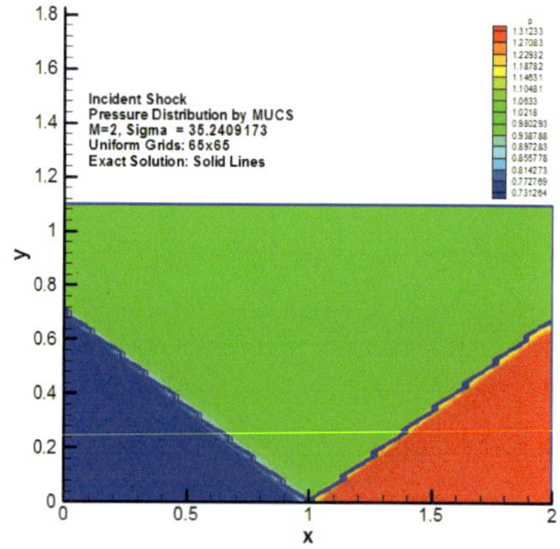

(b)

Fig. (4.28). Numerical test for 2D incident shock on middle grids (**a**) Grids (65×65) (**b**) Pressure contour.

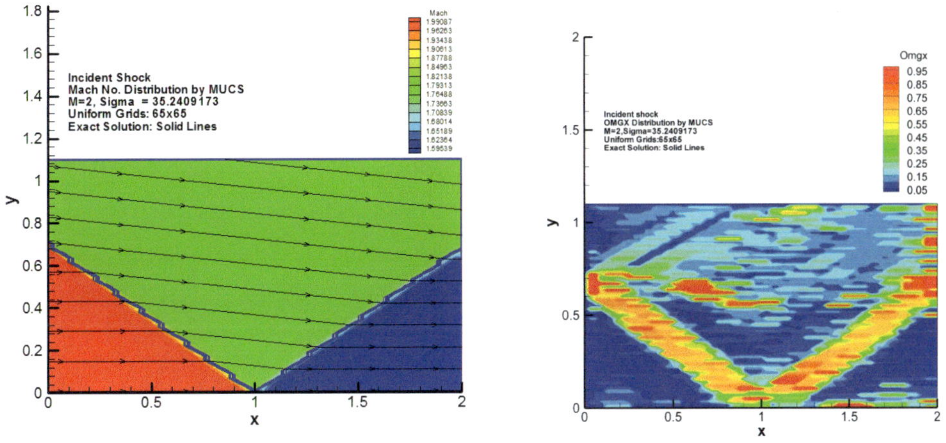

(a) **(b)**

Fig. (4.29). Numerical test for 2-D incident shock on middle grids (**a**) Mach number (**b**) Control function (Red is WENO and blue is UCS).

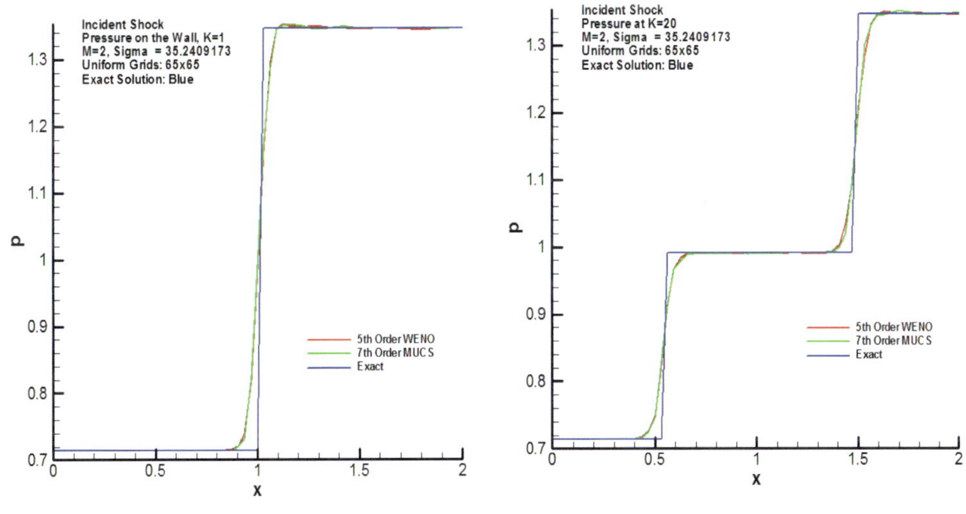

(a) **(b)**

Fig. (4.30). Pressure distribution on middle grids (65×65) (**a**) On the wall k = 1 (**b**) k = 20.

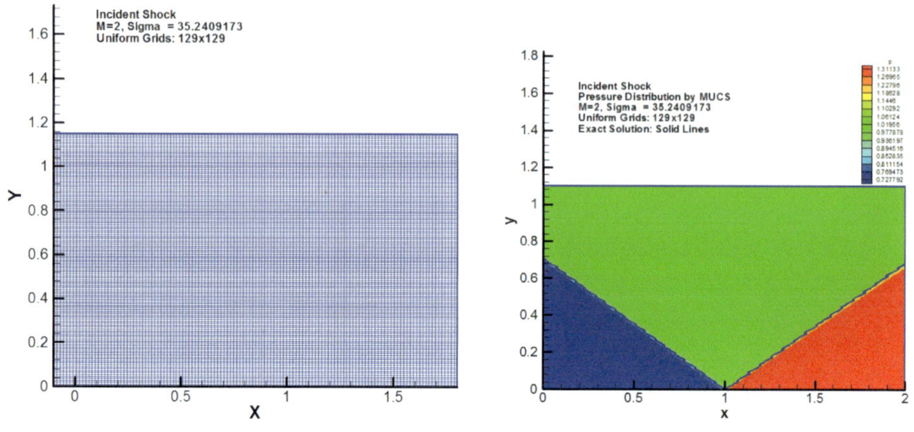

(a) **(b)**

Fig. (4.31). Numerical test for 2D incident shock on fine grids (**a**) Grids (129×129) (**b**) Pressure contour.

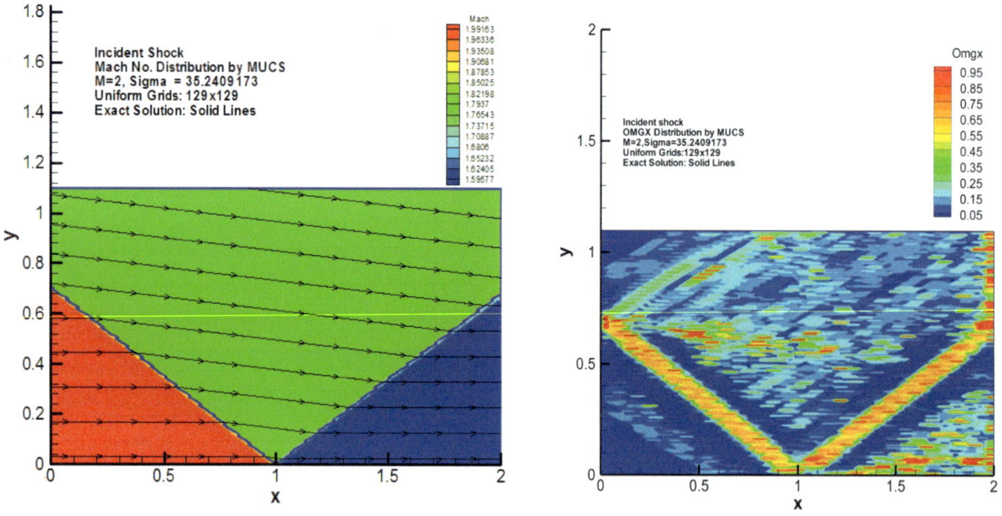

(a) **(b)**

Fig. (4.32). Numerical test for 2-D incident shock on fine grids (**a**) Mach number (**b**) Control function (Red is WENO and blue is UCS).

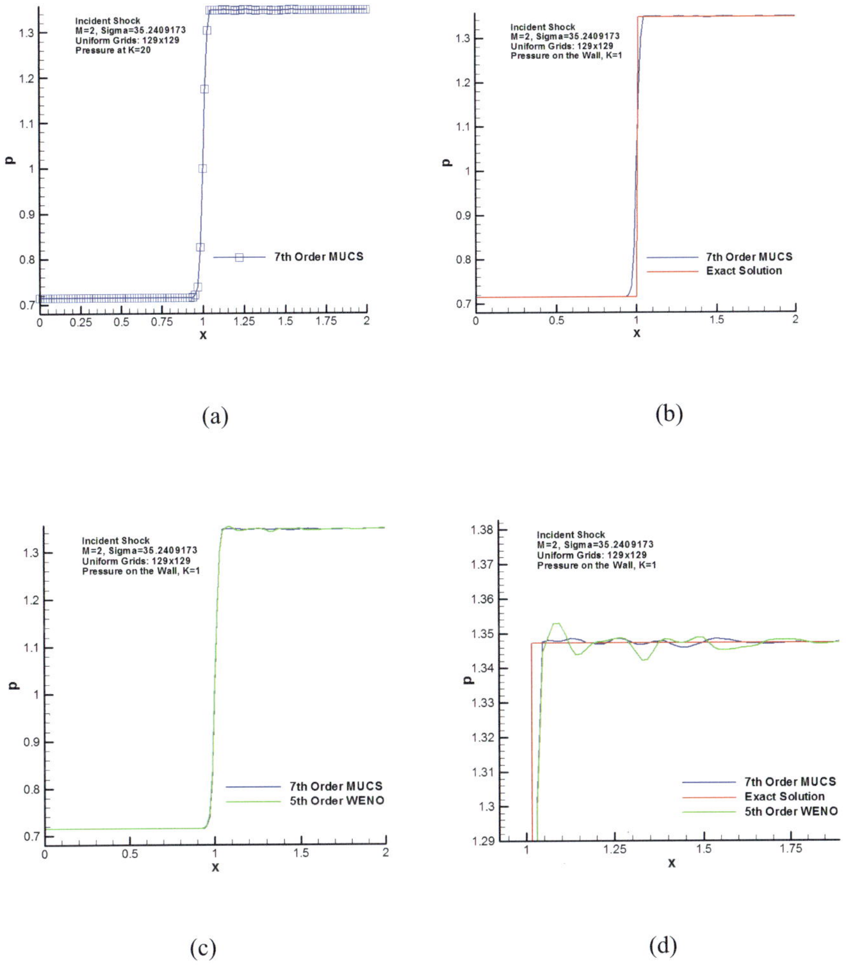

Fig. (4.33). Pressure distribution at wall (k = 1) on fine grids (129 × 129) (**a**) 7th order MUCS (**b**) MUCS and Exact (**c**) 7th order MUCS and 5th order WENO (**d**) Locally enlarged comparison.

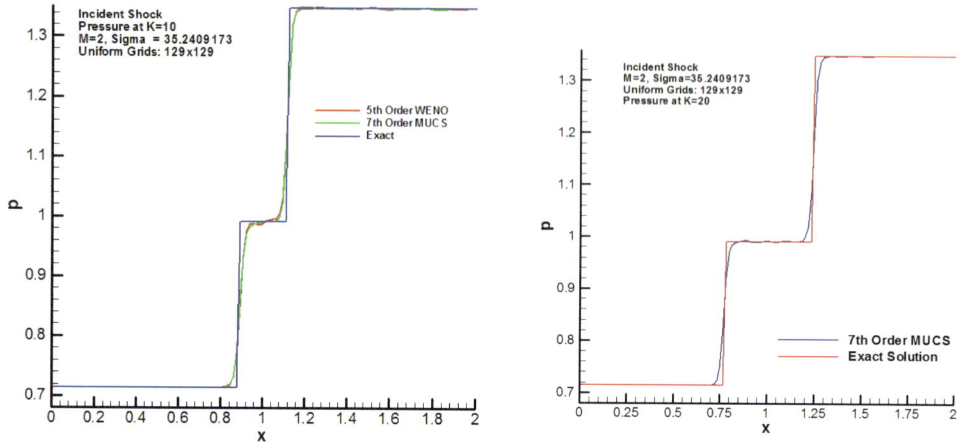

(a) (b)

Fig. (4.34). Pressure distribution on fine grids (129×129) (**a**) k = 10 (**b**) k = 20.

4.6.3. MUCS for 2-D Shock Boundary Layer Interaction

4.6.3.1. 2-D N-S Code Validation

For better testing whether the current code can work for 2 or 3-dimensional shock and boundary layer interaction, a 2-dimensional flow with the incident shock boundary layer interaction (Fig. **4.35**) was investigated by Xie *et al.* [146]. The Reynolds number and the Mach number is respectively set as 10^5 and 2.15. The overall pressure ratio is chosen to be 1.55. The inflow condition was set following the investigation of Degrez *et al.* [147] for the comparison. Their experimental result gave the shock and boundary layer interaction which is laminar and two-dimensional. Therefore, we could simulate a two-dimensional case and compare it with their experimental and computational results. Fig. (**4.36**) shows that the computational grid is 257×257, the grid stretching is set as 1.01 in the streamwise direction and 1.015 in normal direction. A two-dimensional Navier-Stokes equation is solved to be the governing equation. Fig. (**4.37**) gives the computational results, and Figs. (**4.38** and **4.39**) give a comparison between our computational results and the Degrez's results. From Fig. (**4.38**), we can find Degrez's computation did not match his experimental results very well, and our computational results does not either. However, our computation is closer to his experimental results than his computation.

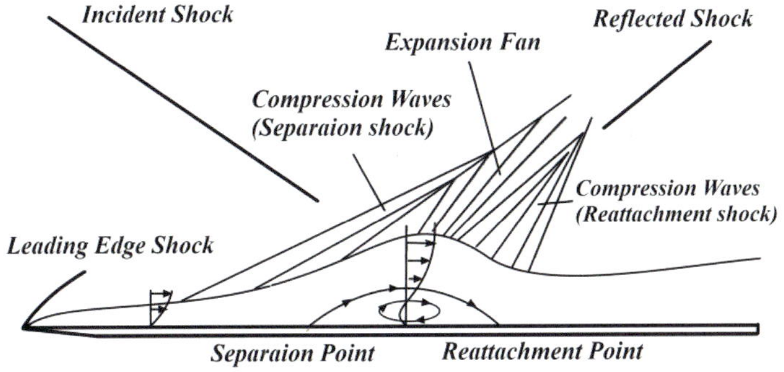

Fig. (4.35). Sketch of incident shock-boundary layer interaction.

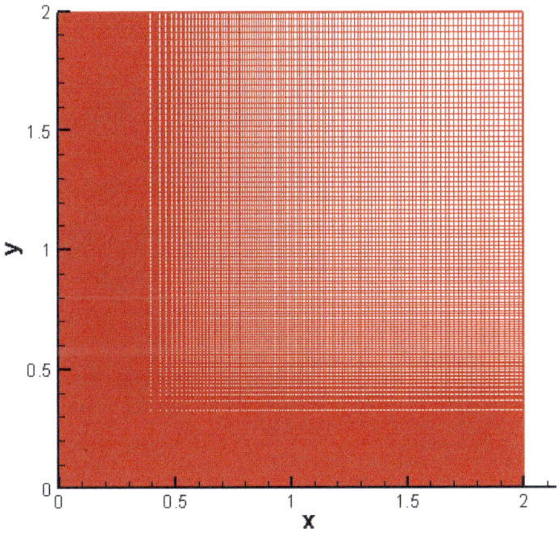

Fig. (4.36). Computation Grids (257×257).

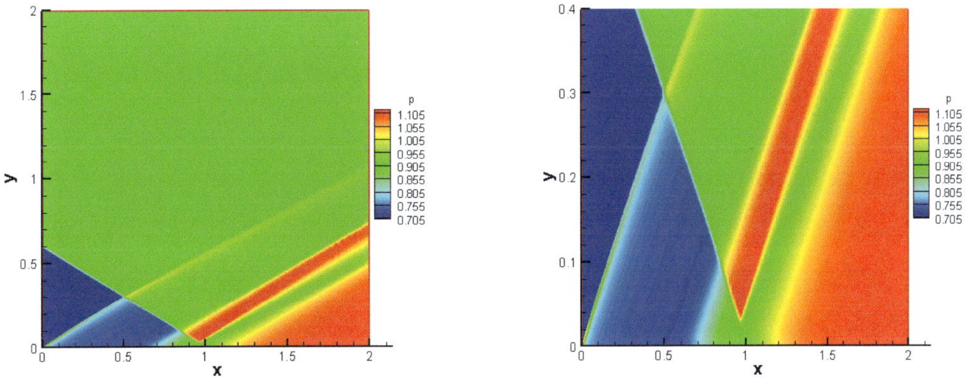

(a) (e)

Fig. 4.37 contd.....

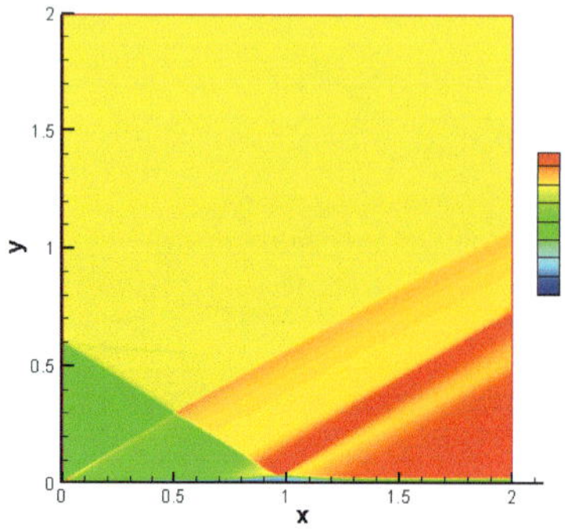

(b)

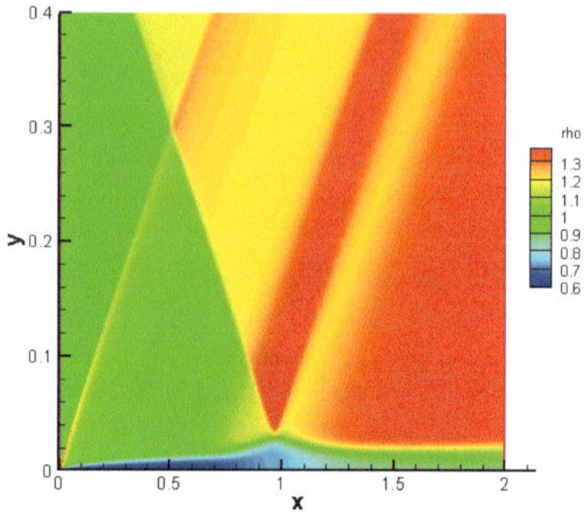

(f)

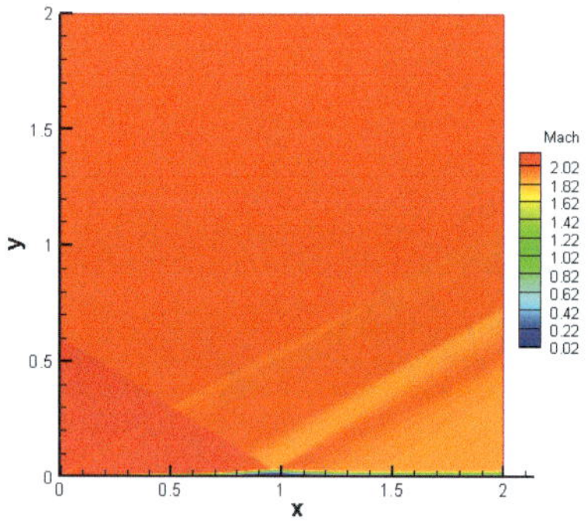

(c)

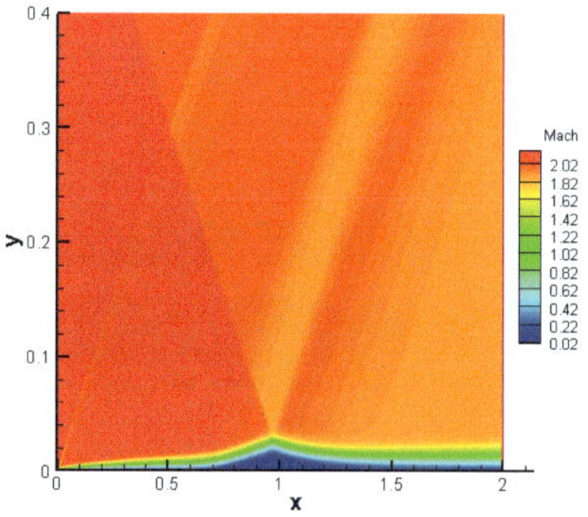

(g)

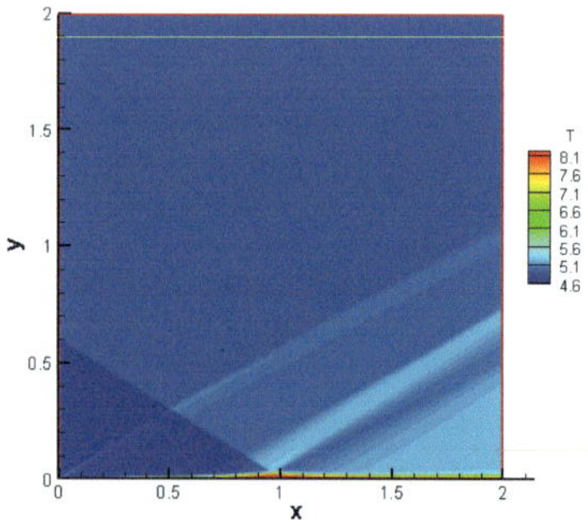

(d)

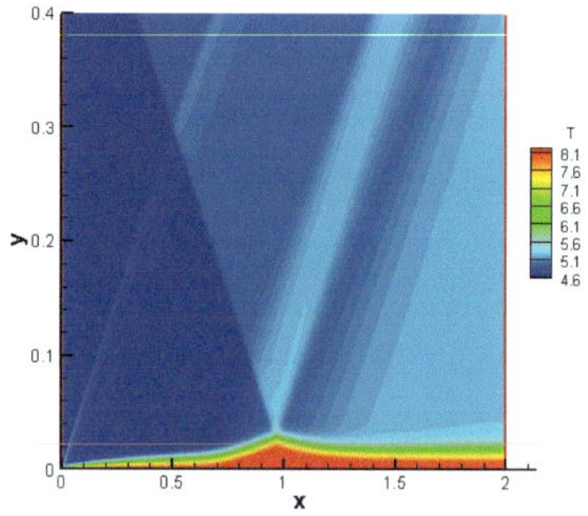

(h)

Fig. (4.37). Left is a normal view. (**a**) Pressure distribution (**b**) Density distribution (**c**) Mach number distribution (**d**) Temperature distribution Right is stretched in the normal direction by a factor of 5. (**e**) Pressure distribution (**f**) Density distribution (**g**) Mach number distribution (**h**) Temperature distribution.

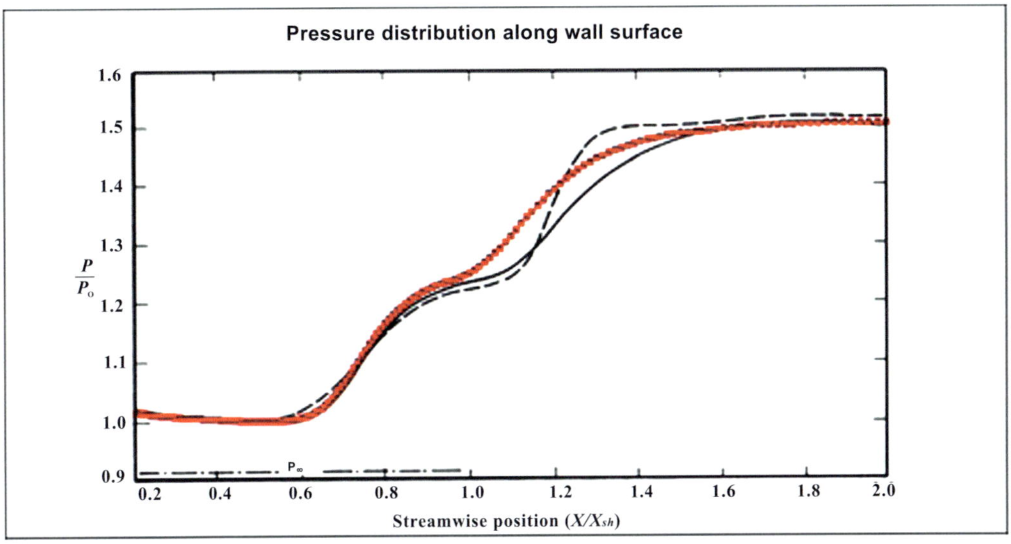

Fig. (4.38). Comparison of pressure distribution on the wall surface. (The red one is our computation, the black dash one and solid one are Degrez's computation and experiment respectively).

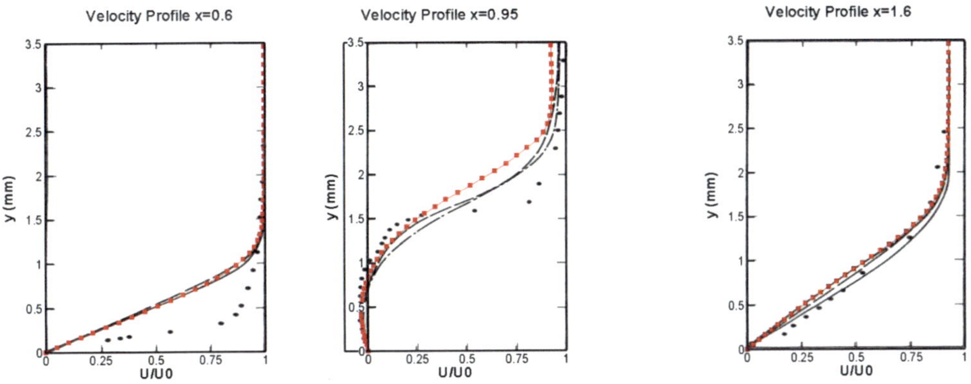

Fig. (4.39). Comparison of velocity profiles at different locations (The red one is our computation, the black solid line and black square are Degrez's computation and experiment respectively).

4.6.3.2. Computation of 2-D Incident Shock-boundary Layer Interaction

For comparing the performance between our MUCS scheme and the standard WENO scheme, a case of incident shock with Ma=3 and Reynolds number Re = $3\times10^4$and the attack angle 35.24° is selected. Three scales of grids were selected: a coarse grid 33×33, a middle grid 65×65, and a fine grid 129×129. The solution with a grid 57x257 is regarded as a reference solution. Fig. (**4.38**) shows a contour distribution of pressure with stream track. Many structures are captured clearly such as the incident shock, leading shock, separation shock, expansion waves and reflecting shock. A separation bubble structure with 5 vertex points is captured. Fig. (**4.40**) shows the control function distribution of OMGX and OMGZ in x direction and z direction. The blue area represents where compact scheme was applied and the red area is the area where the WENO scheme was used. The results of grids 257×257 are used here to be a reference. Figs. (**4.41** to **4.44**) respectively give the grid distribution, Mach number, pressure, control function OMGX, OMGZ obtained by MUCS based on a 129×129 grid. The results based on the grids of 129×129 by using MUCS have good agreement with our reference results. A comparison between seventh order MUCS and fifth order WENO schemes with the reference results on the pressure distribution of the wall surface is given in Figs. (**4.45** and **4.46**). As can be seen, the results obtained by MUCS are more similar to the reference results. The wall surface skin friction coefficients on three levels of grids are shown in Fig. (**4.47**) and the grid convergence is found fine. Table **4.1** provides a comparison of the size of the separation bubbles on three grid levels and it is found that the grid convergence is acceptable. All comparisons are made by a same code at same time step with different schemes and different grids.

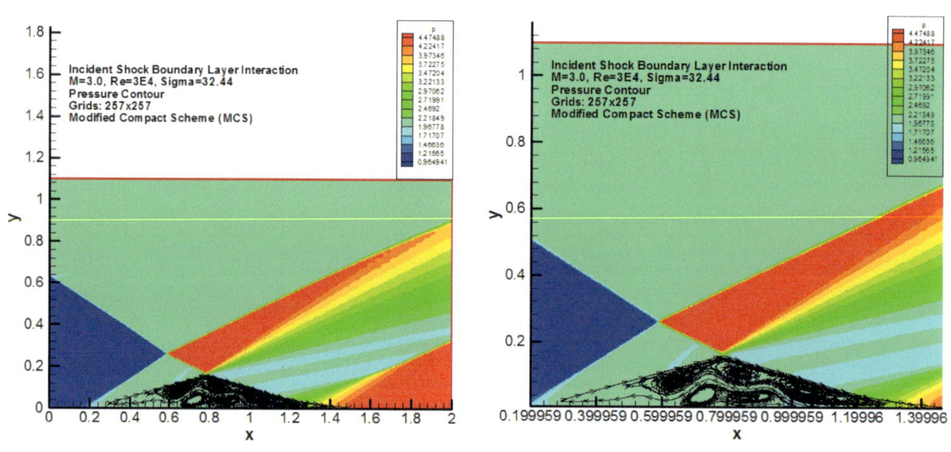

(a) (b)

Fig. (4.40). (**a**) Pressure distribution and stream traces by MUCS on grid of 257×257 as reference (**b**) locally enlarged.

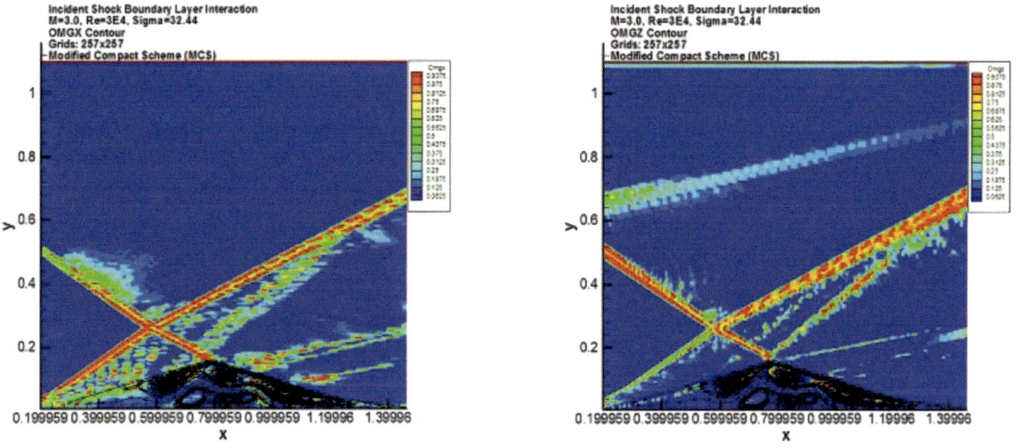

(a) (b)

Fig. (4.41). Control function on girds of 257×257 (**a**) Omega in x-direction (**b**) Omega in z-direction (red area is WENO and blue area is MUCS).

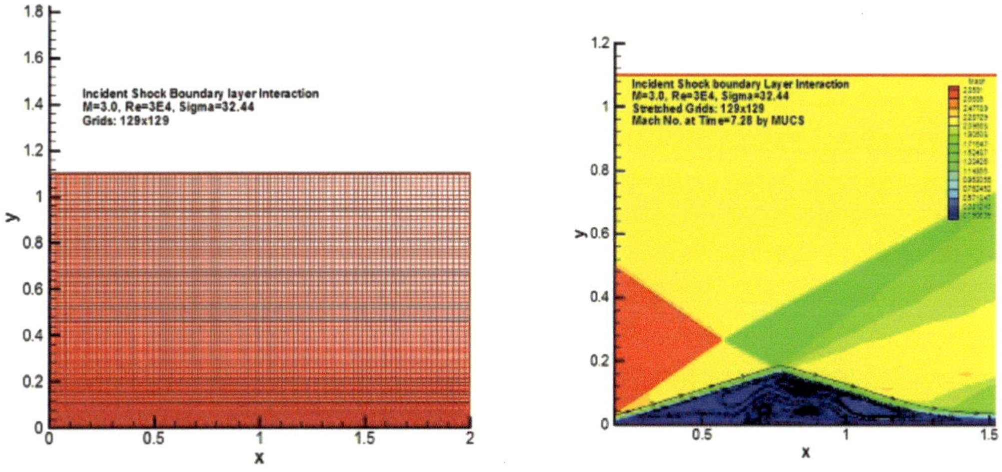

(a) (b)

Fig. (4.42). Case of grids of 129×129 (**a**) Stretched grids (**b**) Mach number distribution.

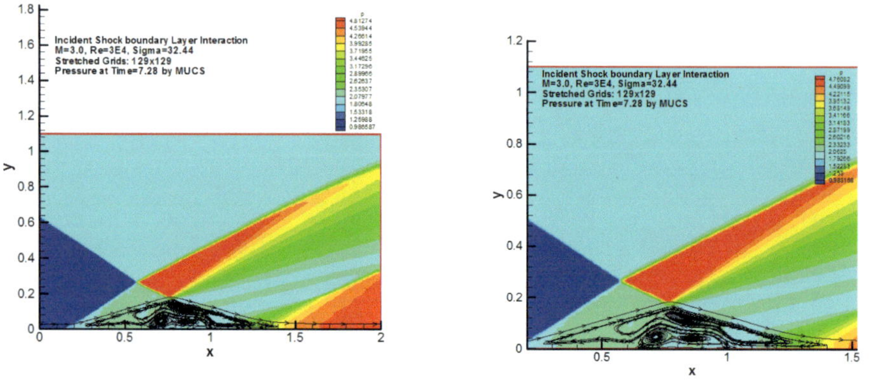

(a) (b)

Fig. (4.43). Pressure distribution obtained by MUCS on grids of 129×129. (**a**) regular (**b**) locally enlarged.

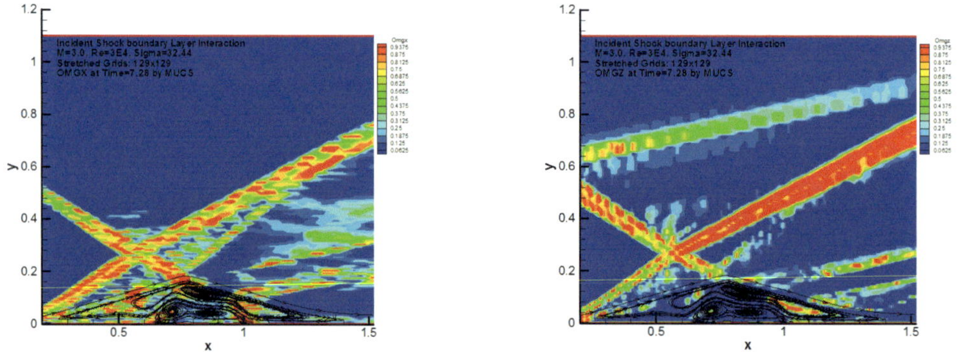

(a) (b)

Fig. (4.44). Control function on girds of 129×129 (**a**) Omega in x-direction (**b**) Omega in z-direction (red area is WENO and blue area is UCS).

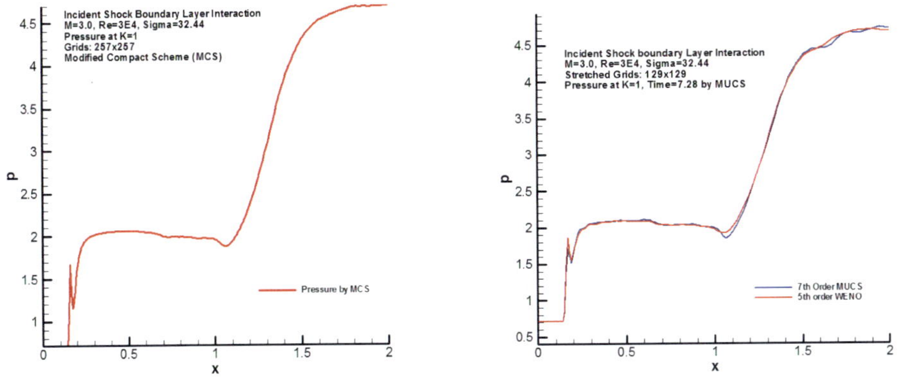

(a) **(b)**

Fig. (4.45). Pressure distribution on the wall (**a**) results on grids of 257×257 as reference (**b**) comparison of 7th order MUCS and 5th order WENO on grid 129×129.

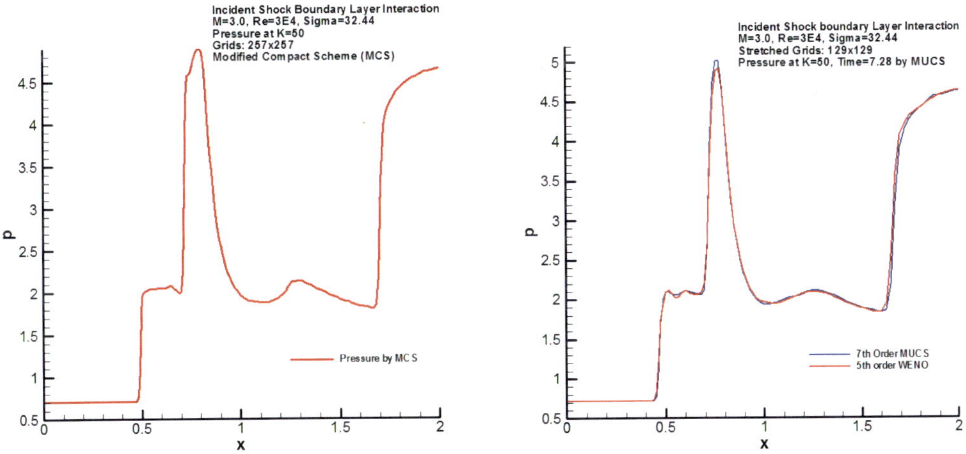

(a) **(b)**

Fig. (4.46). Pressure distribution (**a**) results on grids of 257×257 at k = 50 as reference (**b**) comparison of 7th order MUCS and 5th order WENO on grid 129×129 at k = 50.

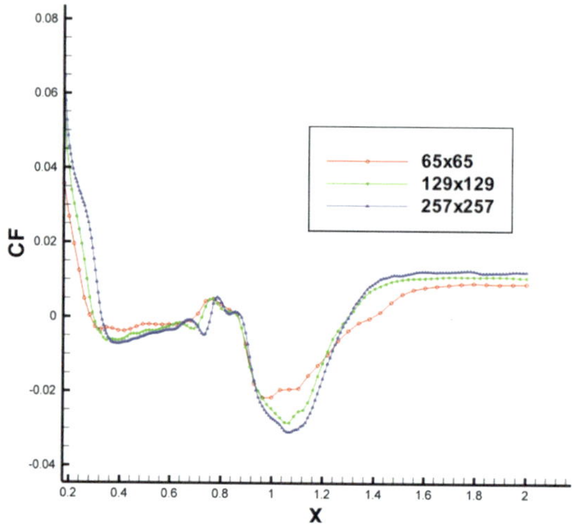

Fig. (4.47). Skin-friction distributions on the wall surface - comparison of three different grids.

Table 4.1. Size of separation bubbles caused by shock/boundary layer interaction.

Grid Number	65×65	129×129	257×257
Size of separation bubbles	1.10	1.01	0.97

4.6.4. CONCLUSIONS

1. Modified up-winding compact scheme (MUCS) with a new shock detector and a new control function, which uses WENO to improve upwind compact scheme, can be adopted for Euler and Navier-Stokes equations for capture of sharp shock and high resolution of small-length scales, and do not have case-related parameters.
2. Modified central compact scheme (MCS) could have better resolution on the interaction between shock and boundary layer than MUCS; however, it cannot pass the test for the fine grid Euler equation.
3. One problem of MCS is that central compact scheme cannot alleviate the numerical oscillations generated by the shock overshooting and does not have dissipation. If the low frequency error is generated by the numerical scheme, CS is not able to remove them since the high order filter can only remove high frequency numerical errors. Thus unfortunately, high order CS with high order filter cannot perform well for DNS or LES simulation of high speed flow within the shock. In order to passing the check of the fine grid for the Euler equations, MCS needs further improvement.

4. MCS, MUCS, MWCS are all fundamentally used for shock boundary and shock acoustic interaction, but still need more modification. They are good in capturing the sharp shock and have higher resolution than WENO.

4.7. THE 5TH ORDER BANDWIDTH-OPTIMIZED WENO SCHEME

4.7.1. The 5th Order Bandwidth-optimized WENO Scheme for the Convective Terms

For integrity, the 5th order *WENO* will be described as follows. Considering the one dimensional hyperbolic equation:

$$\frac{\partial u}{\partial t}+\frac{\partial f(u)}{\partial x}=0 \tag{4.89}$$

The semi-discretized equation can be expressed as:

$$(\frac{\partial u}{\partial t})_j=-\frac{h_{j+1/2}-h_{j-1/2}}{\Delta x} \tag{4.90}$$

Considering the positive flux, the four upwind-biased schemes on candidate stencils can be given as:

$$\begin{cases} h^{+'}{}_0=\frac{1}{3}f_{j-2}-\frac{7}{6}f_{j-1}+\frac{11}{6}f_j \\ h^{+'}{}_1=-\frac{1}{6}f_{j-1}+\frac{1}{3}f_j+\frac{5}{6}f_{j+1} \\ h^{+'}{}_2=\frac{1}{3}f_j+\frac{5}{6}f_{j+1}-\frac{1}{6}f_{j+2} \\ h^{+'}{}_3=\frac{11}{6}f_{j+1}-\frac{7}{6}f_{j+2}+\frac{1}{3}f_{j+3} \end{cases} \tag{4.91}$$

The mark '+' refers to the positive flux after flux splitting. It is easy to know that the third-order accuracy is obtained for each individual scheme.

Weighting and the linear weights to obtain higher order:

$$h^{+}{}_{Linear,\,j+1/2}=\alpha_0 h^{+'}{}_0+\alpha_1 h^{+'}{}_1+\alpha_2 h^{+'}{}_2+\alpha_3 h^{+'}{}_3$$

The optimal order (**Order optimized**) for the weighted scheme is at most $2r$, where r is the number of the stencil. And when the optimal order is realized, the α_i must be determined as:

$$(\alpha_0, \alpha_1, \alpha_2, \alpha_3) = (0.05, 0.45, 0.45, 0.01).$$

The order of the scheme is 6th order, and is **the central scheme**. If the bandwidth-optimized technique is chosen [148] and a fifth-order is desired, the values of α_i can be specified as

$$(\alpha_1, \alpha_2, \alpha_3) = (0.094647545896, 0.428074212384, 0.408289331408, 0.068988910311)$$

The final nonlinear weighted schemes can be expressed as:

$$h^{+}{}_{j+1/2} = w_0 h^{+'}{}_0 + w_1 h^{+'}{}_1 + w_2 h^{+'}{}_2 + w_3 h^{+'}{}_3$$

where w_i is changing from place to place, and

$$w_i = b_i / (b_0 + b_1 + b_2 + b_3),$$

$$b_i = \alpha_i / (\varepsilon + IS_i)^2,$$

and ε is a small quantity to prevent the denominator from being zero, which should be small enough in supersonic problems with shocks ($10^{-6} \sim 10^{-10}$). IS_i is the smoothness measurement.

In order to make the nonlinear scheme still pertain the same optimal order, *i.e.*, 5th order, IS_i should have the property:

$$IS_k = C(1 + O(h^2))$$

where C is the same number for three IS_i.

IS_i has the following form:

$$\begin{cases} IS_0 = \frac{13}{12}\left(f_{j-2} - 2f_{j-1} + f_j\right)^2 + \frac{1}{4}\left(f_{j-2} - 4f_{j-1} + 3f_j\right)^2 \\ IS_1 = \frac{13}{12}\left(f_{j-1} - 2f_j + f_{j+1}\right)^2 + \frac{1}{4}\left(f_{j-1} - f_{j+1}\right)^2 \\ IS_2 = \frac{13}{12}\left(f_j - 2f_{j+1} + f_{j+2}\right)^2 + \frac{1}{4}\left(3f_j - 4f_{j+1} + f_{j+2}\right)^2 \\ IS_3 = \frac{13}{12}\left(f_{j+1} - 2f_{j+2} + f_{j+3}\right)^2 + \frac{1}{4}\left(-5f_{j+1} + 8f_{j+2} - 3f_{j+3}\right)^2 \end{cases}$$

To enforce the upwinding feature or make the scheme stable,

$$IS_r = \max_{0 \le k \le r} IS_k$$

This ensures that the normalized weight ω_r can be no larger than the least of all ω_k. Further improvement for ω_k by Martin *et al.* is:

$$\omega_k = \begin{cases} \alpha_k & if \max(TV_k)/\min(TV_k) < 5 \; and \max(TV_k) < 0.2 \\ \omega_k & otherwise \end{cases},$$

where TV_k stands for the total variation on each candidate stencil.

The scheme for $h^-_{j+1/2}$ has a symmetric form of $h^+_{j+1/2}$ to the point $x_{j+1/2}$.

The large eddy simulation based on the *WENO* scheme was thought to be slightly more dissipative than other implicit LES methods. In order to decrease the dissipation of the scheme, the less dissipative Steger-Warming flux splitting method is used in the computation, not the commonly-used dissipative Lax-Friedrich splitting method.

4.7.2. The Difference Scheme for the Viscous Terms

Considering the conservative form of the governing equations, the traditional 4th order central scheme is used twice to compute the 2nd order derivatives in viscous terms.

4.7.3. The Time Scheme

The basic methodology for the temporal terms in the Navier-Stokes equations adopts the explicit 3rd order TVB-type Runge-Kutta scheme [129]:

$$
\begin{aligned}
u^{(1)} &= u^n + \Delta t L(u^n) \\
u^{(2)} &= \frac{3}{4}u^n + \frac{1}{4}u^{(1)} + \frac{1}{4}\Delta t L(u^{(1)}) \\
u^{n+1} &= \frac{1}{3}u^n + \frac{2}{3}u^{(2)} + \frac{2}{3}\Delta t L(u^{(2)})
\end{aligned}
\qquad \textbf{(4.92)}
$$

CONCLUSIONS

A number of high order shock capturing schemes are introduced. Their truncation errors, dissipation and dispersion are analyzed in details. In our high order code, high order compact scheme is applied to our DNS code to generate the turbulent inflow conditions. The 5th order bandwidth-optimized WENO scheme is adopted in our LES code.

CHAPTER 5

Turbulent Inflow and LES Validation

Abstract: In this chapter, a high order direct numerical simulation (DNS) is applied to generate fully developed turbulent inflow. The fully developed inflow was taken from a case of DNS for flow transition. The inflow condition is carefully checked with the velocity profile and ratio of boundary layer thickness, ensuring that it is a fully developed turbulent flow in order to compare with the wind tunnel test.

Keywords: Boundary layer thickness, Fully developed turbulent flow, High order DNS, Velocity profile.

Generating the same or even similar turbulent inflows as those in experiments is extremely challenging in CFD. In recent decades, there are more and more researches focusing on turbulent inflow boundary conditions for numerical simulation of complex, spatially developing external flows. The most significant one is given by Lund *et al.* [149]. They developed a simplified Spalart method by taking only the transformation on independent variables at two streamwise stations without altering the Navier-Stokes equations. However, due to its semi-empirical nature in turbulent flow studies, even with DNS, it is still challenging to obtain the real turbulent inflows that can compare to the ones in experiments. Therefore, we conducted a DNS on transition flow and investigated the complicated vortical structures in the boundary layer. To generate the turbulent inflow for LES on MVG controlled boundary layer, a number of turbulent profiles are obtained from the DNS data. The inflow conditions are generated by several steps described in the following.

5.1. HIGH ORDER DNS ON BOUNDARY LAYER TRANSITION

In order to get deep understanding on the mechanisms of the late stage of flow transition in a boundary layer and physics of turbulence, recently, a high order direct numerical simulation (DNS) with medium size of simulation with more than 600,000 time steps is conducted by Liu *et al.*, and the mechanism of the late stages of flow transition in a boundary layer at a free stream Mach number 0.5 is studied by Chen *et al.* [150-154], Liu *et al.* [155-164], Lu *et al.* [99, 165-167] and Yan *et al.* [168].

Chaoqun Liu, Qin Li, Yonghua Yan, Yong Yang, Guang Yang, Xiangrui Dong

5.1.1. Case Setup

The computational domain is illustrated in Fig. (**5.1**). The grid level is 1920×128×241, representing the number of grids in streamwise (x), spanwise (y), and wall normal (z) directions. The grid is uniform in the streamwise and spanwise directions and refined at the wall boundary in the normal direction. The length of the first grid interval in the normal direction at the entrance is found to be 0.43 in wall units ($Z^+ = 0.43$). The computational domain is decomposed into multiple sub-domains in the streamwise direction. Table **5.1** shows the flow parameters, including Mach number, Reynolds number, *etc.*. Here, x_{in} represents the distance between leading edge and inlet, Lx, Ly, Lz_{in} are the lengths of the computational domain in x-, y-, and z-directions, respectively, and T_w is the wall temperature.

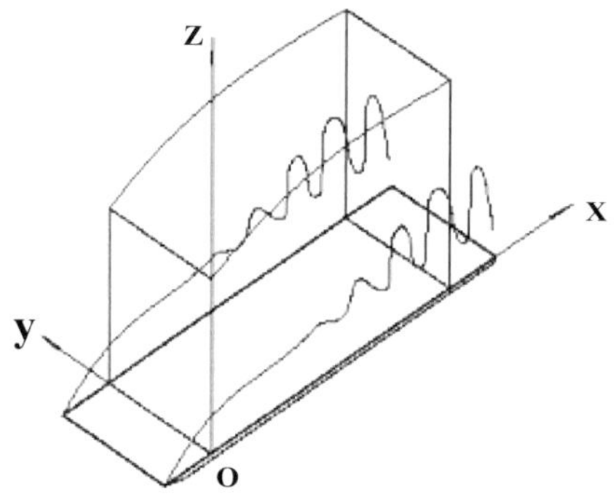

Fig. (5.1). Illustration of the computational domain.

Table 5.1. Flow parameters from the DNS on transition.

Ma	Re	x_{in}	Lx	Ly	Lz_{in}	T_w	T_∞
0.5	1000	$300.79\delta_{in}$	$798.03\delta_{in}$	$22\delta_{in}$	$40\delta_{in}$	273.15K	273.15K

5.1.2. Code Validation

NASA Langley and UTA researchers [155, 166, 169] has carefully validated the DNS code – "DNSUTA" to make sure that the DNS is to be successful. Liu & Chen [159] have reported the details of the DNS on transition flow and the code validations.

5.1.2.1. Comparison with Log Law and Grid Convergence

Fig. (**5.2**) shows the time and spanwise-averaged streamwise velocity profiles at various streamwise locations in two different grid levels (coarse grid and fine grid, to show the grid convergence). The inflow velocity profiles at $x = 300.79\delta_{in}$ is a typical velocity profile for laminar flow. At $x = 632.33\delta_{in}$, the mean velocity profile approaches a turbulent flow velocity profile (log law). We conclude that the velocity profile from the DNS results is turbulent flow and the grid convergence has been realized by this comparison.

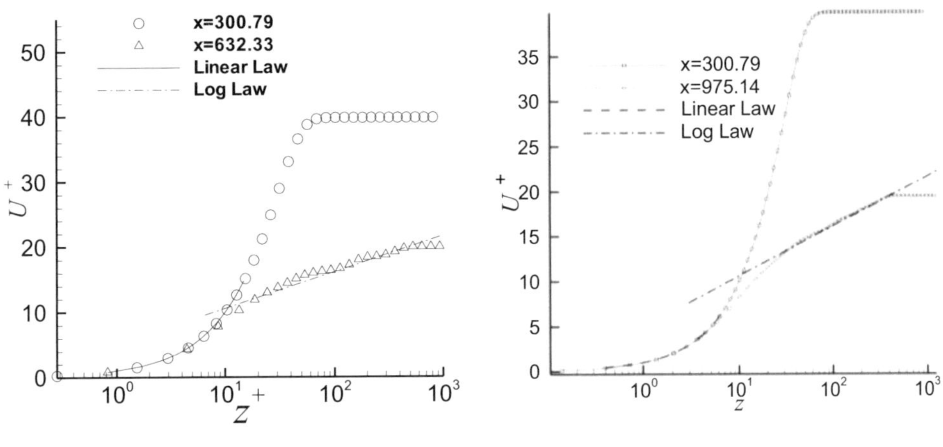

(a) Coarse Grids (960x64x121) (b) Fine Grids (1920x128x241)

Fig. (5.2). Log-linear plots of the time-and spanwise-averaged velocity profile in wall unit.

5.1.1.2. Comparison with Experiment

By using λ - eigenvalue visualization method [86], the vortex structures generated by the evolution of the nonlinear of T-S waves in the transition process are shown in Fig. (**5.3**). After the lambda shaped vortex is formed at t = 6T (Fig. **5.3a**), a ring-like vortex will soon be generated at the tip of the lambda vortex (Fig. **5.3b**). Then, a series of similar ring-like vortices will also appear above the roots of that lambda vortex and in the upstream of the first ring-like vortex. A ring-like vortices chain is thus formed and the flow becomes more and more complicated (Fig. **5.3d**). The formation of ring-like vortices structures is consistent with the experiment results by Lee & Li [170], see Fig. (**5.4**).

Fig. (**5.5a**) shows an experimental result of the vortex structure including ring-like vortex and barrel-shaped head U-shaped vortex by Guo [171]. In Fig. (**5.5b**), our corresponding result from the nonlinear evolution of T-S waves in the transition process given by DNS is also presented. It shows that the experiment and DNS agree with each other in a detailed flow structure comparison.

From the results of comparison between experimental data and our numerical simulation, it can be concluded that the vortical structure is universal for late boundary layer transition. The vortical structure is not influenced by different inflow conditions nor different types of transitions (K type, H type or the others).

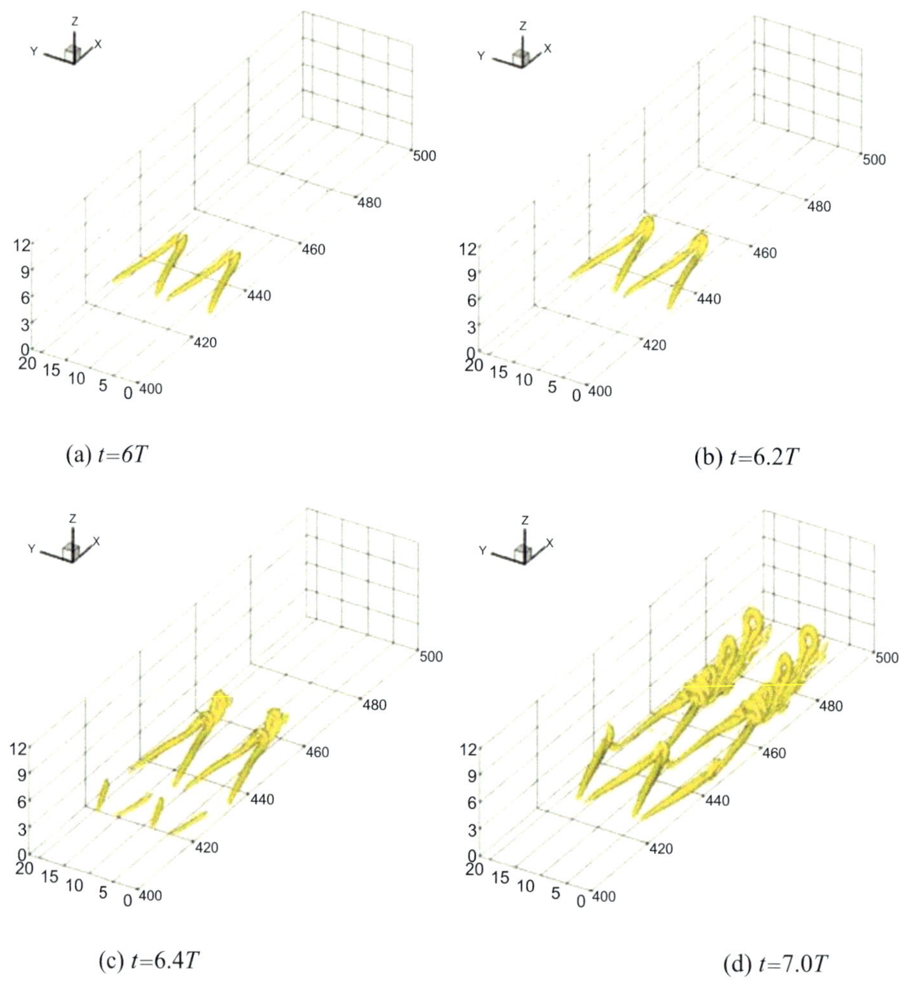

(a) t=6T (b) t=6.2T

(c) t=6.4T (d) t=7.0T

Fig. (5.3). Evolution of vortex structure at the late-stage of transition.

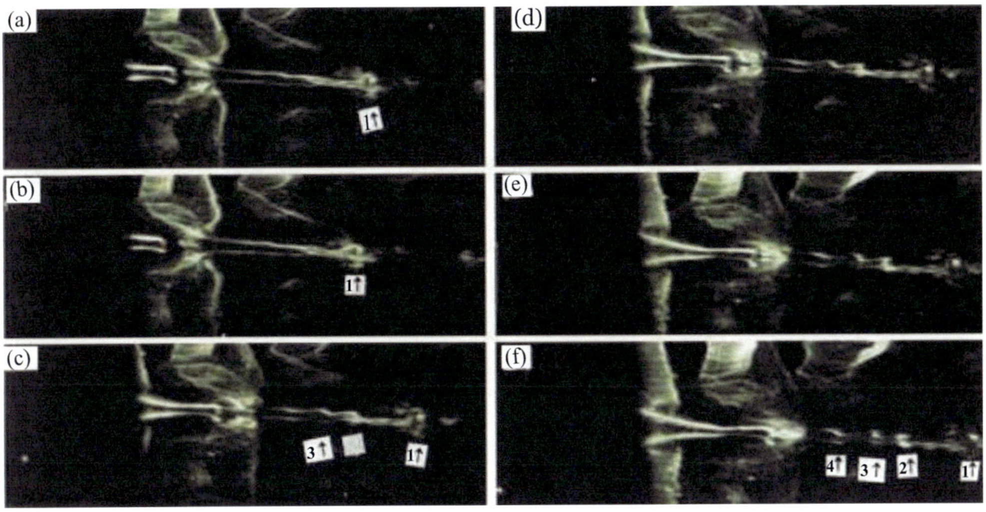

Fig. (5.4). Evolution of the ring-like vortex structure [170].

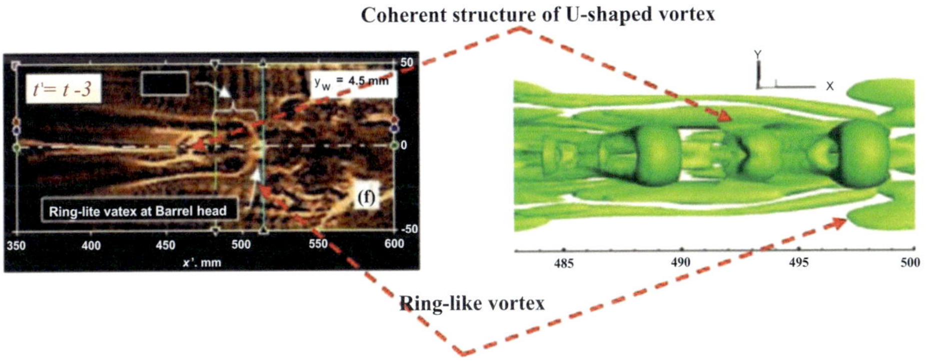

(a). Experiment results given by Guo *et al.* [171] (b). DNS result

Fig. (5.5). Qualitative comparison with experiment on vortex structure.

5.1.1.3. Comparison with Other DNS Results

Our numerical results are also compared with the other DNS results. Fig. (**5.6a**) shows a ring-like vortex along with the vortex filaments inside from our DNS. Fig. (**5.6b**) shows the results with the data set provided by Rist as his personal kindness [172]. The comparison shows both DNS have the similar vortex structure (even the mechanism of generation of the vortical structure).

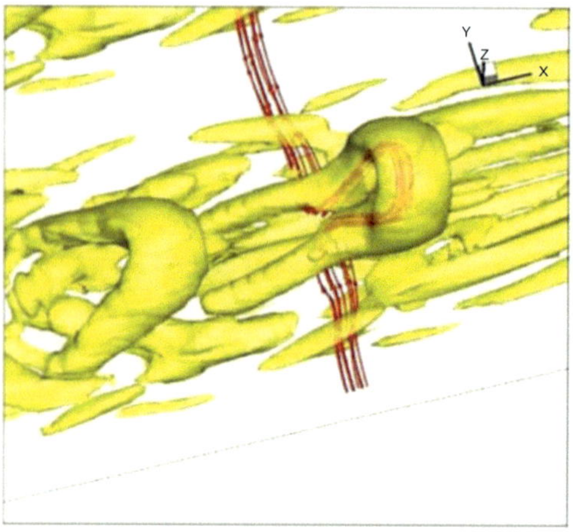

(a) Our DNS

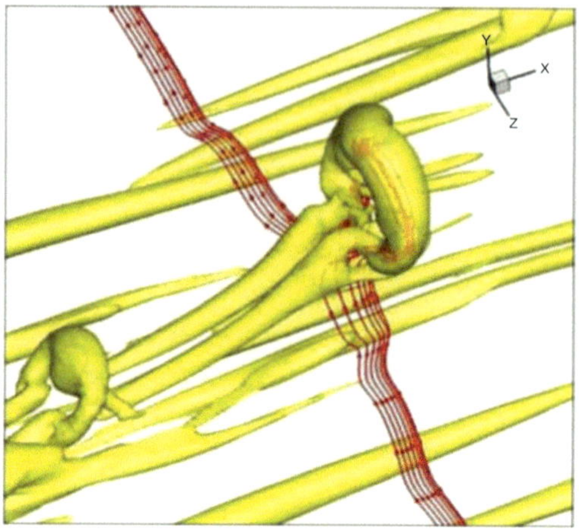

(b) Rist's DNS data

Fig. (5.6). Comparison between our DNS results with Rist's results (on filaments and vortex structures).

5.2. DNS DATA TRANSFORMATION

From the DNS results, 20000 consecutive turbulent profiles are obtained. The instantaneous data were recorded on three sections in the fully developed turbulent zone, which are used as the inflow data for LES on supersonic flow. However, to get a turbulent inflow for the MVG controlled supersonic boundary layer, we still need do some transformation on the flow profile obtained from the DNS on subsonic boundary layer. The assumption of similar perturbation distributions ($U / U_e \sim y / \delta^*$) is used for the transformation of the profiles from the subsonic flow to supersonic inflow. A turbulent mean profile of the streamwise velocity (*w*-velocity) is first extracted from a previous DNS [155, 162]. The smooth mean velocity profiles for the inflow in front of MVG can be obtained by spline interpolation. Its corresponding fluctuating components are then scaled by using the local displacement thickness and free stream velocity. The first grid along the wall normal direction is at $Z^+ = 0.43$.

The pressure is assumed to be constant at the inlet boundary, that is $p = p_\infty$ and $p' = 0$. The temperature profile is obtained by using Walz's equation for adiabatic walls:

$$T/T_e = T_w/T_e - r(\gamma - 1)/2 \times M_e^2 (U/U_e)^2$$

where T_w denotes the adiabatic wall temperature, the subscript "*e*" represents the quantity at edge of the boundary layer and $r = 0.9$ is the recovery factor.

The fluctuating velocity components can be extracted from the instantaneous profiles at each time step obtained in DNS by subtracting the mean values. These fluctuation profiles are rescaled in the same way as the mean profiles. Fluctuation of density is determined by

$$\frac{\Delta\rho}{\bar{\rho}} = -\frac{\Delta T}{\bar{T}} \tag{5.1}$$

Eventually, the transformed parameters become

$$u = U + \Delta u\,,\, v = V + \Delta v\,,\; w = \Delta w\,,\; \rho = \bar{\rho} + \Delta\rho\,,\; p = \frac{\rho T}{\gamma M^2}\,,\;\; T = \bar{T} + \Delta T\,.$$

5.3. FULLY DEVELOPED TURBULENT FLOW IN HIGH SPEED

After doing the transformation, the profiles for all the variables are obtained and they are treated as the inflow condition for the LES of MVG controlled boundary flow. Of course, these new profiles are different from the exact solutions of the Navier-Stokes equations inevitably due to the different Mach numbers of the flows. However, with the evolution under nonlinear Navier-Stokes equations, the flow will soon be adjusted and evolve into a fully-developed turbulent flow while propagating downstream (Fig. **5.7**). In Fig. (**5.7**), it can be found that similar coherent vortical structure which exists in turbulent flow is generated before the MVG.

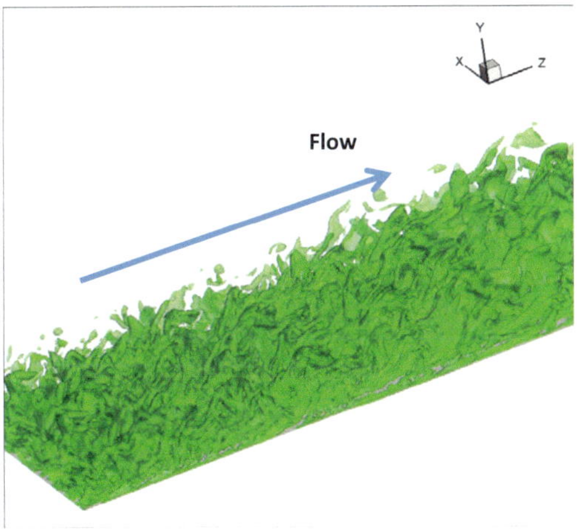

Fig. (5.7). Vortical structure in front of MVG by λ_2.

5.4. TURBULENT INFLOW VALIDATION

5.4.1. Flow Parameters

The flow profiles located at 7.4h (h is the height of the MVG as the reference length) from the inlet are examined carefully in front of the MVG after long periods of computation. It is found that the boundary layer thickness $\delta \approx 2.751h$, the displacement thickness $\delta^* \approx 0.394h$ and the momentum thickness $\theta \approx 0.296h$. The spanwise averaged C_{Ptotrc} is 0.94 and the averaged shape factor H_i is 1.33. The H_i for laminar flow is about 2.6 and about 1.2 ~ 1.4 for a standard turbulent flow. Thus, the results indicate that the methods described above produce a fully developed turbulent inflow in front of MVG.

5.4.2. Boundary-layer Profiles

The inflow boundary layer velocity profile in log - coordinates is shown in Fig. (**5.8**). A good log region can be found in the figure. Overall, the agreement with the analytical profile by Guarini *et al.* [173] is good.

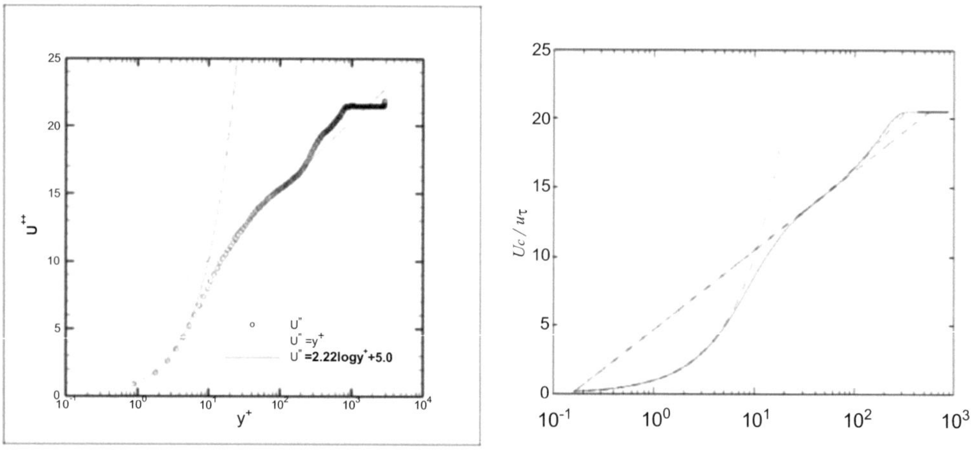

Fig. (5.8). Inflow boundary-layer profile and comparison with Guarini's result [173].

To check influence on the flow in the downstream from the turbulent inflow obtained from DNS as specified above. Profile of $\overline{u'v'}$ behind MVG at z ≈12h, is also compared with experiment data from Sun [174]. Fig. (**5.9**) shows that the profile from LES has a good agreement with the experimental data, especially in the near wall region. A similar peak of $\overline{u'v'}$ is estimated. In the region apart from the wall, the two curves drop following a similar trend. When approaching the top of the boundary layer, $\overline{u'v'}$ approaches zero. The difference between the two curves may be caused by different Mach numbers in our LES (Ma = 2.5) and the experiment (Ma = 2.0).

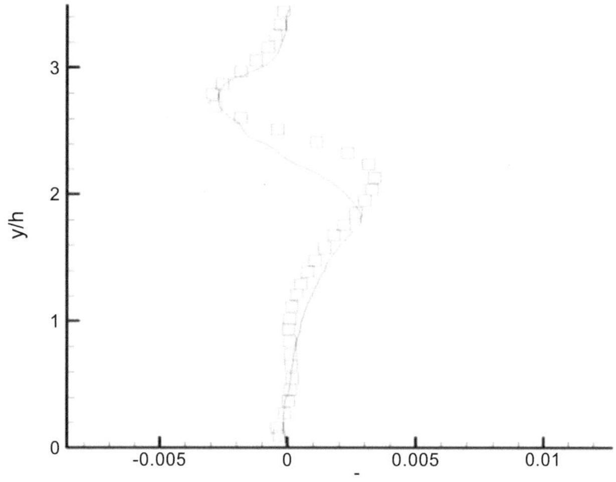

Fig. (5.9). Profiles of $\overline{u'v'}$ along wall normal direction.

CONCLUSION

As a conclusion, the method used in our LES to generate the supersonic turbulent inflow condition is quite successful. A fully developed turbulent flow is generated in front of MVG. The flow is physical and the numerical results in the downstream of MVG are authentic.

CHAPTER 6

Vortex Structure of the Flow Field Around MVG

Abstract: In this chapter, we propose a series of new discoveries from our high order LES, including the spiral point near the leading edge, the origin of the momentum loss, the vortex structure, the topology of the flow separation, and some conclusions of other physics in the flow. Although there are some studies of experiments and numerical simulations on fluid flow controlled by MVG recently, only two-dimensional flow information was provided and confirmed by those studies. The three dimensional flow field behind MVG is still unclear. To better understand the flow structure especially the vortex structure in the downstream of the MVG, the flow field around the MVG and surrounding areas has been investigated in details. In addition, the 3D shock wave structure in the flow field is also studied in this book.

Keywords: Flow separation, Momentum deficit, Spiral points, Topology, Vortex rings.

6.1. RESULTS OF THE SUPERSONIC RAMP ONLY FLOW

In order to compare with the MVG-ramp process, the ramp only flow is studied first.

6.1.1. Flow Structures of the Ramp Only Flow

The flow structure is shown in Fig. (**6.1**) from the numerical instantaneous data. The basic shock wave structure consists of separation and reflection shock waves at the ramp. Behind the shock of separation, continuous interference exists. Many compressed waves or weak shocks are thus induced. The separation shock wave and the potential compression waves eventually converge to the reflection shock above the ramp, as the disturbance propagate to the downstream.

Fig. (**6.2**) shows the streamlines and the distribution of pressure on the central spanwise plane. The cross-sectional pressure profiles in Fig. (**6.2**) and the surface pressure in Fig. (**6.1**) clearly show that the flow exhibits qualitatively turbulence characteristics within and after separation although the inflow is weakly turbulent. This indicates that separation amplifies turbulence. It is observed that the recirculation of the flow causes the waves to reproduce in a limited area and develop non-linearly.

Chaoqun Liu, Qin Li, Yonghua Yan, Yong Yang, Guang Yang, Xiangrui Dong

The instantaneous vortex structure in separation zone appears to contain vortices in multiple scales. It can also be seen from the numerical results that the initial structure is almost quasi-two-dimensional (2-D) in the spanwise direction in the upstream of the separation process. However, the flow quickly becomes three-dimensional as the three-dimensional interference in upstream arrives.

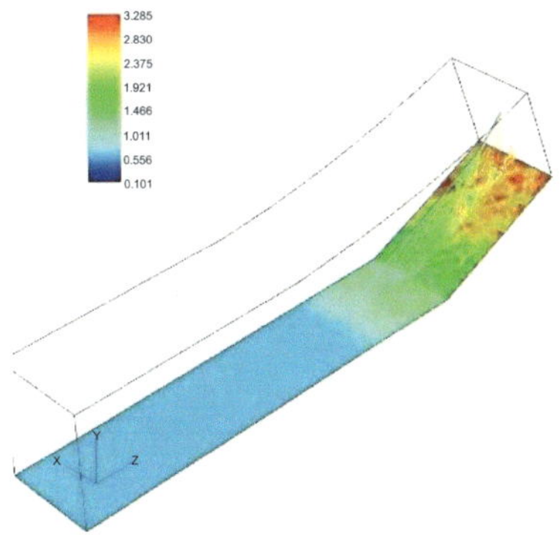

Fig. (6.1). Pressure distribution at the central plane and wall surface.

Fig. (6.2). The streamlines at the central plane.

Fig. (6.3). The instantaneous digital schlieren at central plane.

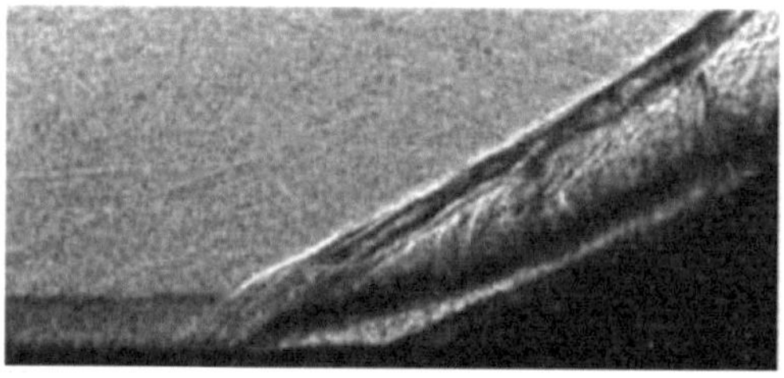

Fig. (6.4). The schlieren image from the experiment by Zheltovodov [5].

Fig. (**6.3**) shows the instantaneous image of the distribution of $|\nabla\rho|$ on the center plane. For a qualitative comparison, Fig. (**6.4**) shows an experimental plot of the ramp only flow with a larger Reynolds number, M = 2.9, from Zheltovodov's experiments [5]. Fewer tiny structures can be observed in Fig. (**6.4**) since the experimental images were averaged span wisely and temporally over the film exposure. As can be seen from the two figures, the separation shock wave has a bevel almost the same as the inclination angle of the ramp. There produced a complex series of compression waves in the boundary layer under shock waves. However, the separation shock is completely separated from the reflection shock and it would have a much smaller angle than in the case of laminar flow.

In order to study the vortex structures in the separation at the ramp corner, the visualization technique - λ_2 method [86] is used. By choosing a small negative value for iso-surface of λ_2 (-0.001), vortex structure are illustrated in Fig. (**6.5**). A large number of vortices with different length scales are observed in the separation zone. In addition, hairpin vortices with annular heads are also found. It must be pointed that the vortices in the upstream are not shown since they are much weaker. This provides an indirect proof of the mechanism on amplification of disturbance of the flow separation.

In Fig. (**6.6**), Two instantaneous figures are plotted by using the distribution of streamwise velocity in front of the shock waves at two x-z planes with y= $0.01\delta_0$ and $0.07\delta_0$ respectively. The common stripes are found which are similar to those found in experiments and the other numerical simulations [175]. As shown in the figure, the strip consists mainly of alternating fast and slow flow. These structures are considered to be coherent structures that may lead to unstable movements of shock waves [25, 176]. The other planes are also examined at different heights and the stripe structure is not found (or very weak), it indicates that the strip structure is only present in certain areas with a specific height.

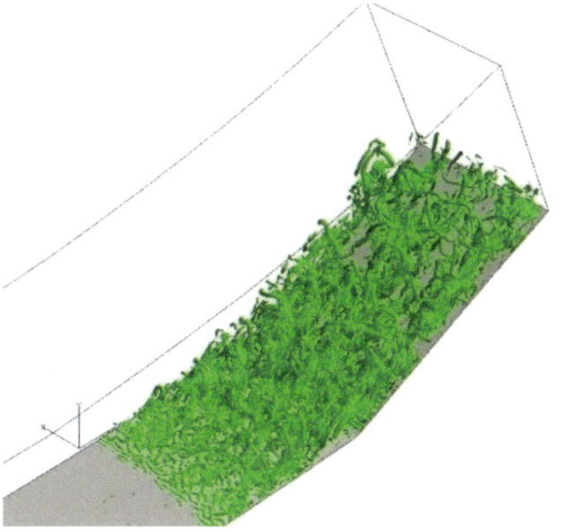

Fig. (6.5). Vortices shown by iso-surface of λ_2.

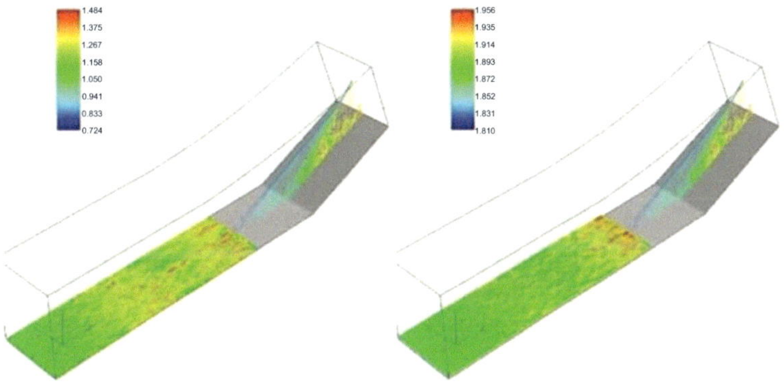

a) The plane at y=0.01δ_0 **b) The plane at y=0.07δ_0**

Fig. (6.6). The stripe structures on two planes at different heights.

6.1.2. The Separation Length at the Ramp Corner in Ramp only Flow

The size of the separation is measured on the wall surface in front of the ramp. The separation length on the ramp is ignored since the reattachment position is scattered. Both stream lines and profile of pressure distribution along streamwise are used to determine the starting point of separation. As a result, the separation length is estimated at about $4.10\delta_0$ to $4.20\delta_0$ which is in consistent with the experimental result - $4.20\delta_0$ [177].

6.2. RESULTS OF THE MVG CONTROLLED RAMP FLOW

As previously described, numerical simulations are made at $M = 2.5$ and Re_θ =

5760, and Babinsky's experiment has a much larger Re_θ, about 28800. Therefore, the comparison between the computations and experiments conducted in our study is mainly qualitative.

6.2.1. The Three Dimensional Shock/Expansion Wave System Around the MVG

Due to the inflow of supersonic boundary layer, it is expected that MVG will cause strong interference such as shock waves. Based on the experiment images, Babinsky [51] proposed a structure wave system, *i.e.*, a first reflection shock, expansion and re-compression shock wave system, as shown in Fig. (**6.7**). Since "the change in Mach number", the first shock will bend within the incoming boundary layer and "a second shock wave then turns the flow back to horizontal". At the foot of the picture's re-compression shock, a subtle oblique "λ" structure is observed. In Fig. (**6.8**), we show the numerical distribution of the time-averaged flow field on the central plane. As can be seen from the figure, the two shock waves are captured. The bending of the main shock to the wall can be found; the oblique "λ" structure can also be found at the foot of the second shock. Babinsky measured the shock angles: the first shock is 26.869°, and the re-compression shock is 21.93°. In contrast, the calculated value is: the first shock is 26.98°, and the re-compression shock is 24.65°. Taking into account the declining angle of the trailing-edge is 70°, rather than 90° in the experiment, the computational results show that the experimental and numerical results are reasonably consistent. It must be pointed out that there was no ramp in the experiment.

Fig. (6.7). The schlieren picture from Babinsky[7].

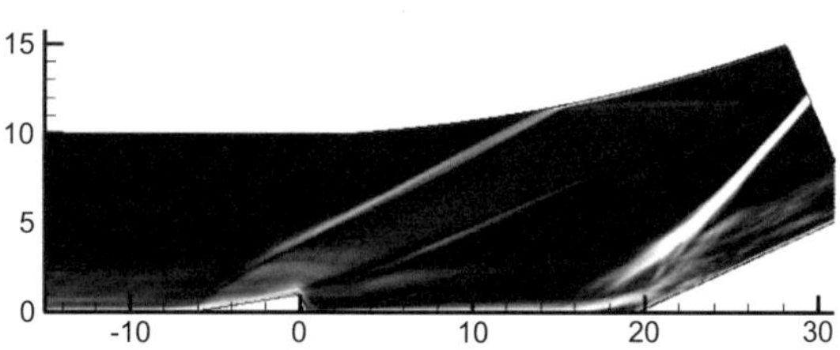

Fig. (6.8). The numerical schlieren picture at the central plane.

While the clear 2-D views are given experimentally and numerically, there is no information on the possible 3-D structure for the shocks given in the past. In Fig.

(**6.9**), the instantaneous cross-sectional pressure contours give the spatial structures of the wave system. As can be seen in the first cross-section in Fig. (**6.9**), the wave system is consisted of the main reflection shock which is above the MVG and the expansion wave which is emanated from the edge of the MVG.

The other sections show an arc-like structure of the re-compression shock wave after the first reflection shock. The size of the arc keeps growing as it moves downstream. The re-compression shock wave would cause the expansion flow satisfy the "virtual" boundary condition which generated by the streamwise vortices, at least at the initial stage of the shock wave. Detailed studies have shown that, the head and feet of the arc-like shock are separated from each other at the initial stage (see Fig. **6.10**). They begin to connect in a downstream position and form a complete arc curve, as shown in Fig. (**6.11**). In the next sub-section, the mechanism for this phenomenon will be discussed.

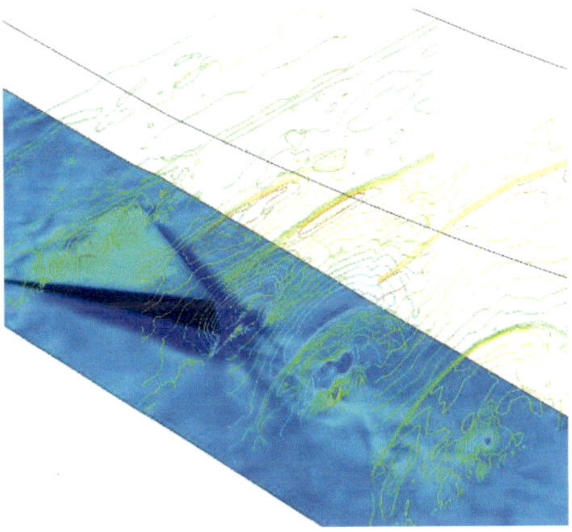

Fig. (6.9). The 3-D structure of the shock waves on cross-sections.

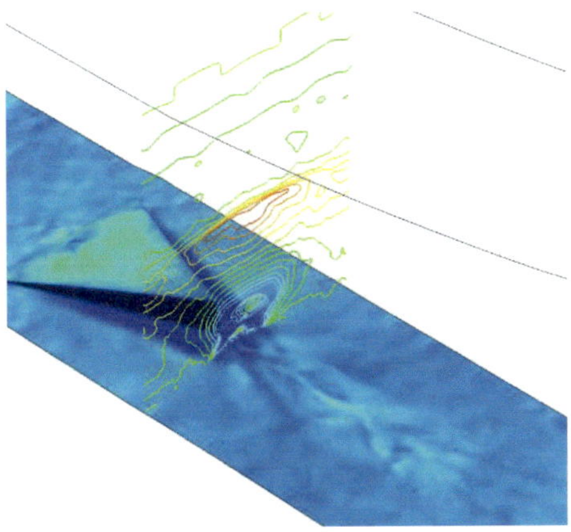

Fig. (6.10). The 3-D structure at the inter-mediate section 1.

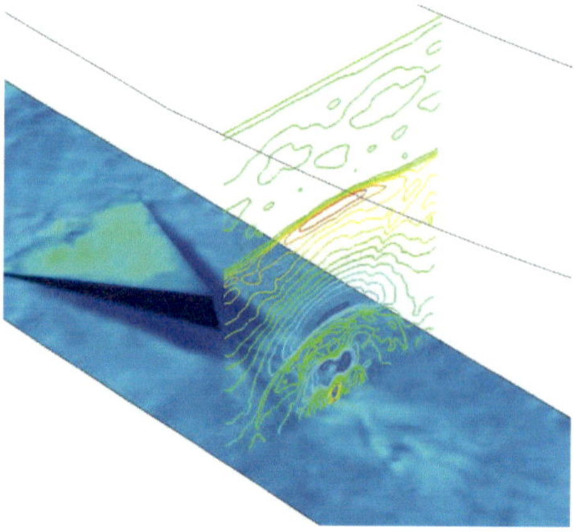

Fig. (6.11). The 3-D structure at the inter-mediate section 2.

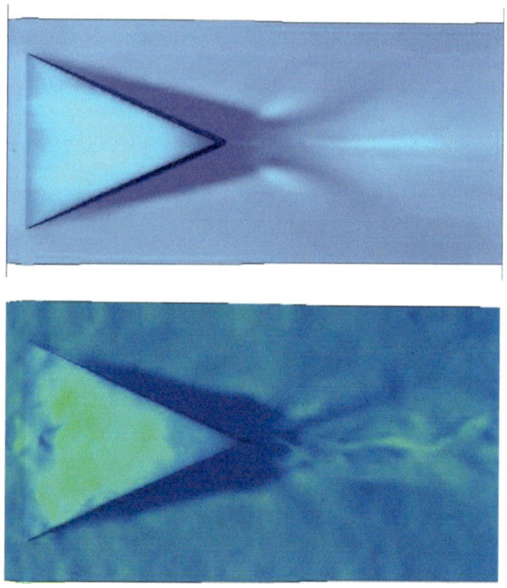

Fig. (6.12). Time-averaged (upper) and instantaneous (lower) pressure distribution on the wall.

Fig. (**6.12**) shows the time-averaged and instantaneous surface pressure contours of the surface pressure structure. As can be seen from these figures, the surface pressure structure can be described as follows: MVG has low pressure regions on both sides, which should be caused by the flow expansion on sides of the MVG. Behind the foot of the trailing-edge, there is an isolated high pressure region corresponding to the stagnation point near the flow on the wall. After that there is a pair of oblique high pressure slots, followed by low pressure slots. Another area of high pressure (averaged) appears again in the area around center line. The mechanism for the structures will then be discussed.

6.2.2. The Topology of the Separation Around MVG

As can be seen from Babinsky'sexperiment [51], useful information can be derived from the surface oil flow. For comparison, Figs. (**6.13** and **6.14**) shows the flow of the surface oil from experiment and the limiting streamlines from the computation. The topology structure of the two graphs is basically the same. The separation lines of the horseshoe vortex, the secondary separation lines next to the MVG and the ones after the MVG are clearly described by computation. Due to the deposit of the oil, such lines are the most obvious traces found in the experiment. There are some differences between the computation and the experiment. In computation, the horseshoe vortex has a stable or a slightly shrinking area, whereas in experiment, the area next to the MVG seems to increase. The reason for this difference may be the mirror-symmetric state of the two lateral sides of the computational domain in the spanwise direction.

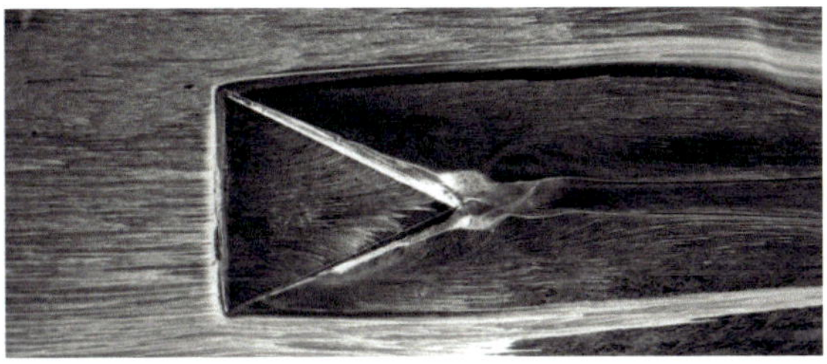

Fig. (6.13). The surface oil flow from Babinsky.

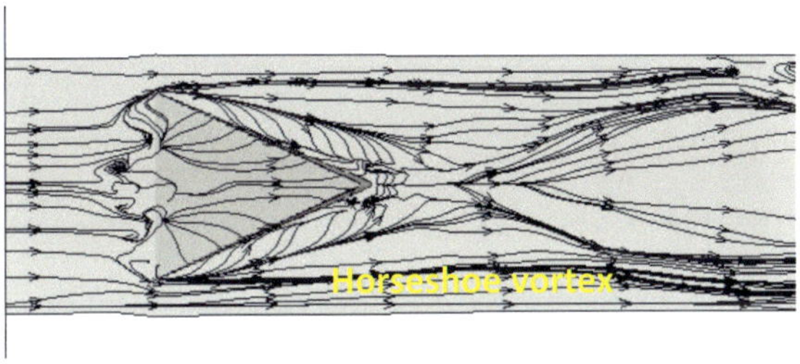

Fig. (6.14). The streamlines on the wall surface from our simulation.

In order to explore the mechanism of separating topology structure, the separation mode of the flow on the wall surface around MVG is further studied. Surface limiting streamlines are depicted on the wall in Figs. (**6.15** and **6.16**). It can be seen that the separation mode of the main streamwise vortices is open separation (see Fig. **6.16a**). The separation lines are located on the upper edge of the MVG and end with a degraded nodal point (NP) near the top of the MVG. This indicates that the separation completely leaves the surface and becomes completely three-dimensional separation. Besides the MVG, there are two pairs of secondary separations, one occurs on the wall surface behind MVG, and the other one occurs on the sides of the MVG. The two separations are open dominant ones ending with spiral points (SPP). This also means the end of the surface separation and the rise of secondary vortices. Thereafter, a pair of secondary separation are newly created on the wall surface in the downstream of the MVG, starting from a pair of saddle points (SDP). After the trailing-edge, there is a source-typed nodal point (NP) on the wall corresponding to the stagnation point in the back dead-water region. There are other small separations as well, just as starting form saddle, ending with a spiral points. They usually have a fairly small length scales and are considered to have played a small role.

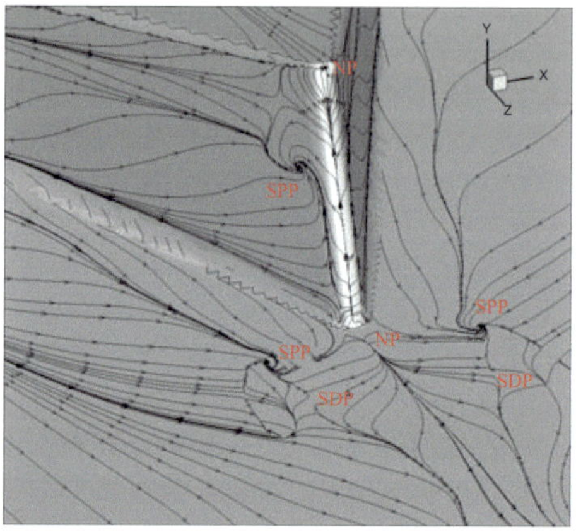

Fig. (6.15). The separation pattern behind the MVG.

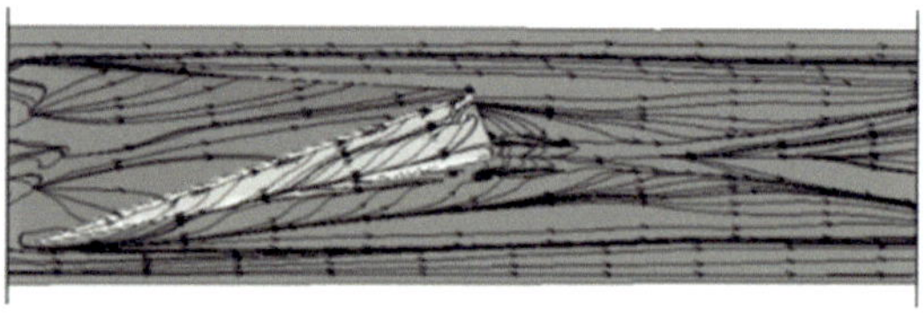

a) Side view of MVG

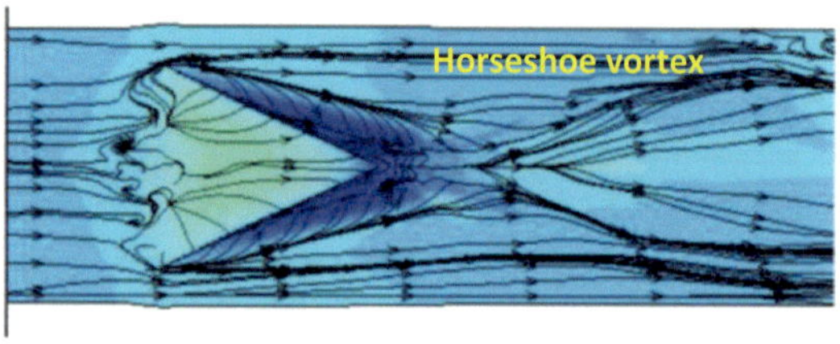

b) Top view with the background by using the pressure contour

Fig. (6.16). Surface separation pattern.

By giving the contours of the streamwise velocity component, u, at different heights of the domain in Fig. (**6.17**), the flow structure is more obvious on the wall surface close to the MVG.

Fig. (**6.17a**) shows the distribution at $y/h = 0.018$. Two obvious regions with back flow can be found and their locations are similar to those observed in the visualization of the oil stream. The back flow achieves a speed of about $0.2U_\infty$ and has extended about $2.0h$ to the downstream. A shear layer is formed at the boundary of the back flow region. By comparing with the stream limits on the wall surface, three vortices (V1, V2 and V3 in the figure) are distinguished at the edges of the back flow regions.

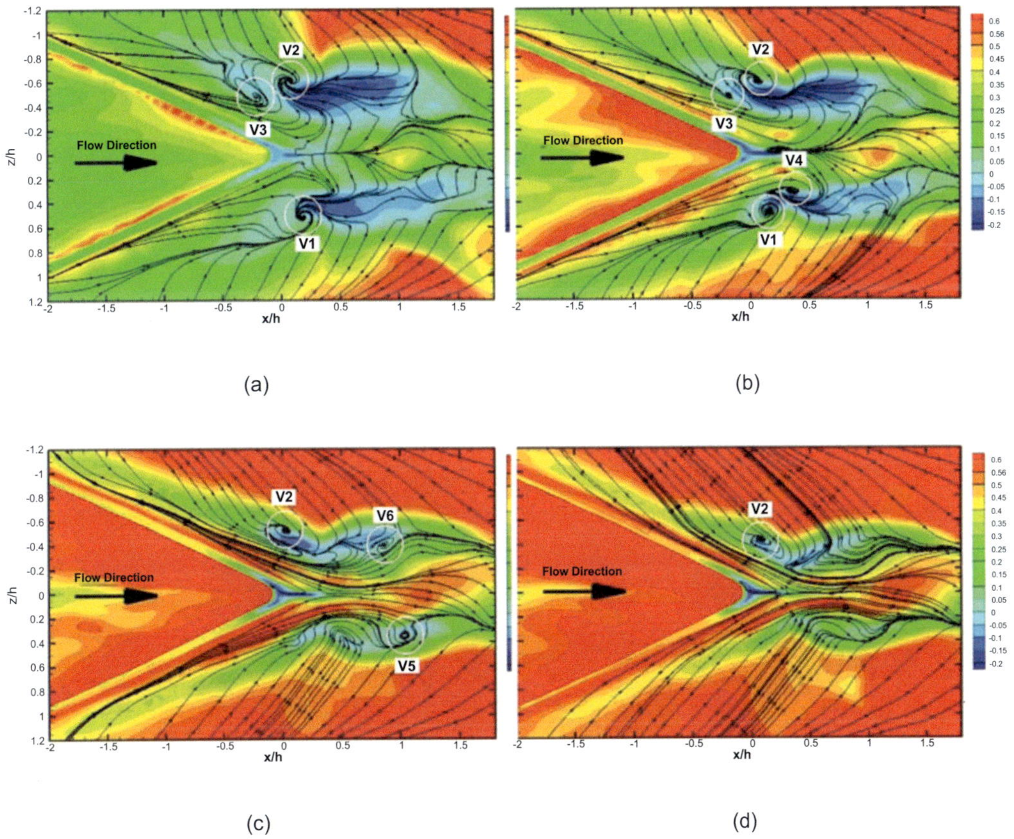

Fig. (6.17). Distribution of streamwise velocity at four heights with projected streamlines.

Fig. (**6.17b**) gives the distribution at *y/h* = 0.036. The back flow regions in this figure is narrowed. The vortices V2 and V3 observed in Fig. (**6.17a**) still exist at the boundary. The locations of these vortices are slightly offset and can be compared in Table **6.1**.

Figs. (**6.17c** and **6.17d**) show the distributions at *y/h* = 0.047 and 0.100 respectively. The back flow is further weaken and narrowed. Besides the vortex V2, two additional vortices, V5 and V6, are generated at the boundary of the back flow region in the downstream. The locations of all the vortices mentioned above can be found in Table **6.1**.

Table 6.1. The coordinates of the revealed vortices in Fig. (**7.42**).

	***y/h*=0.018**	***y/h*=0.036**	***y/h*=0.047**	***y/h*=0.100**
V1 (*x/h,z/h*)	(0.192,0.514)	(0.154,0.489)	-	-
V2 (*x/h,z/h*)	(0.085,-0.600)	(0.057,-0.600)	(0.036,-0.529)	(0.075,-0.438)
V3 (*x/h,z/h*)	-0.197,-0.473	(-0.191,-0.471)	-	-
V4 (*x/h,z/h*)	-	(0.351,0.318)	-	-
V5 (*x/h,z/h*)	-	-	(1.033,0.349)	(0.864,-0.401)
V6 (*x/h,z/h*)	-	-	-	-

By giving the iso-surface of the streamwise velocity at u = $-0.01U_\infty$ in Fig. (**6.18**), a three-dimensional back flow zone can be found. An additional back flow zone (marked as reverse region 3) which is attached on the trailing edge of the MVG is observed.

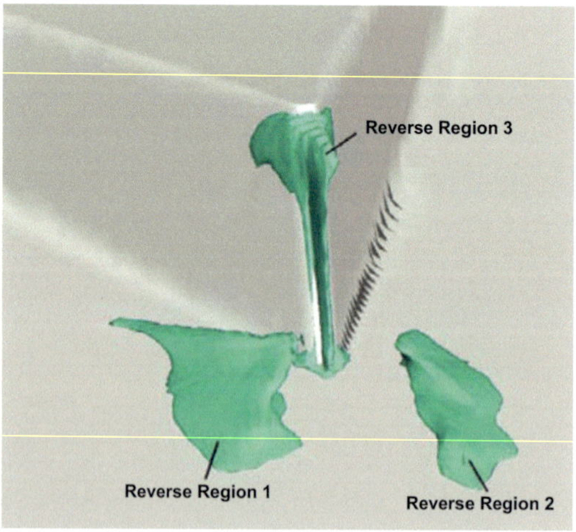

Fig. (6.18). The iso-surface of $u = -0.01U_\infty$.

Fig. (**6.19**) gives five vortex filaments around the iso-surface of a back flow zone. These vortex filaments are initiated from the local maximums of vorticity on wall surface. Once these vortex filaments leave the wall surface, they follow the sloping surface of the separation region, and become omega shaped.

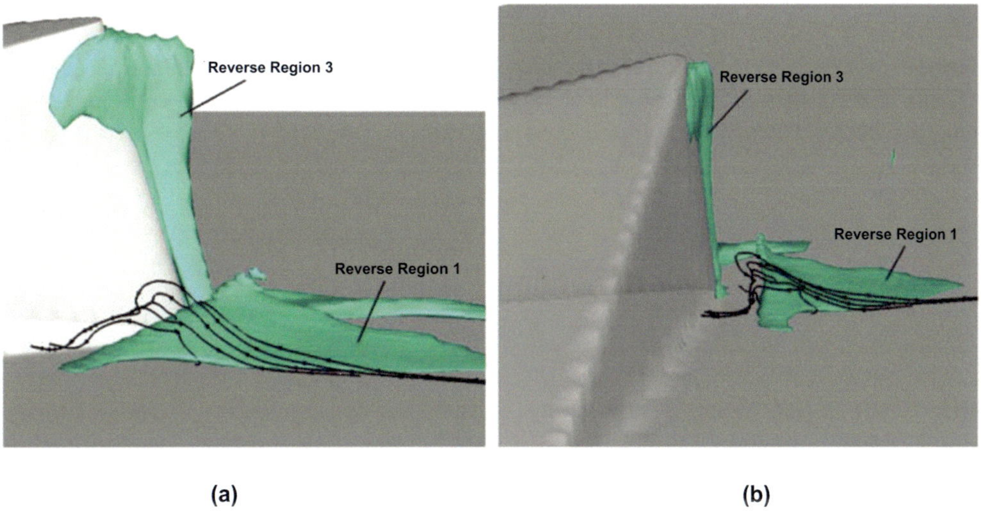

Fig. (6.19). The Ω-shaped vortex lines around the back flow zone.

According to the above analyses, we can get a dynamic vortex model as shown in Fig. (**6.20**) in half of the domain: there are mainly 5 pairs of vortices in the flow field - one pair of horseshoe vortices; one pair of primary vortices by inflow separation next to the MVG; two pairs of secondary vortices at the sides of MVG and below the streamwise vortices, which will leave the wall surface at spiral points and then become completely 3-D separations. The secondary vortex will interact with the primary vortices when propagating downstream. A pair of newly created secondary vortices above the wall surface in the downstream of the back flow region behind the MVG are also included, they are induced by the primary vortices in the momentum deficit.

It is worthy of mention that the five pairs of vortex model near MVG is different from those reported by Babinsky [57]. The difference is mainly in the secondary vortex structures. In their report, the first secondary beside the MVG and the new secondary vortex under the primary streamwise vortices were considered to be the same vortex. Actually, the first secondary vortex pair will end with a pair of spiral points, then be lift up and mixed with the streamwise vortices in the momentum deficit. Thus, the key to verifying the current model is that there is a spiral point due to surface separation. In April 2010, the experimental team at the University of Texas at Arlington repeated the experiment with the same Mach number after being told to predict and need to verify the spiral points [75]. Instead of only checking the oil flow after the wind tunnel was running, they watched a video from the top, recorded the process of the MVG oil flow. A pair of unique oil accumulating points was found in the video (see the experimental snapshot in Fig. **6.21**). These points just correspond to the spiral points, and this verification proves the vortex model proposed from the computations. A more comprehensive analysis can be found in the report by Lu *et al.* [76]. This dynamic vortex model is also mostly confirmed by the experiment work of Saad [178] recently (see Fig. **6.22**).

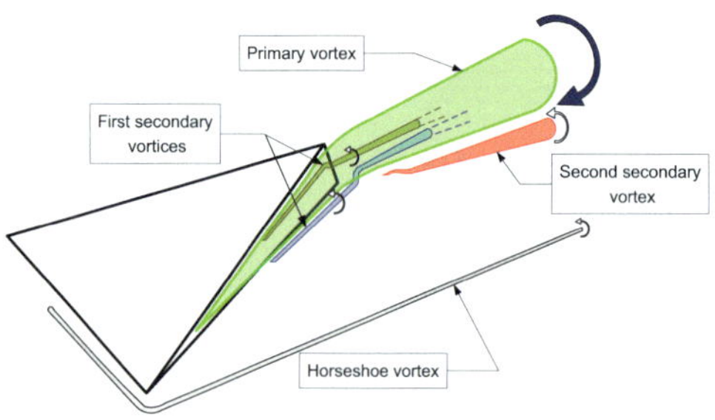

Fig. (6.20). The dynamic vortex model.

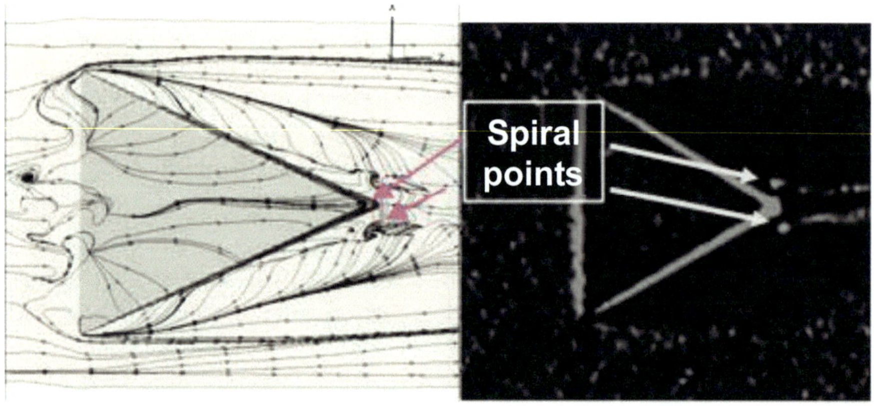

Fig. (6.21). The comparison between the computation and the validating experiment.

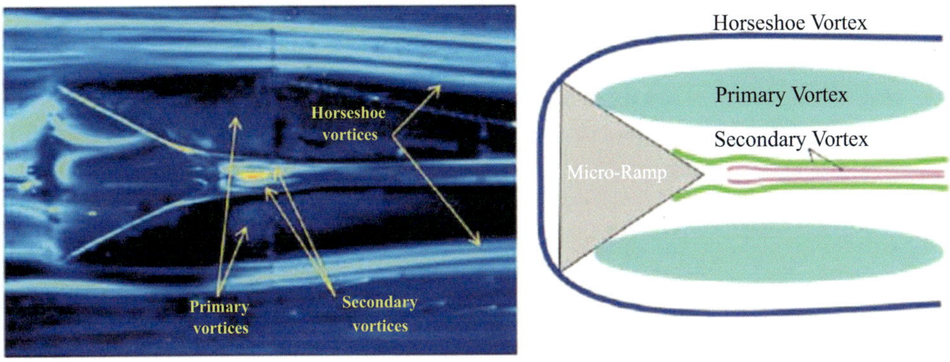

(a) (b)

Fig. (6.22). Visualization of surface flow and the vortex model given by Saad *et al.* [178].

The surface separation mode can explain the surface pressure structure to a certain extent. Around the stagnation point behind the MVG, the flow moves to downstream and decelerates to the wall. This process will transfer the kinetic energy to the potential energy of the fluid, thus, the decrease in velocity leads to an increase in pressure. That's why there is a high pressure point on the center line behind the MVG (see Fig. **6.12b**). The existence of the secondary separations behind the MVG with spiral points indicate the secondary vortexes will rise and interact with the external high velocity flow. This interaction will slow down the fast flow and cause compression. In the supersonic flow, the disturbances will produce a shock wave that propagates substantially along the Mach line, and produces a pair of high pressure slots. The re-acceleration of the compressed flow produces the accompanying low pressure slots. The distribution of the surface pressure present a reasonable correspondence with the flow structure.

In order to study the three-dimensional structures of the vortex structure, the cross-sectional streamlines are plotted in Figs. (**6.23-6.28**). The time-averaged velocity field is used to generate streamlines with the contour for streamwise velocity on these corss-sections. It can be seen from the figures that the so-called momentum loss is related to the streamwise vortices, which will be discussed late. The streamwise vortices are initially located at a lower position but then gradually rise in the downstream. Secondary vortices are observed under the momentum deficit. The results from our numerical simulation are consistent with the experimental ones from the references. The streamwise vortices are found asymmetric to the central plane in the downstream position. In fact, the asymmetric flow occurs elsewhere in fluid dynamics. Its mechanism is considered to be the structural instability of the cross-section topology. In the downstream, the streamwise vortices become weaker and further reduced within separation zone at the ramp corner (see Fig. **6.27**). After that, there exists a recovery on the vortex structure (see Fig. **6.24**).

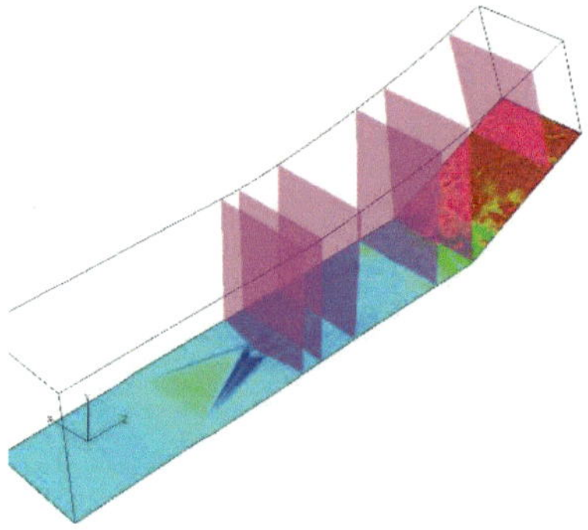

Fig. (6.23). The locations of the cross sections.

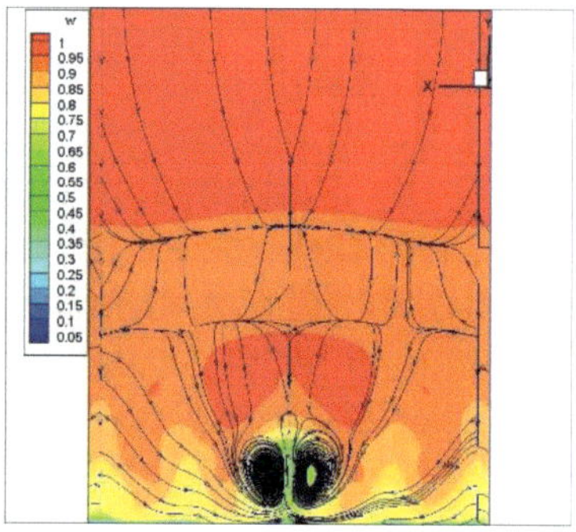

Fig. (6.24). The streamlines at cross section 2.

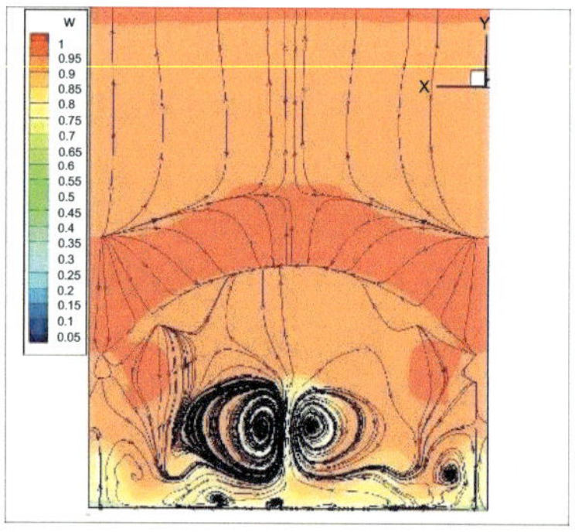

Fig. (6.25). The streamlines at cross section 3.

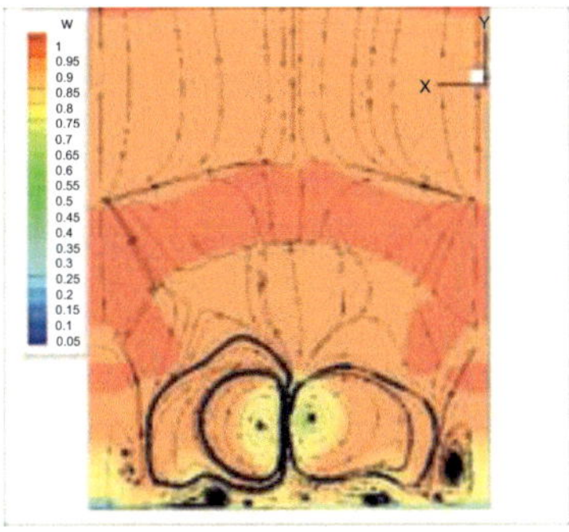

Fig. (6.26). The streamlines at cross section 4.

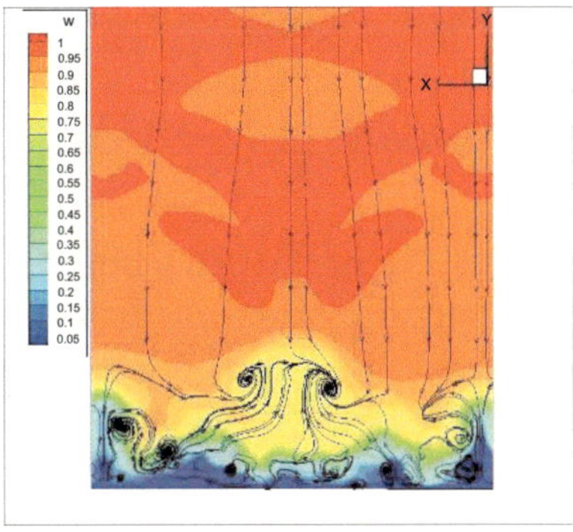

Fig. (6.27). The streamlines at cross section 5.

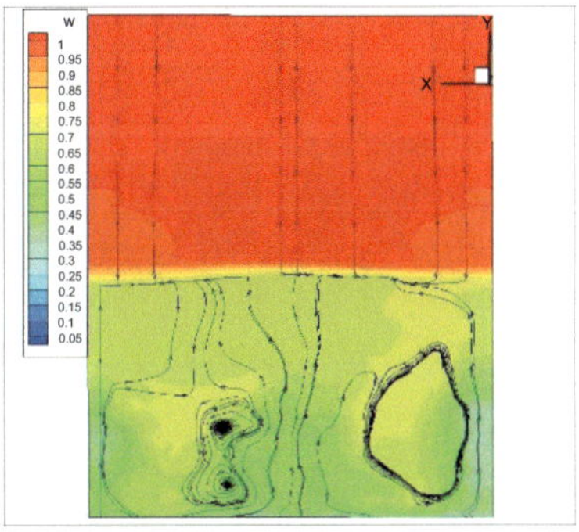

Fig. (6.28). The streamlines at cross section 6.

6.2.3. The Formation of the Streamwise Momentum Deficit after the MVG

The momentum deficit which was later confirmed by the simulation by Ghosh *et al.* [55] and Lee *et al.* [62] is a unique phenomenon first observed by Babinsky [51] in the experiment. Fig. (**6.29**) shows the typical structure of the deficit using the averaged streamwise velocity in section 2 of Fig. (**6.23**). A review of the history of exploration has been described in Chapter 1. The existing explanations does not provide a clear mechanism for the formation of deficits. In our simulation, we studied the relationship between the deficit and the flow structure and the origin of the low-speed flow.

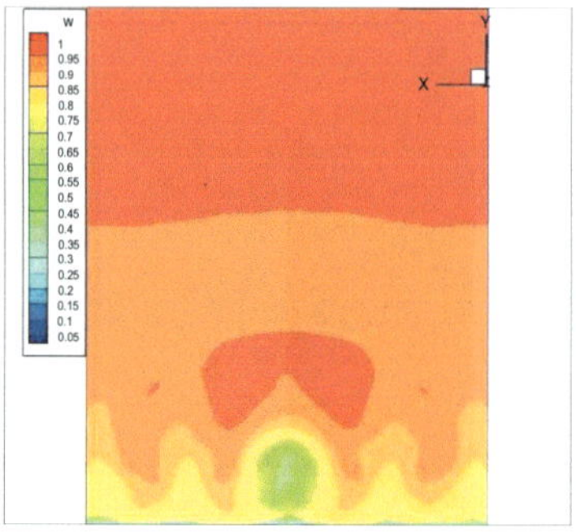

Fig. (6.29). The contour of the averaged streamwise velocity at section 2.

In order to investigate the generation of the momentum deficit, a heuristic analysis is conducted by investigating specific streamlines associated with certain cross-section (see Fig. **6.30**) across the deficit zone. The momentum deficit which is the green circular zone on the cross section is shown by using the contour of the instantaneous stream wise velocity. To initiate the streamlines in the momentum deficit, some seeds around the boundary of the deficit are generated in Fig. (**6.26**). The three-dimensional streamlines initiated from these seeds are shown in Fig. (**6.31**). The distribution of the backward streamlines qualitatively reflects the origins of the deficit. As can be seen from Fig. (**6.31**), all backward streamlines revolve around the streamwise vortices in the deficit indicating that the formation of the deficit is caused by streamwise vortices. All backward streamlines are from the upper surface of the MVG which indicates the main source of the deficit is the shedding of the upstream low-speed boundary layer, rather than the boundary layer flow around the MVG and behind the MVG.

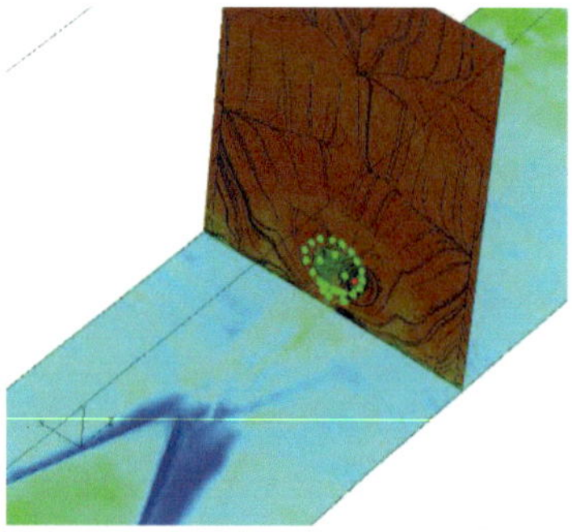

Fig. (6.30). The cross-section.

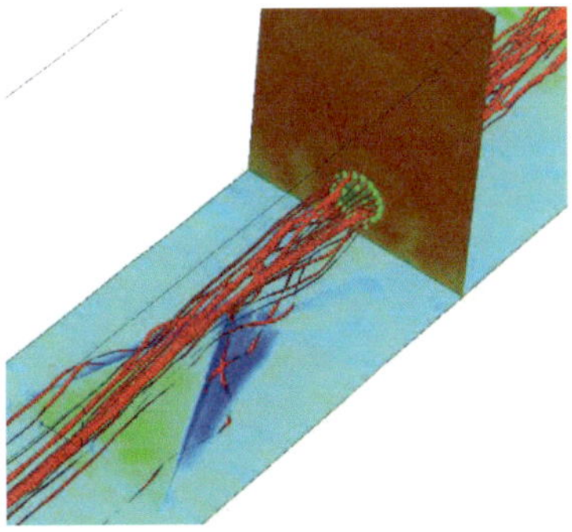

Fig. (6.31). The spatial stream-lines passing through the boundary of deficit.

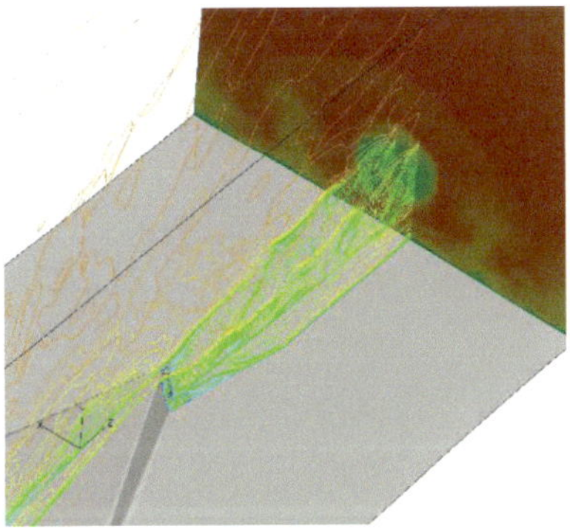

Fig. (6.32). The streamwise velocity contour at two streamwise computational planes.

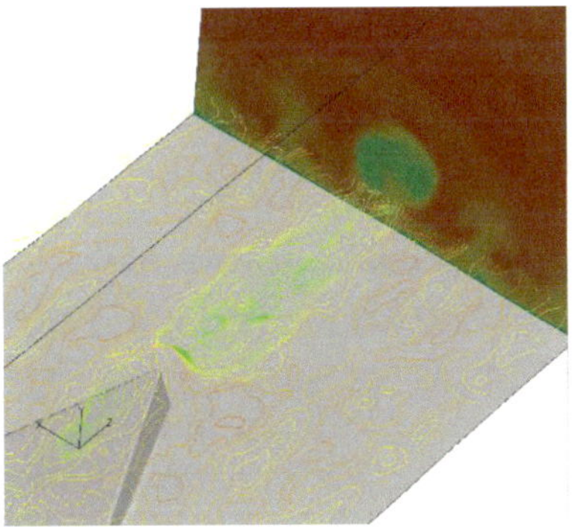

Fig. (6.33). The streamwise velocity contour at a horizontal computational plane.

The streamwise velocity distribution on the streamwise and horizontal cross-sections are also shown in Figs. (**6.32** and **6.33**), respectively. It can be seen from these figures that the slow fluid in deficit is mainly the fall of the upstream boundary layer. The boundary layer in outer spanwise position will be carried along by vortices to the area near the center line. That is why the figure shows streamwise velocity contour seems to start from a limited distance after MVG in Fig. (**6.33**). This phenomenon is consistent with the conclusion made from spatial streamlines in Fig. (**6.30**). Finally, an iso-surface of streamwise velocity is shown in Fig. (**6.34**). The iso-surface which can be considered as the three-dimensional surface of the deficit wrapped the MVG and forms an irregular cylinder.

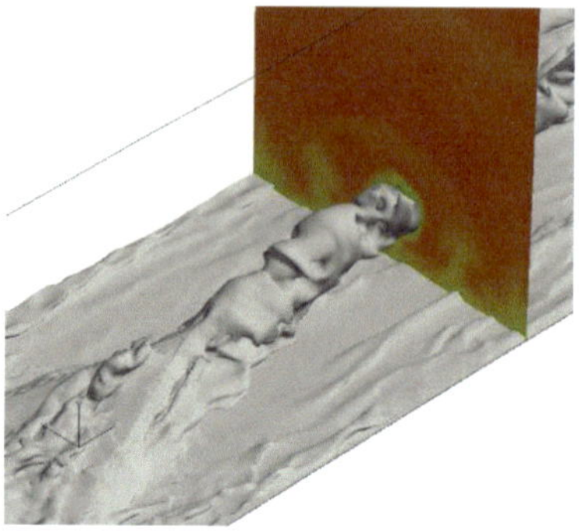

Fig. (6.34). The iso-surface of the streamwise velocity.

6.2.4. The Characteristics of the Separation and the Comparison with the Ramp only Flow

Fig. (**6.35**) shows the streamlines on the wall boundary at the ramp corner where flow separation occurs from time-averaged data. The length of the separation zone is measured from the starting position of the separation to the ramp corner. The separation length at the two sides of the domain is around 6.9*h* but it is shorten to 4h in the middle of the domain. It shows the separation zone is "V" shaped in the MVG controlled ramp flow. The separation length in ramp only flow is measured as about 8.3*h*. Thus, it shows clearly that MVG can significantly reduce the flow separation in the ramp flow. The spiral points on the wall boundary are from the interaction between the horseshoe vortex and ramp separation. The separation of the horseshoe vortex starts on these spiral points. Fig. (**6.36**) shows the time-averaged streamwise velocity profile from both the ramp only flow and MVG controlled ramp flow at the streamwise location $4.5663\delta_0$ in front of the ramp corner. With the existence of momentum deficit, the profile in MVG controlled ramp flow involves a dent. However, there generates a higher shear layer at the sub-layer compared to the profile in the ramp only flow.

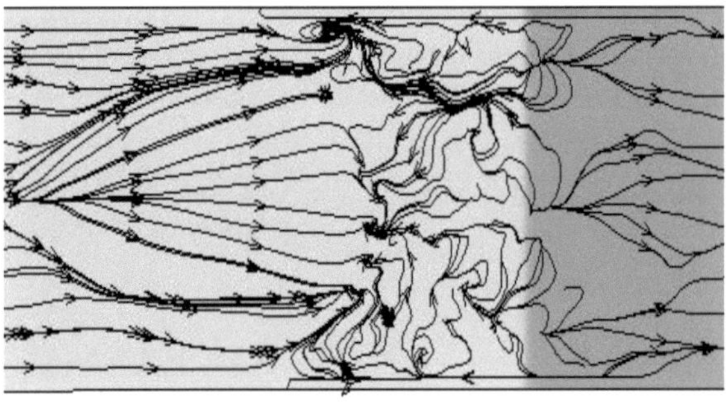

Fig. (6.35). Streamlines at the ramp corner.

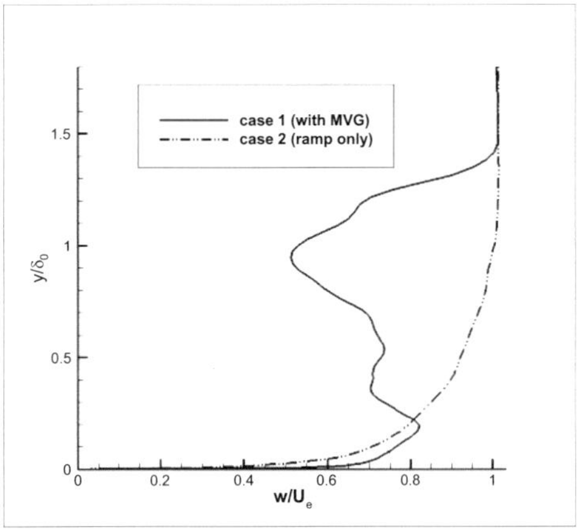

Fig. (6.36). Profiles of the time-averaged streamwise velocity.

6.2.5. Conclusion

Through the discussions above, we can make the following conclusions:

1. From the numerical results of ramp-only flow, the major structures of such as the separation and re-compression shock waves at the ramp corner are captured. The flow separation and the SBLI are investigated. The results obtained for ramp-only flow are very helpful to study the MVG-ramp flow.
2. For the MVG controlled ramp flow, compared to the experimental results of Babinsky's, same wave system is obtained in our numerical simulation. Agreements have been make on the shape of shock waves around MVG and the velocity profiles in the downstream of MVG.
3. The structures of flow separation besides the MVG are investigated in detail. Both 2D and 3D topology of the separation is given. A pair of spiral points are

found on wall boundary behind the MVG which are confirmed by experiments. A new and detail vortex model around the MVG is proposed which is different from the ones proposed by previous investigations. The circular momentum deficit behind the MVG is found that generated by the shedding of the upstream boundary layer.

4. Compared with the results from ramp-only flow, the existence of MVG can significantly reduce the separation zone at the ramp corner and improve the shape factor of the boundary layer.

6.3. RING-LIKE VORTICES - A NEW MECHANISM IN MVG-RAMP FLOW CONTROL

6.3.1. LES Observation and Comparison to the Experimental Results

A pair of counter-rotating streamwise vortices and the secondary vortices underneath are considered as the mechanism for flow control based on MVG from the latest results of experimentation and numerical simulation. These two counter-rotating prime vortices brings high speed flow to the lower boundary layer which is conducive to resisting the adverse gradients of pressure [60]. All the existing results [179], including our numerical result (see Fig. **6.37**), found that the prime vortices mentioned above would be weaken quickly and the make the oscillation on the streamlines in the downstream due to the shear layer instability around the circular momentum deficit behind the MVG. This suggests that there may be some other mechanisms in the MVG control process, except that momentum exchange induced by the primary counter-rotating streamwise vortices.

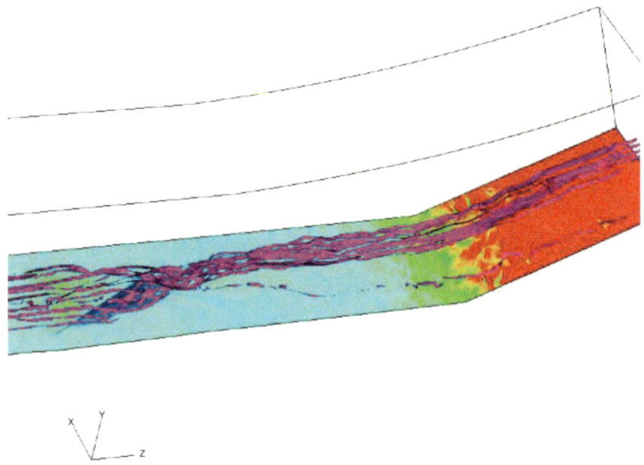

Fig. (6.37). Instantaneous spatial streamlines.

Since the vortex cores usually locate at the positions with the local minim of pressure, the pressure iso-surfaces with a small pressure value is given in Fig. (**6.38**) from an instantaneous data of our numerical results. Besides the shock wave and complicated vortex structure near the trailing edge of the MVG, a series of circular structures can also be observed in the downstream region. These circular structures would become larger and more irregular when moving to the downstream. In the meanwhile, they would also be weaken and less observable

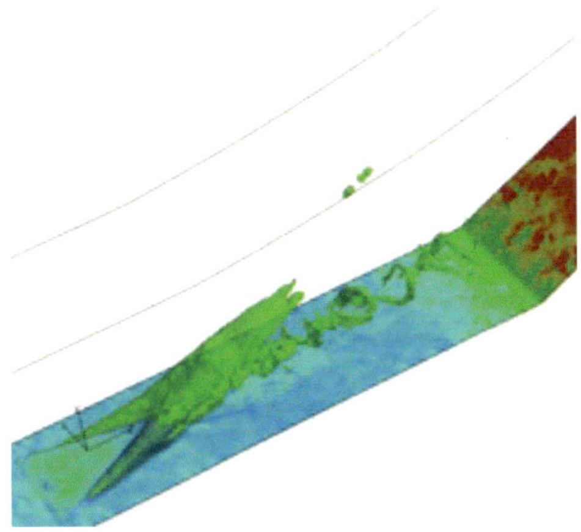

Fig. (6.38). The iso-surface of instantaneous pressure.

Fig. (6.39). Ring-like vortices shown by the iso-surface of λ_2.

To better investigate the coherence structures in the flow, iso-surface of λ_2 = -0.0006 is given in the Figs. (**6.39** and **6.40**) which can be used to illustrate the vortex structure. It clearly shows the existence of a ring-like vortex structure in the downstream of the MVG trailing edge. These vortices are generally perpendicular to wall surface. Then they would be badly distorted and enlarged when propagating downstream. They will eventually impact the separation shock at the ramp corner and interact strongly with the ramp shock. The generation of the ring-like vortex structure is not influence by the inlet condition. The ring-like vortex structure remains in the numerical results of MVG controlled flow with very weak turbulent inlet [73]. These ring-like vortices first found in our LES are considered as the mechanism of MVG in control of shock boundary layer interaction.

Fig. (6.40). The side view of ring-like vortices.

6.3.2. Confirmation of the Existence of the Ring-Like Vortices and Comparison to the Experimental Results

An instantaneous 2D structure of the vortices on the central plane from our numerical results is given in Fig. (**6.41**) by using the quantity of $|\nabla\rho|$ (which can be considered as the numerical schlieren on that plane). We can see many circular structures on that plane, which are the ring-like vortices mentioned in the context. To confirm the existence of these ring-like vortices, Lu conducted an experiment to validate the numerical results [75, 76]. In his experiment, both PIV (the particle image velocimetry) and acetone vapor screen visualization method are used to illustrate the structure in the flow. Flash of the laser is used to provide exposure at microsecond intervals. Fig. (**6.42**) shows the experimental result on the vortex structure on the central streamwise plane. It is clearly confirmed that a similar series of ring-like vortical structure exists in the downstream of the MVG (see Fig. **6.41**).

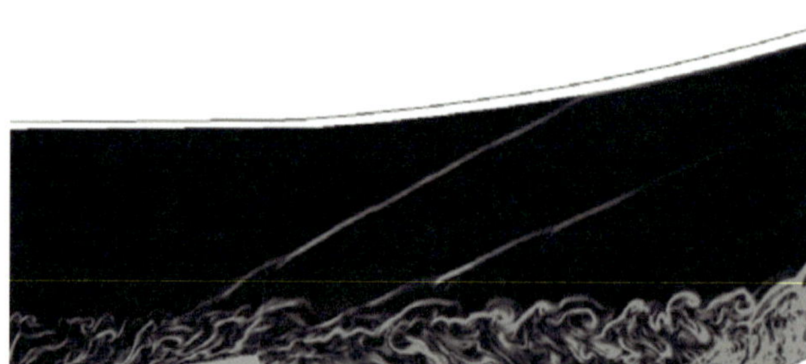

Fig. (6.41). The numerical shilieren on the central streamwise plane.

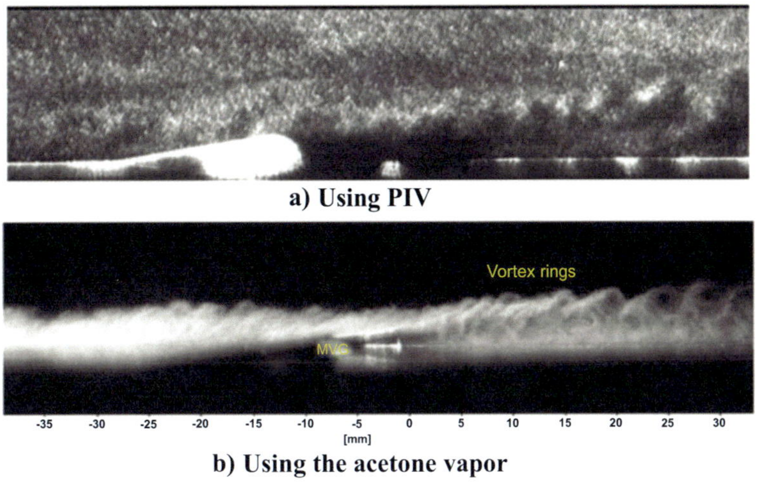

Fig. (6.42). The laser-sheet flash image on the central plane [75].

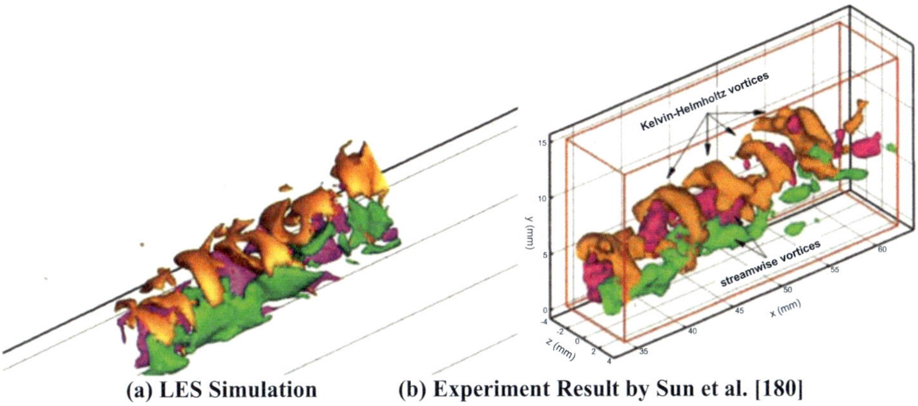

Fig. (6.43). Distribution of vortices behind the MVG.

Our numerical discovery on the vortical structure are also confirmed by Sun's 3-D PIV experiment by Sun *et al.* [180]. Fig. (**6.43**) shows the 3-D distribution of the three components of vorticity from both numerical and experimental results. Compared with the experimental results, we can find similar distributions of streamwise and lateral vorticity components. The experimental result also firmly proves the existence of 3-D ring-like vortical structure. The Kevin-Helmholtz vortical component in Fig. (**6.43b**) corresponds with the top of ring-like vortices. The component underneath is the streamwise component which forms the two counter rotating primary vortices in the momentum deficit. The streamwise vortical component is considered to be the main source of the energy for generating the ring-like vortices which will be explained in the following.

6.3.3. Study on the Origin of Ring-Like Vortices

The wall normal component of vorticity - ω_y is not shown in Fig. (**6.43**). By adding this missing component, we can see the whole ring-like vortical structure clearly as shown in Fig. (**6.44**). From the 3-D distribution of vorticity, we find that a ring-like vortex is mainly composed by ω_x and ω_y. ω_z is definitely the main source of two counter-rotating streamwise vortices inside the ring-like vortical structure. ω_x constitutes the top and the bottom of the ring-like vortices and ω_y forms the two sides.

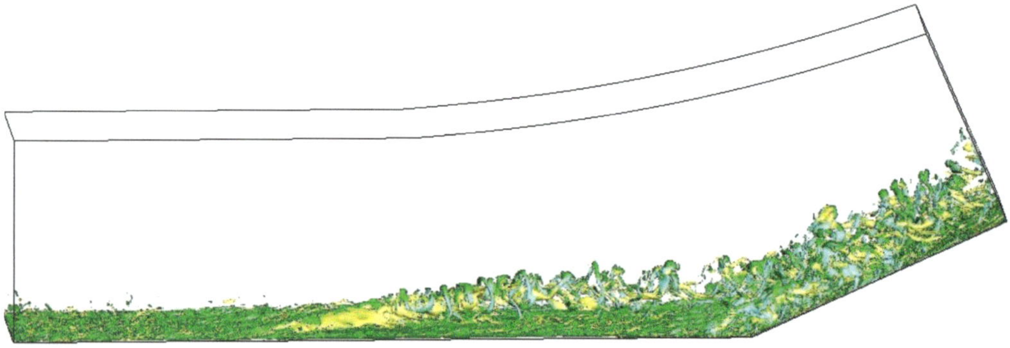

Fig. (6.44). Vortical structure shown by the components of vorticity.

Fig. (**6.45**) shows the iso-surfaces of the vortex structure by both the vorticity and λ_2 , it shows that both visulization methods can illustrate the vortex structure very well and the results are consistent. However, by using iso-surface of λ_2, the counter-rotating streamwise vortices are also illustrated which make the vortex structure to be more complicated. Besides, by using a smaller value (<-0.0006) to show the iso-surface of λ_2, the vortical structure with weaker rotation will be ignored. Fig. (**6.46**) only uses two vorticity components - ω_x and ω_y, to show the ring-like vortices. The roots of actual ring-like vortices seem to be connected to the streamwisely rotating vortex tubes inside the momentum deficit [181]. It shows that the vortices are not exactly circular. When they are just formed behind the MVG in the initial stage, their bottoms are open. Of course, this doesn't mean the breakdown of these ring-like vortices generated by the MVG. Fig. (**6.47**) shows the vortex filaments initiating from the tops of ring-like vortices. It shows the circular parts of these ring-like vortices are the vortex tubes where concentration of the vortex filaments occurs. However, these vortex filaments eventually scatter into the flow field in the upstream or at the two sides of the flow. As these vortices spread, the corresponding vorticity distribution becomes more uniform, and the corresponding vortex structure disappears in isochromatic visualization with a certain vorticity.

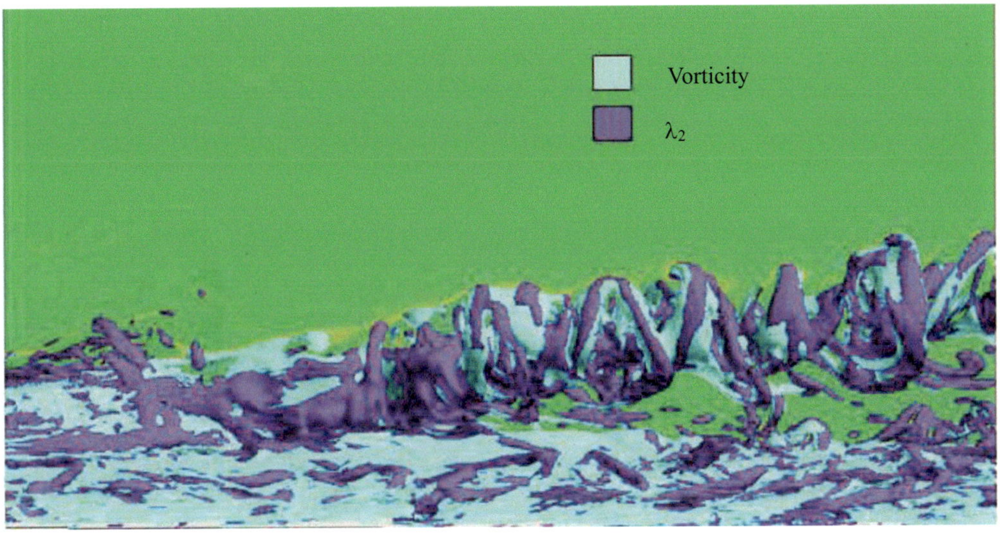

Fig. (6.45). Ring-like vortices shown by the components of vorticity and λ_2.

(a) global view

Fig. 6.46 contd.....

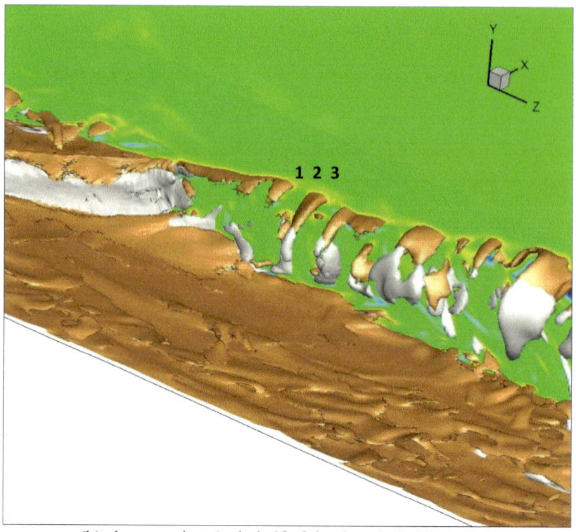

(b) close-up view (only half of the domain is illustrated)

Fig. (6.46). Ring-like vortices shown by ω_x and ω_y (only positive values are used).

Fig. (**6.47**) also shows the two-dimensional distribution of the spanwise vorticity component on the central flow plane. The red spots which locate at the upper bound of the momentum deficit on the figure indicate the tops of the ring-like vortices on the plane. However, in the lower limit of insufficient momentum, no such vorticity intensity was found, which also confirms that these ring-like vortex structures are not closed.

From the 2-D and 3-D vorticity distributions behind the MVG, it appears that these ring-like vortex structures come from the breaking down of the velocity shear layer generated behind the MVG and at the upper bound of the momentum deficit. In fact, the momentum deficit behind MVG is a very important source of vortex structure vortex [104, 181-183]. The shear layer at the upper bound is unstable and is considered to be a key factor in the formation of ring-like vortices [181]. However, compared with the vortex structure [73] in previous studies, with a relatively low turbulence intensity given in front of the MVG, more vortices are generated in the lower boundary layer in the present results of the simulation. The vortex structures in the upstream and in the lower boundary layer may also affect the formation of the ring-like vortices.

To trace the origin of these ring-like vortices, the vortex filaments of the first three distinguishable ring-like vortices in Fig. (**6.46b**) are given in Fig. (**6.47**). Those vortex filaments are initiated from local maximum regions of these ring-like vortices on the central streamwise plane.

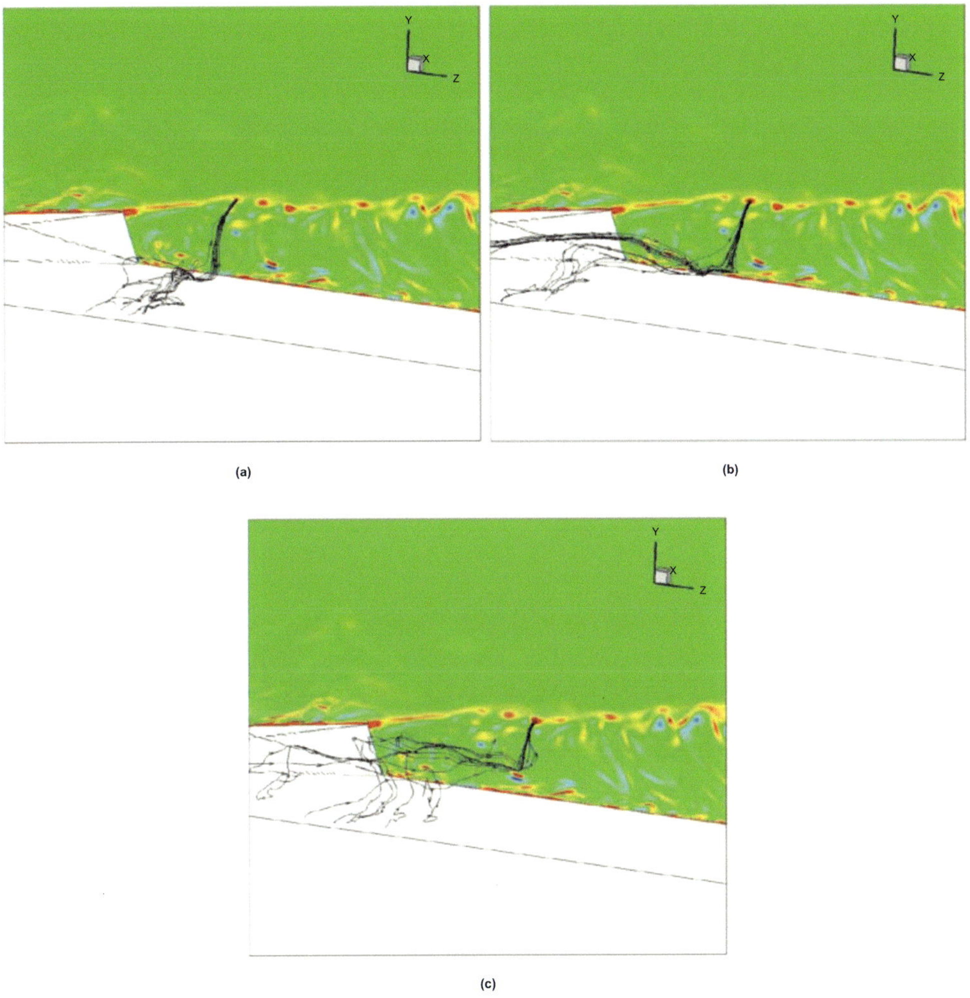

Fig. (6.47). Vortex filaments in the (a) first, (b) second, (c) third ring-like vortices.

In Fig. (**6.47a**), it shows that the vortex filaments are almost predominantly from both sides in the spanwise direction of the plane. The distribution pattern of vortex filaments indicates that the origin of the ring-like vortices may come from the unstable shear layer at the boundary of momentum deficit due to shear layer instability. That is, the circular momentum deficit which comes from the boundary layer around the MVG plays a very important role on the flow structure behind MVG.

Fig. (**6.48**) gives the time and spanwise averaged velocity profiles in the downstream of MVG from our numerical simulation and the experiment conducted by Babinsky [51]. In both cases, a strong circular shear layer can be observed by Yan *et al.* [184]. This is in consistency with other computational and experimental results [179, 59].

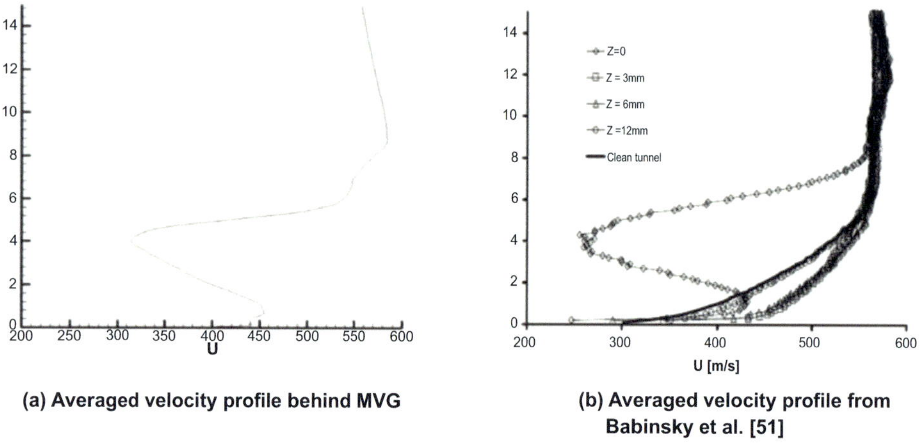

Fig. (6.48). Comparison of averaged velocity profile behind MVG with experiment.

Fig. (**6.49**) also gives a qualitative check on the shear layer which is performed by using the instantaneous data from the simulation. In this figure, the distribution of streamwise velocity is shown on the central plane and a spanwise cross-section. The green zone on both planes illustrates the momentum deficit. The distribution of pressure is also contoured on the central plane by lines. It shows that the cores of the ring-like vortices on the plane are located at the upper bound of the deficit. This location is exactly the same as the high shear layer.

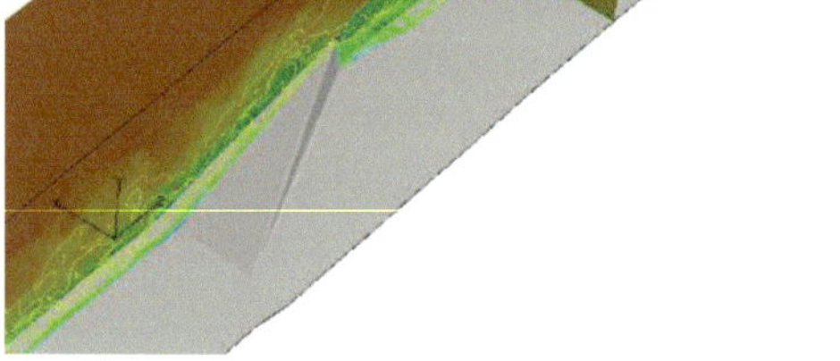

Fig. (6.49). The instantaneous pressure and streamwise velocity contour on different sections.

Fig. (**6.50a**) shows the time-averaged streamwise velocity profile at four different positions behind the MVG. In each velocity profile, a high velocity shear layer at the upper limit of the momentum deficit can be observed. Fig. (**6.50b**) shows the profile of second derivative of the first velocity profile in Fig. (**6.50a**). It is easily to find an inflation point at the upper limit of the momentum deficit. The presence of inflection points in the shear layer will cause the flow to be unstable and produces vortices through the Kelvin-Helmholtz (K-H) instability mechanism. The loss of stability in the shear layer will cause the aggregation of vorticity and forms spanwise vortices like the tops of the ring-like vortices. Thus, the mechanism by which the ring-like vortices are generated may be caused by K-H instabilities.

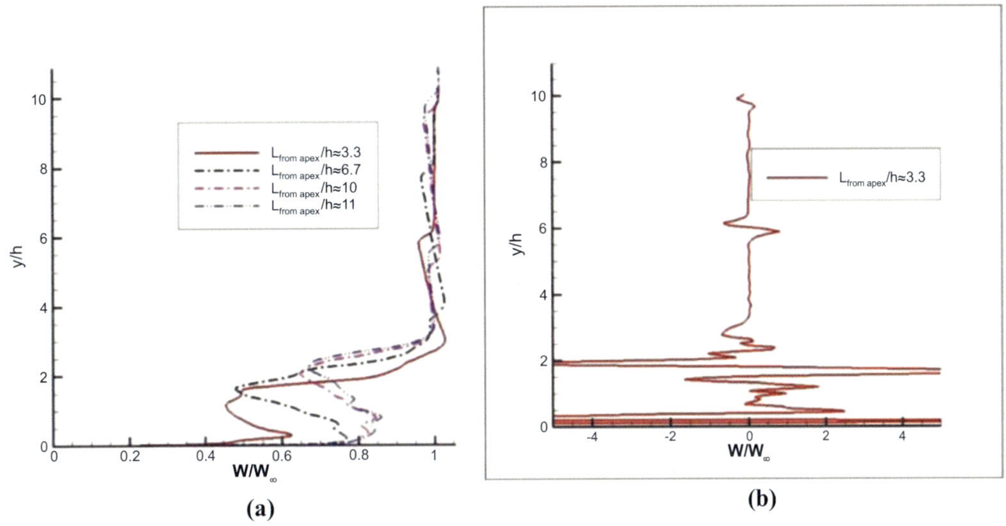

Fig. (6.50). (a). Averaged streamwise velocity at different sections (b). Second derivative of the velocity profile.

In our previous work by Yan *et al.* [185], the velocity profile was analyzed for stability, including the effect of fluid viscosity. The results show that the velocity profile from our numerical result is unstable. Fig. (**6.51**) shows the 3D iso-surfaces of the three vorticity components right behind the MVG. It shows that spanwise vorticity is first aggregated at the upper bound of the momentum deficit (illustrated by the streamwise vortices in side). In the downstream, the sides of the ring-like vortices are formed immediately on two sides of the circular momentum deficit. A whole ring-like vortex shows up soon due to the evolution of this unstable shear layer around the circular momentum deficit. This confirms the above assumptions. The velocity profile at the upper bound of the momentum deficit has the most unstable shear layer, but there is no such strong shear layer observed in Fig. (**6.50a**) at the bottom of the momentum deficit. As a result, the ring-like vortex is not closed at the bottom, the vortex lines for the ring-like vortex must come from the two sides instead.

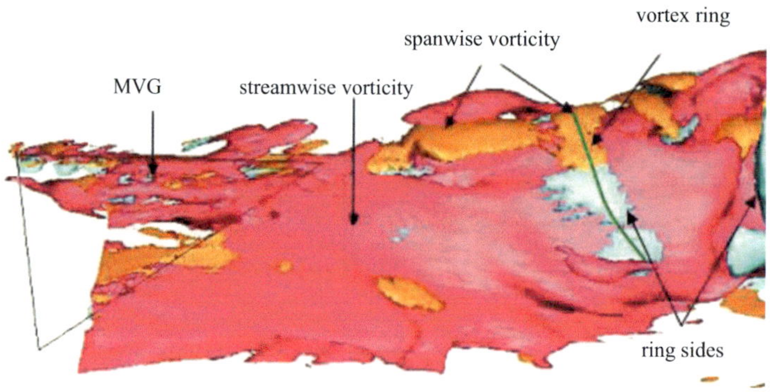

Fig. (6.51). The distribution of instantaneous vorticity.

Fig. (**6.47b**) shows the vortex filaments in the second ring-like vortex in Fig. (**6.46**) initiating from the top of this vortex. It is different from the distribution observed for the first ring-like vortex. Many of the vortex filaments in Fig. (**6.47b**) are from the upstream in front of the trailing edge of MVG. This vortex filament diagram shows that the second ring-like vortex is mainly derived from the upstream vortex. We tracked the evolution of the second ring-like vortex in Fig. (**6.52**) starting from the structure in the upstream (which is circled) by using the iso-surfaces of spanwise vorticity. It clearly shows this vortex structure from the upstream which is carried to the upper bound of the momentum deficit and eventually becomes part of the second ring-like vortex. It supports our conclusion above. In fact, this indicates that the second ring-like vortex consists mainly of upstream vorticity.

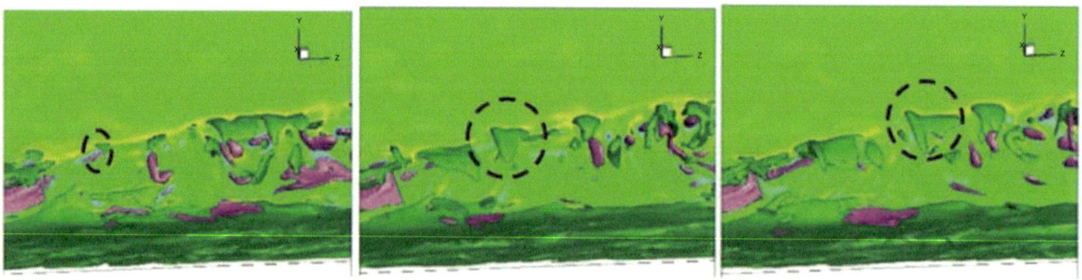

Fig. (6.52). Distribution of spanwise vorticity at three sequential times.

Fig. (**6.47c**) shows the vortex filaments in the third ring-like vortex in Fig. (**6.46**). The distribution of vortex filaments of this vortex seems more complicated. It shows that many of these filaments from the lower boundary layer in the upstream and also the region near the side surfaces of the MVG. In order to explain the difference in vortex filament distribution, we reviewed the evolution of this vortex as well.

Fig. (**6.53**) shows the spanwise vorticity distribution for three consecutive periods on the central plane and in the 3D space (iso-surfaces). It indicates that the body of third ring-like vortex is mostly initiated at the lower sub-layer and then it is pushed up to the upper bound of the momentum deficit. This vortical structure finally forms the third ring-like vortex ring behind the MVG. However, during such process, the vortical structure circled in Fig. (**6.53**) would inevitably mix with other vortex structures. In the meanwhile, the vortex filaments of the third ring-like vortex are stretched and pushed up gradually as well. In Fig. (**6.53**), it can also be found that the vorticity is continuously aggregated into these ring-like vortex structures during the process of forming the tops of these vortex at the upper limit of the defect. Thus, the structure of the third vortex in Fig. (**6.46**) is much more complex than the structure of the other two vortices in the upstream. In addition, the vortex filaments of the third vortex can also be derived from the rolling effect of the upper shear layer and the ones in the lower sub-layer. Eventually, the vortex line pattern shown in Fig. (**6.47c**) forms.

From previous study [184], we got the conclusion that the ring-like vortices are merely formed by the shear layer instability at the upper bound of the momentum deficit. However, some parts of the vortex structure can also be produced by the vortex structure in the lower boundary layer if the flow in front of the MVG is a fully developed turbulent boundary layer. The vortex structure in the lower sub-layer can also be raised to the upper bound. Of course, this only occurs in the initial stage of the generation of ring-like vortices. By carefully tracking the evolution of the ring-like vortices, the lower boundary layer does not produce a large structure and forms a large scale vortex in the downstream.

The vortex filament distribution for the other ring-like vortices are too complicated to analyze. Actually, the generation of ring-like vortices in the downstream comes from all the three situations mentioned above. The vortex structures in the downstream interact with each other as well.

Besides the clockwise rotating vortex structures shown in the Fig. (**6.47** and **6.52**), we can also see the counter rotating structures in Figs. (**6.52** and **6.53**). The spanwise tops of ring-like vortices are mostly positive rotating vortex tubes. However, there also exists counter rotating structures (with negative spanwise vorticity) in the flow. Fig. (**6.53**) shows these structures on both 2-D plane (in blue color) and 3-D space (in purple color). These 3D structures are mainly from the lower sub-layer under the momentum deficit and the shredding on the side surfaces of the MVG. By tracking the evolution of these structures, it shows none of them will evolve to counter rotating ring-like vortices although they can also be carried up to even the upper bound of the momentum deficit.

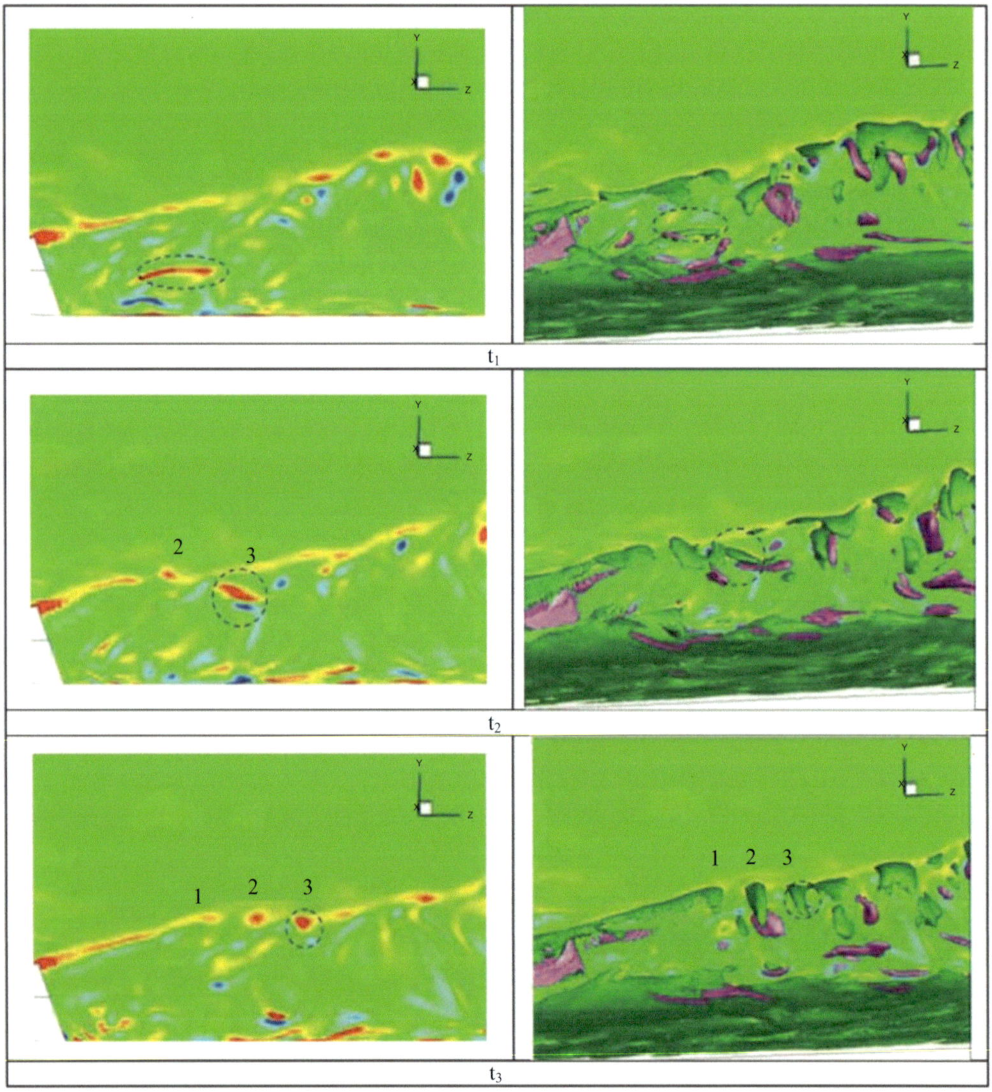

Fig. (6.53). Distribution of spanwise vorticity at three sequential times on the central plane (left) and 3D space (right).

On the other hand, the existence of the anti-rotating structure does affect the evolution of the ring-like vortices. In Fig. (**6.54**), we tracked some of the vortex structures in the circular area, which shows a reduction between the vortex structures with different spanwise vorticity. The complex effects of these vortex structures may continue to distort the ring-like vortices. This can also be considered to be the influence on the ring-like vortex structure from the lower boundary layer.

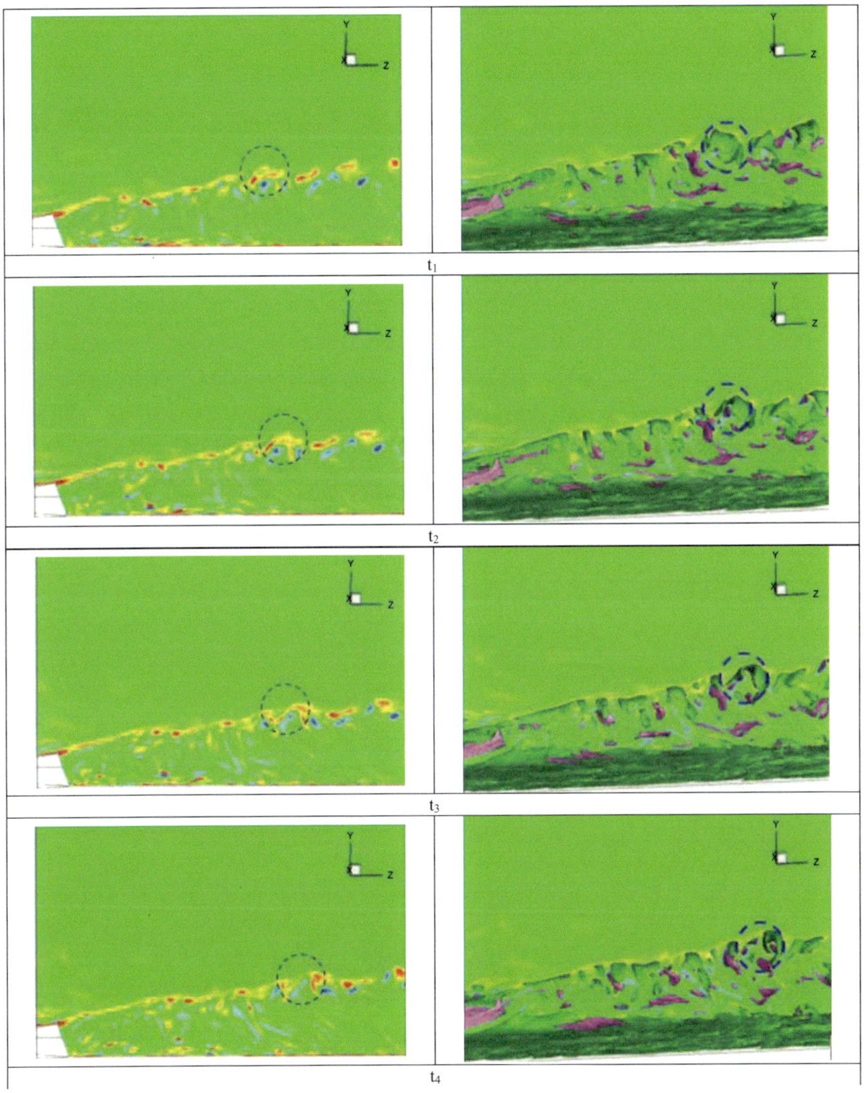

Fig. 6.54 contd.....

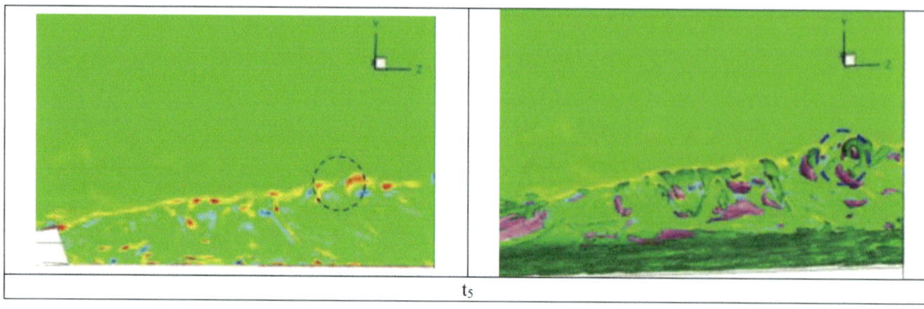

Fig. (6.54). Distribution of spanwise vorticity at five sequential times on the central plane (left) and 3D space (right).

CONCLUSION

As a conclusion, it shows that the major mechanism of the generation of ring-like vortices is the K-H instability of high shear layer at the upper bound of the momentum deficit behind the MVG. However, it has also been found that some of the vortex structures in the lower boundary layer below the momentum deficit can be lifted and evolve to ring-like vortices. The fully developed turbulence in front of the MVG strongly influences the formation and development of the ring-like vortex structure as well. The generation of ring-like vortices is a true 3-D process. The interaction between vortex structures is a new and interesting topic. Finally, a conceptual vortex model for the ring-like vortices in the wake of MVG is sketched in Fig. **(6.55)**.

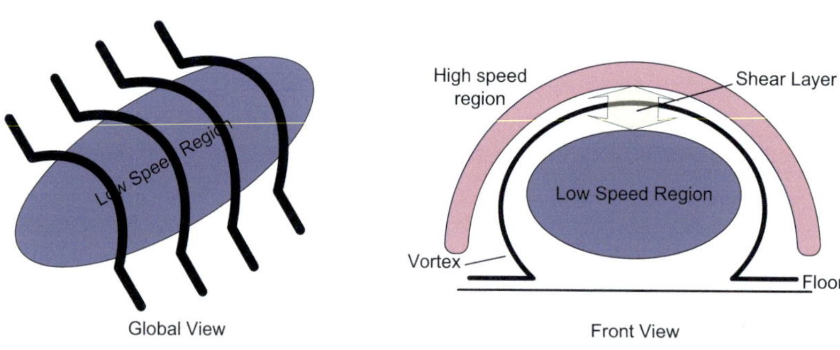

Fig. (6.55). The conceptual model of the generation of ring-like vortex.

CHAPTER 7

Interaction Between Ramp Shock Wave and Ring-like Vortices

Abstract: In this chapter, interaction of shock and vortex rings which were generated by MVG is described and analyzed in details. The change of vorticity, vortex, shock strength and location is described in details. The physics of shock and vortex interaction and its influence on the flow structure is also discussed.

Keywords: Flow separation, Shock, Shock and vortex interaction, Vortex ring, Vorticity.

7.1. INTRODUCTION

The ring-like vortices discussed in Chapter 6 play a very important role in the shock wave boundary layer interaction (SWBLI) at the ramp corner. These special ring-like vortex structures are also the mechanism of the flow separation reduction in MVG controlled ramp flow.

Once generated by MVG, the ring-like vortices would soon be carried into the downstream and interact with the strong ramp shock wave induced by the inclined ramp. The shock wave-boundary layer interaction (SWBLI) at the ramp corner thus involves interaction between ramp shock wave and ring-like vortices. There are some studies on shock wave-boundary layer interaction in recent decades [186-188]. These studies mainly focus on the following topics: the oscillations of the shock wave; the multi-stage characteristics of the impact of the vortex with the primary and the reflection shocks; the acoustic properties, including near- and far-field of acoustic, the dipole and quadruple acoustic pressure structure. However, the interaction between the ramp shock wave and ring-like vortices generated in the MVG controlled supersonic flow is completely different and should be considered as a new topic of the study. The differences include: 1) this interaction is completely three dimensional and very complicate since both of the ring-like vortices and the oblique ramp shock wave are three dimensional; 2) The interaction occurs in or near the upper boundary layer of the separation region in which there exists other flow structures, such as small scale vortices in the turbulence; 3) The interaction with ring-like vortices is a continuous process.

Chaoqun Liu, Qin Li, Yonghua Yan, Yong Yang, Guang Yang, Xiangrui Dong

Obviously, the previous studies on the standard vortex-shock wave interaction can still provide some hints and suggestions for our research.

7.2. INFLUENCE ON THE RING-LIKE VORTICES FROM THE INTERACTION

For further analysis, it is necessary to obtain some kinetic information about the ring-like vortices first. According to the results of our numerical simulation, those ring-like vortices appear to be irregular behind the MVG. When ring-like vortices travel to the downstream, they keep twisting. The ring-like vortices are badly deformed before they reached at the ramp and interacted with the ramp shock wave. The kinetic information of the first three ring-like vortices marked in Fig. (**7.1**) is collected because they are relatively regular at this stage. It is found that they move almost at the same speed and the averaged parameters are shown in Table 7.1.

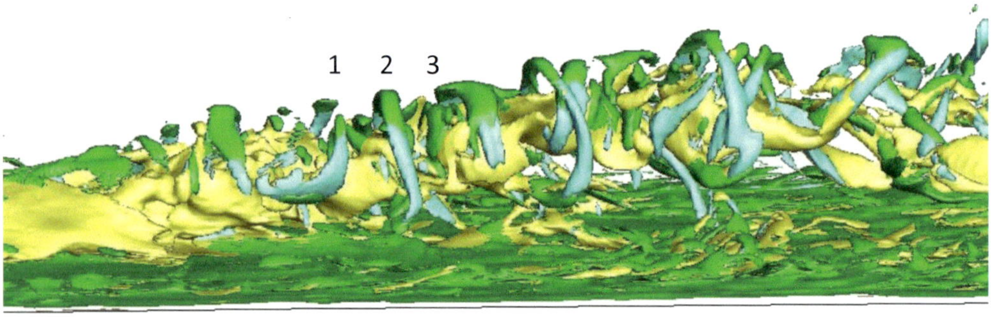

Fig. (7.1). Vortex structure plotted by iso-surfaces of the vorticity.

Table 7.1. Kinetic parameters of the first three ring-like vortices in Fig. (**7.1**).

Δs_{ring}	V_{ring}	T_{ring} (h/U_∞)	St_h
$1.21h$	$0.78U_\infty$	1.55	0.26

In Fig. (**7.2**), two ring-like vortices at the two sides of ramp shock wave (marked as 1 and 2) are investigated. The streamwise velocity of them are: $V_{ring,1}\approx 0.77U_\infty$, $V_{ring,2}\approx 0.47U_\infty$ respectively. The total velocity of the second vortex along the ramp direction is about $0.68U_\infty$. These values are reasonable, since the convection structure travels at a speed of about $0.7U_\infty$ normally.

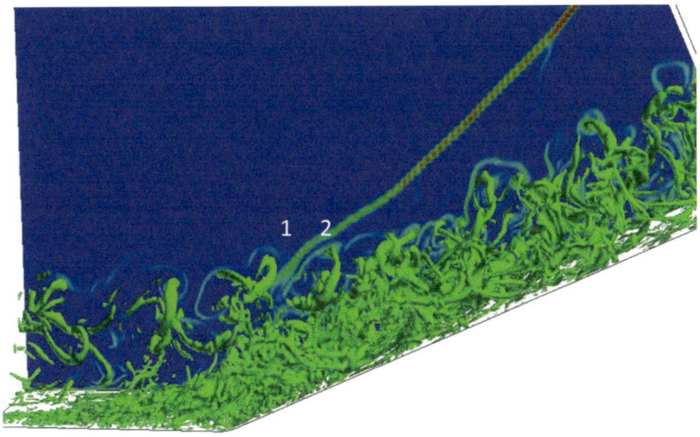

Fig. (7.2). Two ring-like vortices besides the shock wave.

It can be found that the total velocity of the ring-like vortex does not vary greatly when penetrating the shock wave. A slight decrease in velocity may be caused by the interaction between the ring-like vortex and the viscous sublayer. The ring-like vortices would eventually merge into the sublayer after they penetrate the ramp shock and travel above the ramp.

In Fig. (**7.3**), a ring-like vortex is tracked before and after it penetrates the ramp shock wave. It can be found that the vortex structure is slightly distorted but it does not break down during the interaction with the strong ramp shock wave without losing its initial topology. Since the ring-like vortices are kept distorting after they are created and carried to the downstream, the deformation of the vortex may be caused by the sublayer or compression effect at the ramp corner rather than the interaction with the ramp shock wave.

Fig. (**7.4**) shows the 3-D vortex structure above the ramp. Fig. (**7.5**) shows the 2-D distribution of the gradient of both density and pressure on the central streamwise plane when a ring-like vortex is passing through the ramp shock wave. The pressure and density gradient of the vortex increases immediately after the impact, however, it clearly shows that those ring-like vortices survived after the interaction with the shock wave and they never break down even as they travel to the end of the domain. The vortex in Fig. (**7.5**) is not even distorted. Fig. (**7.6**) shows the streamwise vorticity distribution on the central plane right before and after the ring-like vortex in Fig. (**7.3**) passing through the ramp shock. The distribution of streamwise vorticity is not very much influenced by the shock wave. Fig. (**7.7**) shows the vortex filaments contained in the ring-like vortex in Figs. (**7.3** and **7.6**) which had just impacted the ramp shock wave. A strong shock wave produces a strong discontinuity for speed, density and pressure. However, the vortex filaments of the vortex are not broken during the interaction. They are connected to both sides of the flow field that are cut by the shock wave.

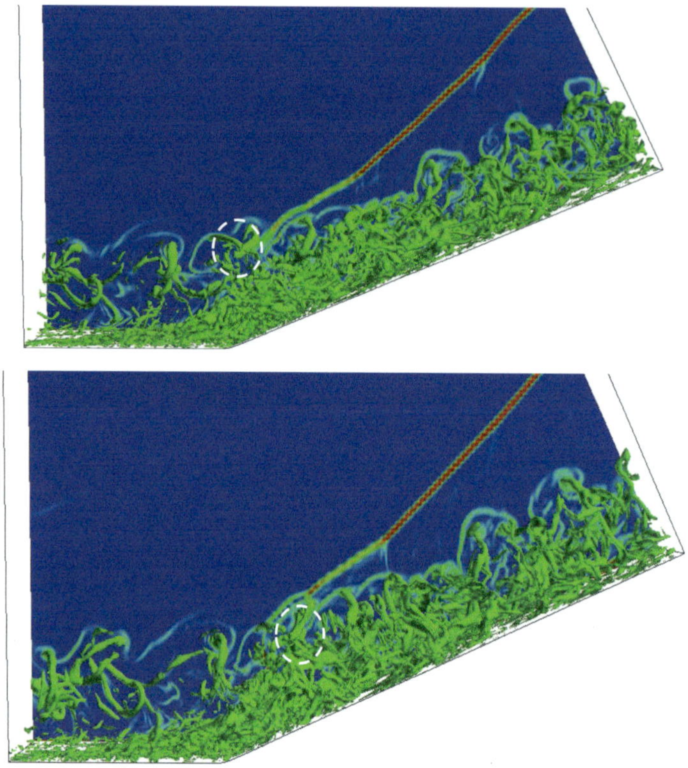

Fig. (7.3). The vortices at the ramp corner at two consecutive time steps.

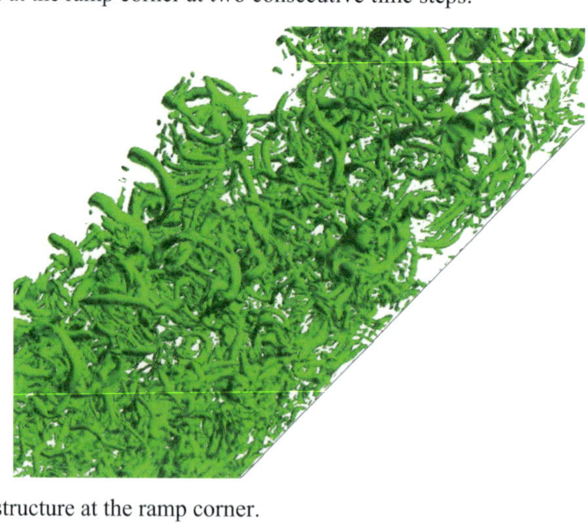

Fig. (7.4). The vortex structure at the ramp corner.

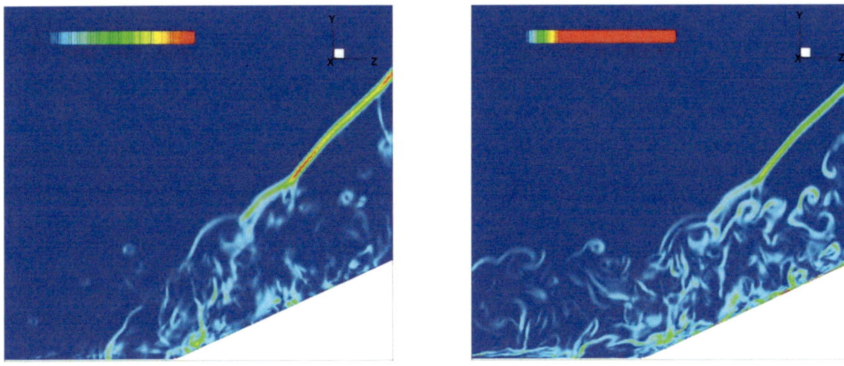

(a) (b)

Fig. (7.5). Distribution of (a) pressure gradient and (b) density gradient on the central plane.

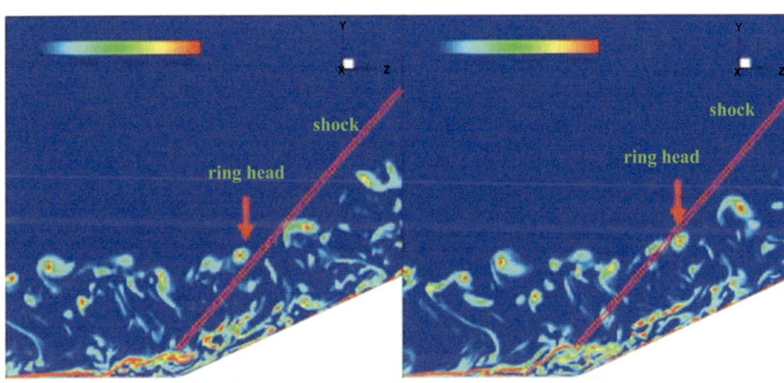

Fig. (7.6). Distribution of streamwise vorticity at two different time steps.

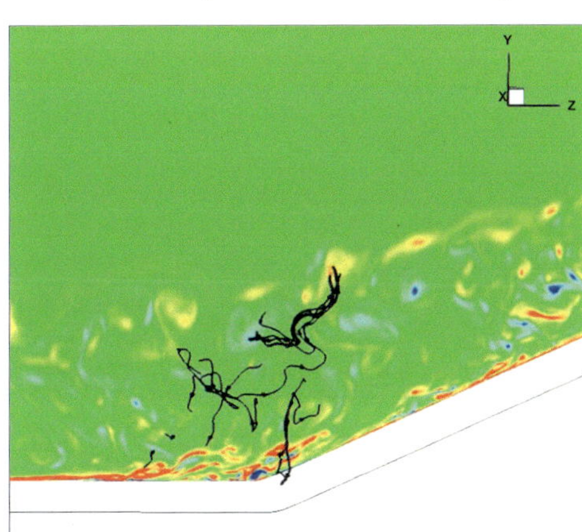

Fig. (7.7). Vortex filaments initiated from the tops of ring-like vortex behind the shock wave.

From all the observation mentioned above, it shows the ramp shock rarely determines the evolution of the ring-like vortices during the interaction. These ring-like vortices pass smoothly through the strong inclined ramp shock wave. They keep their shape, the number of vortices, and move without hindrance. The interaction is a continuous process for the ring-like vortex structure produced by the MVG. The ring-like vortex structure is quite robust and does not decompose during the interaction process with ramp shock wave.

To explain why ring-like vortices are not influenced by the interaction with ramp shock wave, the two leading factors -- shear layer instability and Richtmyer-Meshkov instability(RMI) which have impact on the vortical structures in the flow field are examined.

The velocity shear layer in the vicinity of shock wave could be a source of vorticity to alter the evolution of ring-like vortices. High speed shear layer may be unstable. The unstable shear layer will redistribute the vorticity in the flow field and form rotating vortex structures.

We know that there is a discontinuity in velocity distribution at the shock wave. However, the discontinuity in velocity distribution shown in Fig. (**7.8**) does not form a high shear layer of the streamwise velocity.

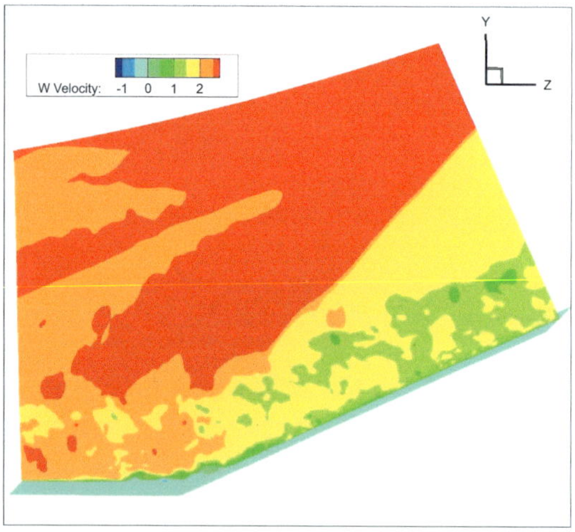

Fig. (7.8). Distribution of time-averaged streamwise velocity on the central plane.

In order to assess the discontinuity caused by the shock, we should consider the velocity distribution of the components tangent to the shock, rather than in the direction of flow. Fig. (**7.9**) shows that when the velocity component tangent to the shock wave is used, the velocity shear layer on the central plane disappears. In addition, Fig. (**7.10**) gives the time and spanwise averaged velocity distribution on a spanwise cross section which is perpendicular to the shock wave. It indicates that there is no shear layer at the ramp shock wave. As a conclusion, the shock-induced discontinuity does not form an unstable high shear layer of velocity, and therefore it does not affect the vorticity of the ring-like vortex structure due to shear layer instability.

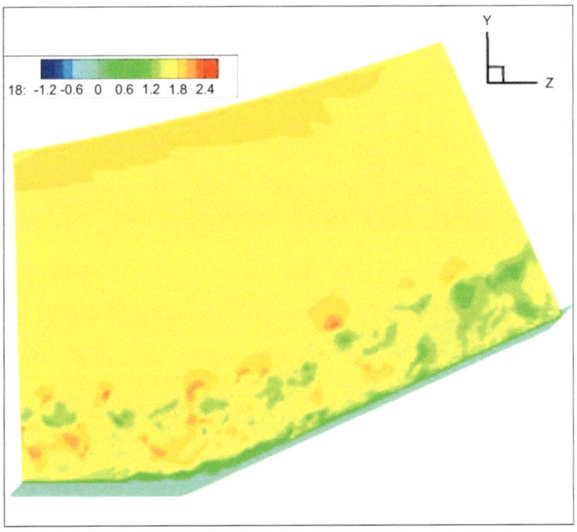

Fig. (7.9). Distribution of time-averaged velocity (tangent to the ramp shock wave).

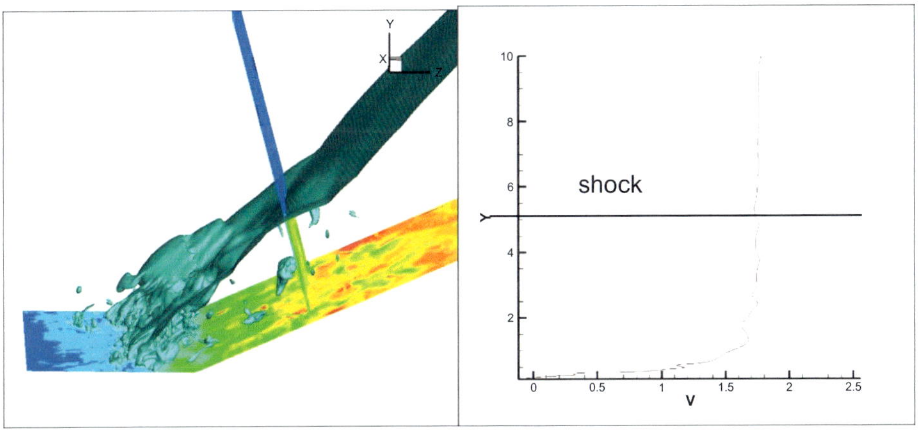

Fig. (7.10). Spanwise cross section (left) perpendicular to the shock wave and the averaged velocity profile on it (right).

The other factor that can dominate the interaction between the ring-like vortices and ramp shock wave is the Richtmyer-Meshkov instability (RMI). The mechanism of RMI is the baroclinic effects between $\nabla\rho$ and $\nabla\rho$. When an interface between fluids of differing density is impulsively accelerated, for example, by the passage of a shock wave, Richtmyer-Meshkov instability occurs.

The vortex structure is a different density wave. The process of its interaction with shock waves is the same as what RMI theory indicates.

$$\frac{d\Omega}{dt}-(\Omega\cdot\nabla)\mathbf{v}+\Omega(\nabla\cdot\mathbf{v})=\nabla\times\mathbf{F}+\frac{1}{\rho^2}\nabla\rho\times\nabla p+\nabla\times(\nu\Delta\mathbf{v})+\frac{1}{3}\nabla\times(\nu\nabla(\nabla\cdot\mathbf{v})) \quad \mathbf{(7.1)}$$

The mechanism of RMI can also be revealed by the vorticity equation (Eq. 7.1) derived from NS equations. In the supersonic ramp flow, the most important source term of vorticity comes from the quantity of $|\nabla\rho\times\nabla p|$ which plays a very important role on the generation and convection of vorticity in the flow. Fig. (**7.11**) shows the 3-D distribution of $|\nabla\rho\times\nabla p|$ with some specified values. It can be found that the quantity of $|\nabla\rho\times\nabla p|$ reaches a larger value only in a limited small zone, and the smaller value of $|\nabla\rho\times\nabla p|$ happens in the lower boundary layer. Thus, the flow field, especially where exists the ring-like vortex structure, is rarely influenced. As a result, the shock wave has little influence on the ring-like vortices much since that they only exist at the upper boundary layer.

7.3. INFLUENCE ON OBLIQUE SHOCK WAVE

As mentioned above, the interaction between the ramp shock wave and ring-like vortices has different features from classic 2D regular shock-boundary layer interaction. The series of ring-like vortices generated by the MVG are in large scale and have 3-D structures. The ramp shock is in fully 3D geometry as well. Although the ring-like vortex structure is not influenced, the discontinuity induced by the shock wave seems to be blurred at the ramp corner a lot.

Fig. (**7.12**) shows the 2-D distribution of the gradient of density and pressure on three different spanwise cross-section above the ramp. Fig. (**7.13**) shows the gradient distribution of these variables on the central streamwise plane at two sub-sequential time steps when a ring-like vortex passes through the ramp shock. From these two figures, it can be observed that there is no obvious shock wave at the corners of the ramp, and the original shock wave returns to the downstream position of the ramp. Although the upper part of the shock wave keeps well in the shape, the bottom portion is twisted, and its strength is also reduced significantly and even be eliminated at the lower boundary layer at the corner. This is mainly caused by the interaction with the ring-like vortices and the viscous sub-layers. This kind of phenomenon is a kind of a complicated shock wave boundary layer interaction.

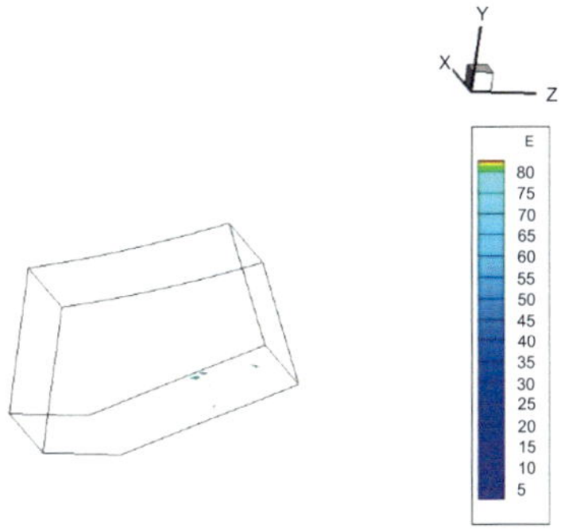

$|\nabla\rho \times \nabla p| = 80$

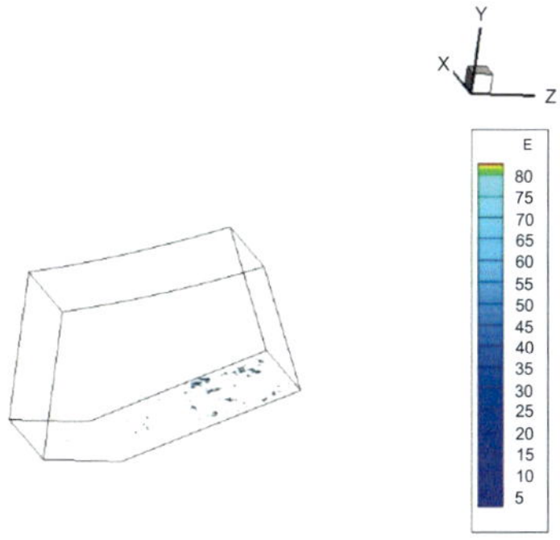

$|\nabla\rho \times \nabla p| = 40$

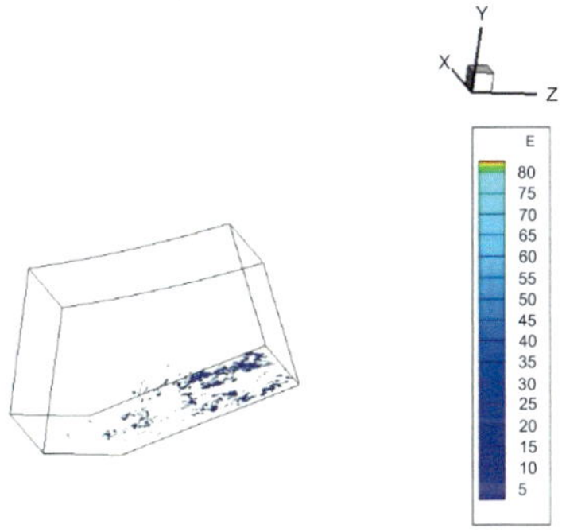

$|\nabla\rho \times \nabla p| = 20$

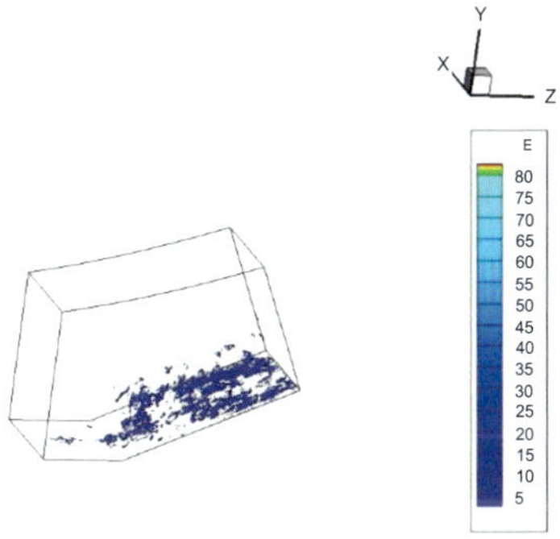

$|\nabla\rho \times \nabla p| = 10$

Fig. (7.11). Iso-surfaces of $|\nabla\rho \times \nabla p|$.

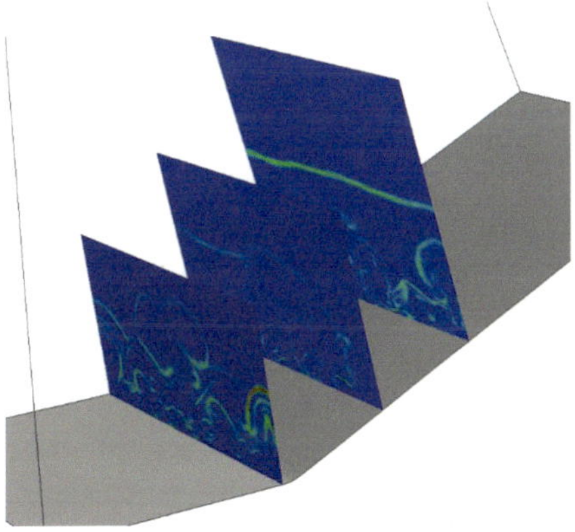

(a) $|\nabla\rho|$

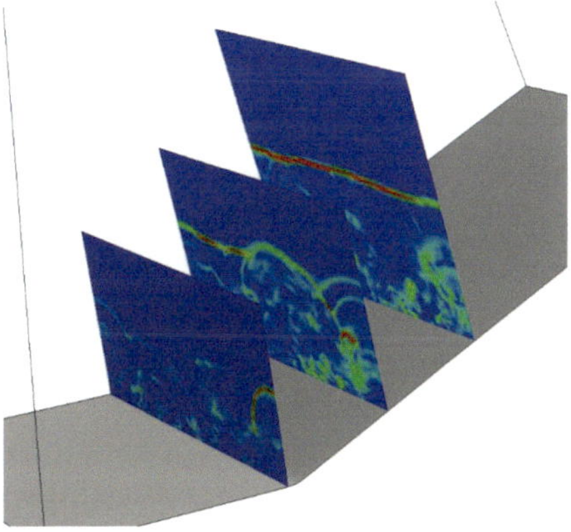

(b) $|\nabla p|$

Fig. (7.12). The contours of $|\nabla\rho|$ and $|\nabla\rho|$ on three different spanwise cross sections.

In Fig. (**7.13**), the structure of multiple layer shock wave is found. When ring-like vortices pass through the shock wave, this typical multiple layer shock structure will be formed continuously. The number of the shock layers changes during the interaction, usually forming 2-4 layers. Fig. (**7.14**) gives the 3-D shock wave structure at 6 sub-sequential time steps. It is observed that a particular bump shape will be formed during the interaction with a ring-like vortex.

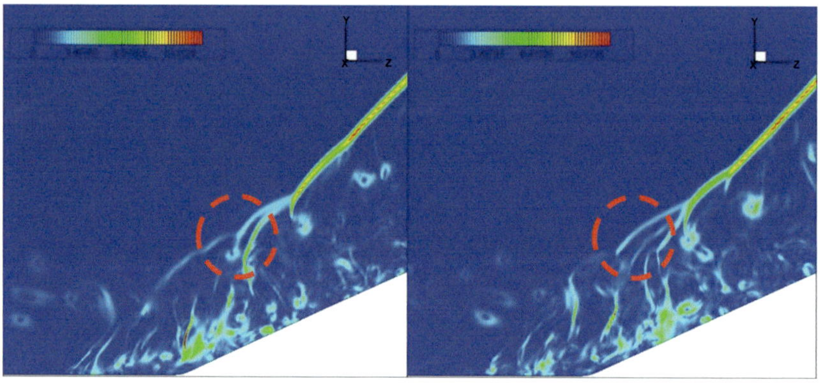

(a)

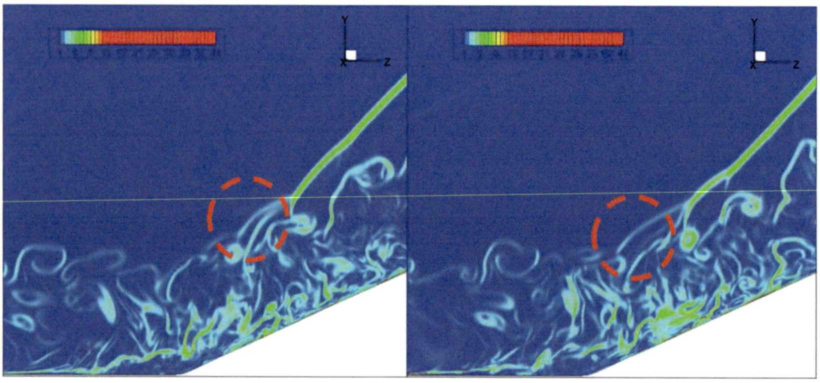

(b)

Fig. (7.13). Distribution of pressure gradient (a) and density gradient (b) at two sub sequential time steps.

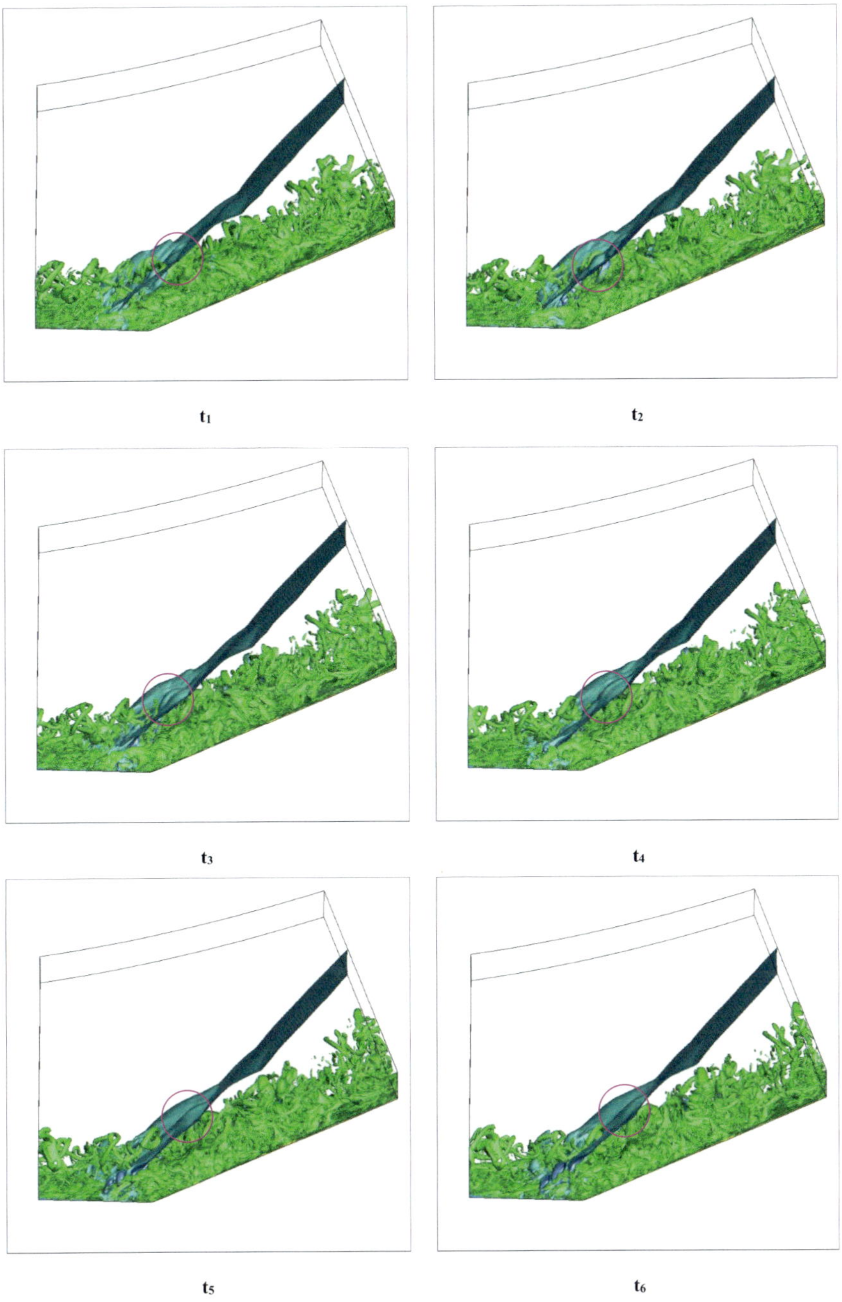

Fig. (7.14). The shock wave presents by the iso-surface of pressure and ring-like vortex structure.

To explain the mechanism of the multiple layers and bump shape of the shock wave, in Fig. (**7.15**), the second ring-like vortex in Fig. (**7.2**) is taken out andthe velocity filed on a spanwise cross section, which passes through the center of this vortex, is provided. The velocity field shown in the figure is the tangent component of the total velocity distributed on the spanwise cross-section, and the normal component is ignored. It is interesting to note the upward ejection at the center of the ring-like vortex. Such up-wash flow will impact on the surface of the ramp shock wave and bend it to be an arc-like shape which is the mechanism of the multiple layers and bump on ramp shock wave. The upper layer is the original but rather weak separation / reflection wave; the other one is created by the strong interaction between the upwash flow induced by the ring-like vortex and the shock wave. After interacting with a ring vortex, the multi-layer structure will disappear. With the movement of the ring-like vortex, the bump will also be carried to the downstream. It will eventually disappear when the ring-like vortex moves away from the shock wave. The process is repeated the next ring-like vortex passes through the shock wave, a new layer is thus formed.

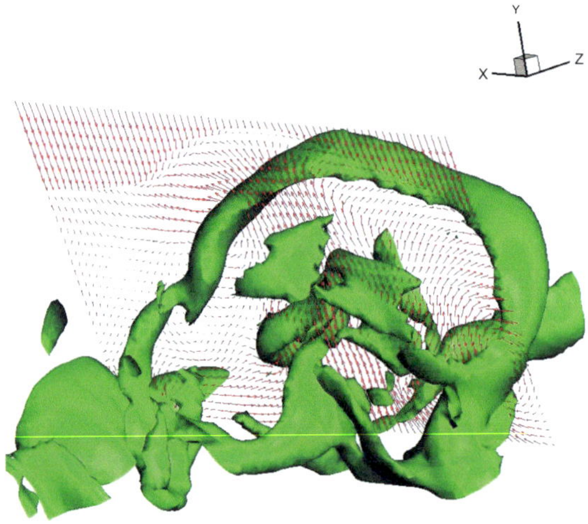

Fig. (7.15). The vector field at a cross section across the ring-like vortex.

The shock wave boundary layer interaction studied in our book is entirely 3D. The shapes of the shock wave in Figs. (**7.12 and 7.14**) clearly state the asymmetry of the flow.

In order to better study the effect of the SWBLI at the ramp corner, the interaction process on the central plane is compared with the same process occurs on a streamwise cross section that is away from the ring-like vortex structure (see Fig. **7.16**). On this spanwise cross section, there's no interaction between the shock wave and ring-like vortices. The contours of the pressure gradient and vorticity are compared in Fig. (**7.17**) on the two sections mentioned above. It shows that the shock wave structure of the central part changes much more significantly. The larger ring-like vortices made a dramatic influence on the shock wave, and it seems to be eliminated during the interaction at the ramp corner. The shock wave would reconnect quickly after the interaction, but its portion below the interaction area is bent and almost perpendicular to the wall boundary. We think this is the mechanism on the reduction of the separation at the ramp corner.

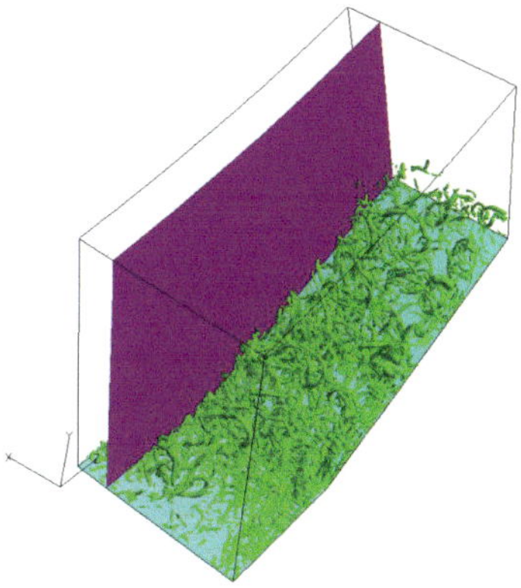

Fig. (7.16). The streamwise section beside the ring-like vortices.

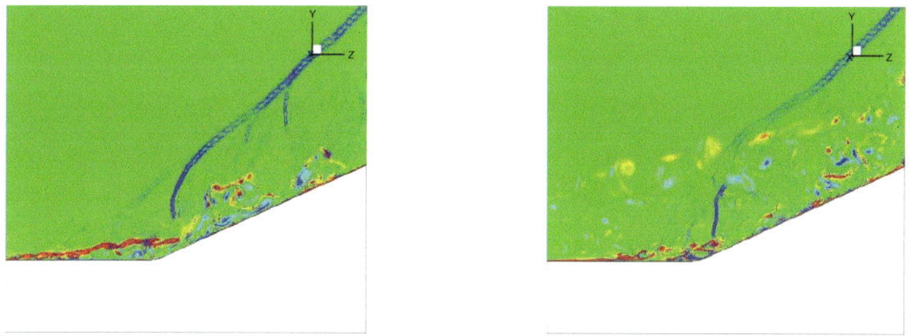

t_1

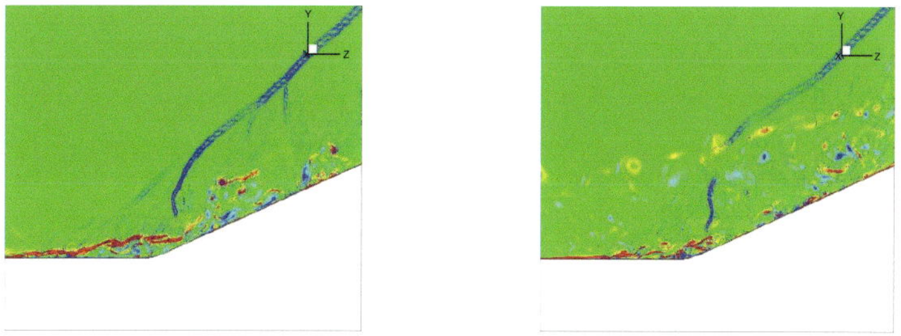

t_2

Fig. 7.17 contd.....

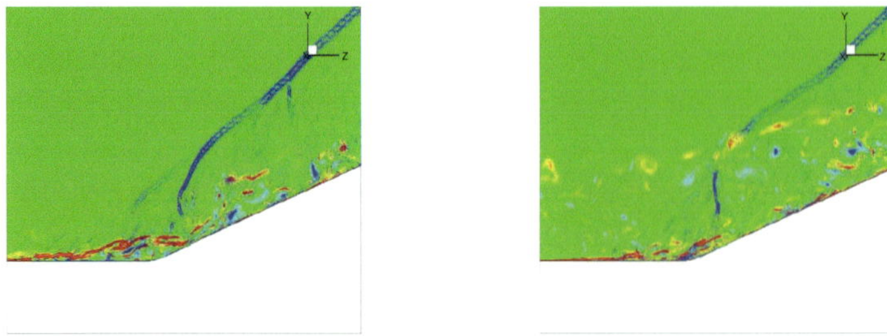

t3

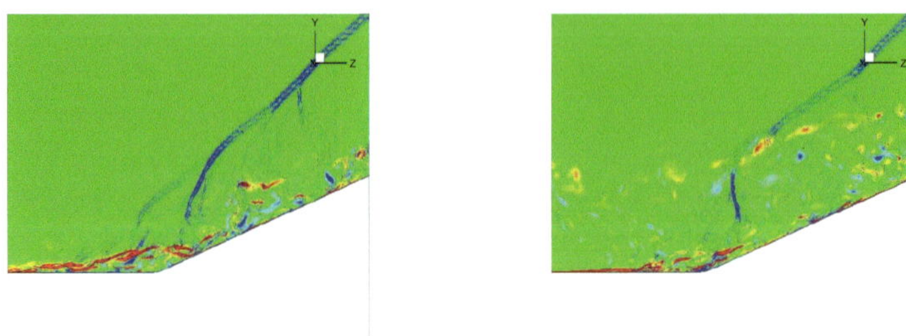

t4

Fig. (7.17). Contour of pressure gradient and the distribution of vorticity on two streamwise sections at four subsequential time steps (a) on the section beside the ring-like vortices (b) on the central plane.

The distributions of pressure and temperature on two streamwise cross sections are also given in Figs. (**7.18** and **7.19**). If there exists the ring-like vortex structure, the temperature boundary layer becomes thicker. In addition, the pressure distribution is also found to be disturbed a lot by these ring-like vortices. As a conclusion, the boundary layer becomes thicker with the presence of the ring-like vortex structure. Thus the special SWBLI involving ring-like vortices influence the ramp shock wave more to a severer extend.

Finally, the time-averaged distribution of pressure on the two streamwise cross-sections and the time-averaged three-dimensional iso-surfaces of pressure gradient are shown in Figs. (**7.20** and **7.21**), respectively. It clearly can be found that the multi-branches of the ramp shock wave are eliminated by the interaction with the ring-like vortices. The shock wave is also much shortened. Thus, the flow separation induced by the ramp shock wave at the ramp corner is significantly reduced. The three-dimensional shock wave remains the two sides where there is no ring-like vortices which shows the significant role of the ring-like vortices on reducing the intensity of the shock wave and the flow separation.

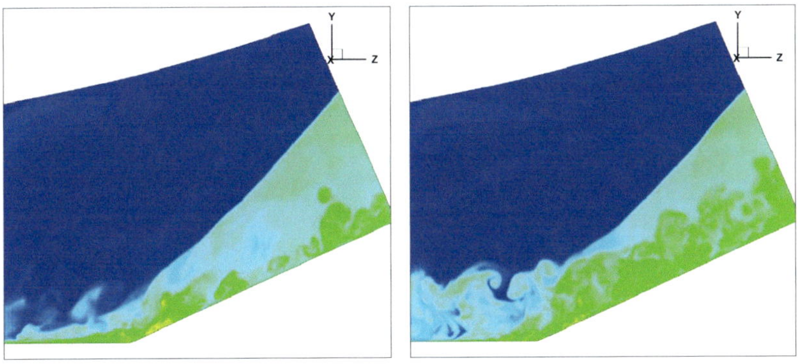

(a) (b)

Fig. (7.18). Temperature boundary layer (a) on the streamwise section beside the ring-like vortices and (b) on the central plane.

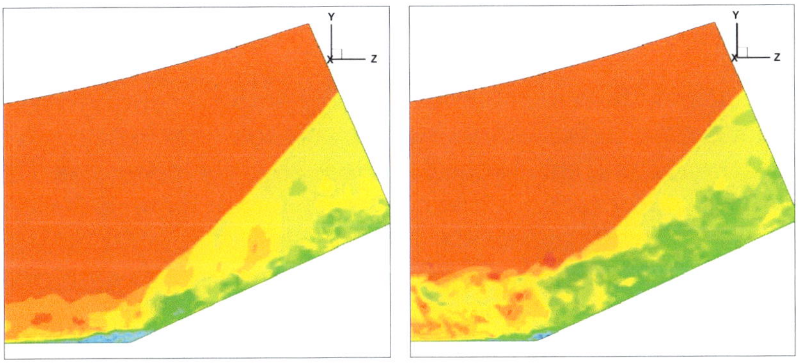

(a) (b)

Fig. (7.19). Distribution of pressure on (a) the streamwise section beside the ring-like vortices and (b) on the central plane.

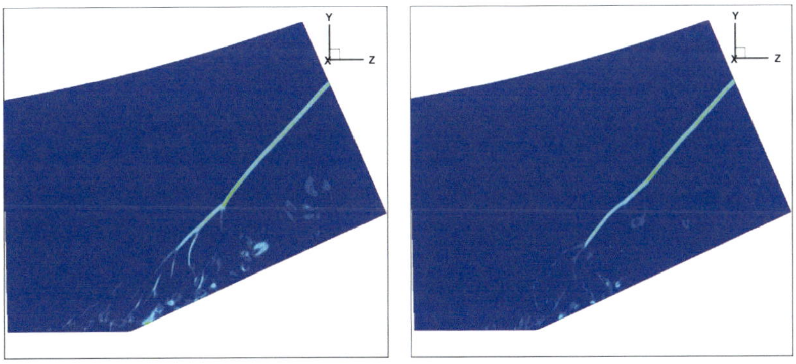

(a) (b)

Fig. (7.20). Distribution of time-averaged pressure gradient (a) on the streamwise section beside the ring-like vortices and (b) on the central plane.

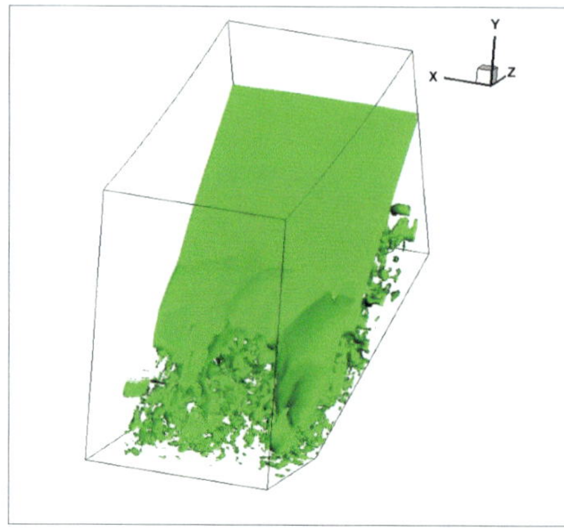

Fig. (7.21). Iso-surface of time-averaged pressure gradient at the ramp.

CONCLUSION

The investigation on the interaction of ring-like vortices generated by MVG and the ramp shock wave indicates that such interaction does not greatly affect the ring-like vortex structure, but the shock wave is severely influenced. The ring-like vortices structure never ruptures, even when they pass through the strong shock wave. Conversely, the instantaneous and time averaged results both showed the decomposition or reduction shock wave in the process. It is confirmed that the existence of the ring-like vortices played the most important role on the shock wave boundary layer interaction (SBLI) at the ramp corner.

CHAPTER 8

Mechanism of MVG for SBLI Control

Abstract: In this chapter, a new mechanism of SBLI control by MVG is presented. Traditionally, people consider the reduction of shock-induced separation is caused by streamwise vortex which mixes the boundary layer flow and makes the velocity profile to be more able to withstand the flow separation. The new mechanism shows the spanwise vortex rings are critical which are quickly moving and fast rotating to destroy or undermine the shock and ease up the shock-induced flow separation significantly. Potentially, this study could contribute to some technology revolution in control of SBLI, flow separation and interference reduction.

Keywords: Flow control, Flow separation, Noise reduction, Pressure fluctuation, Shock, Shock and vortex interaction, Vortex ring, Vorticity.

8.1. MVG REDUCE SEPARATION OF SHOCK INDUCED FLOW

8.1.1. Ring-like Vortices Undermine Shock Waves

Liu *et al.* [164, 189] proposed a new vortex identification criterion, called Ω-method, which is given by:

$$\Omega = \frac{\mathrm{Trace}\left(B^{\mathrm{T}}B\right)}{\mathrm{Trace}\left(A^{\mathrm{T}}A\right)+\mathrm{Trace}\left(B^{\mathrm{T}}B\right)+\varepsilon}, \tag{8.1}$$

ε is a small positive number to avoid division by zero. A and B are the symmetric and anti-symmetric part of the velocity gradient tensor $\nabla \boldsymbol{V}$, respectively, as follows,

$$A=\frac{1}{2}\left(\nabla V+\nabla V^{\mathrm{T}}\right), \tag{8.2}$$

$$B=\frac{1}{2}\left(\nabla V-\nabla V^{\mathrm{T}}\right). \tag{8.3}$$

Chaoqun Liu, Qin Li, Yonghua Yan, Yong Yang, Guang Yang, Xiangrui Dong

Per Liu *et al*. [164, 189], a connected region where $\Omega > 0.5$ identifies a vortex and the iso-surface of $\Omega = 0.52$ can be utilized to demonstrate vortex structures. Fig. (**8.1**) reveals the position of the ring-like vortex chain by $\Omega = 0.52$ iso-surface. The surface color shows the value of X-coordinate. According to Fig. (**8.1**), the chain of ring-like vortices is located between $X = -1.5h$ and $X = 1.5h$, which means vortex rings appear mainly near the center. And no large-scale ring-like vortex is shown near the spanwise boundary. Two planes, $X = -3.45h$ (side one) and $X = 0$ (central one), are chosen for comparison (Fig. **8.2**), based on the distribution of vortex rings.

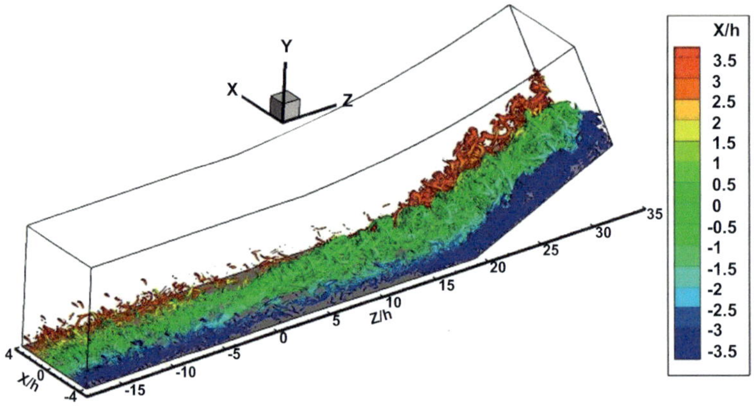

Fig. (8.1). Iso-surface of $\Omega = 0.52$.

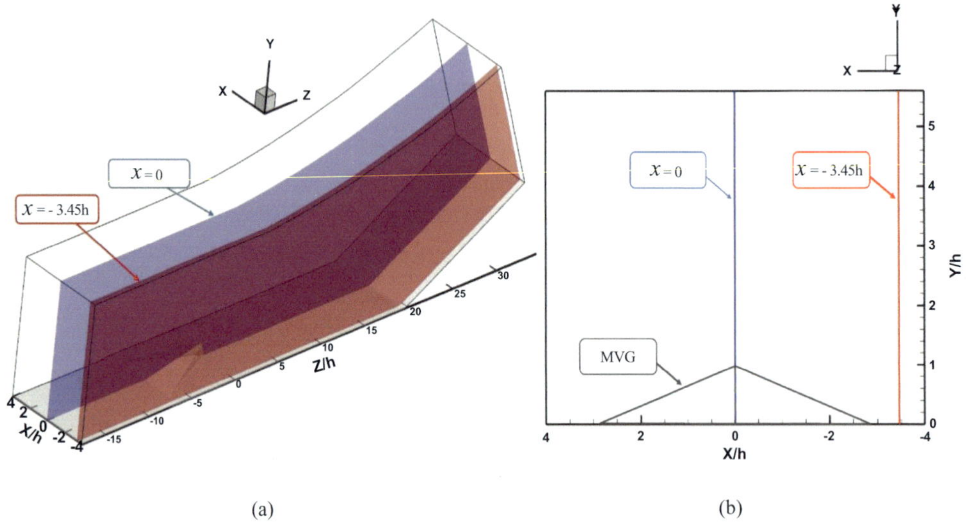

Fig. (8.2). Position of two planes in different views.

Fig. (**8.3**) shows the distributions of instantaneous pressure gradient magnitude ($\|\nabla p\|$) on these two planes, which indicates shock waves. On plane $X = -3.45h$, shock waves show up in branches and the multiple layer shock structure is observed. Yan *et al.* [190] gave the mechanism of the multiple-layer shock waves. Relatively, the central plane $X = 0$ shows shock waves are distorted and undermine, as a result of shock waves and the ring-like vortices interaction. On the side plane, shock begins at = $18h$, while $Z = 21h$ on the central plane. Fig. (**8.4**) gives the time-averaged streamwise velocity w and contour lines of $\frac{\partial w}{\partial y} = 0$ near the wall boundary. The gray curves are the contour lines of $\frac{\partial w}{\partial y} = 0$, where y is normal direction. The front contour line lying at the region where $w = 0$ in Fig. (**8.4b**) indicates, on this line, the velocity is zero and the derivative of velocity is also zero. Thence it demonstrates the flow separation position. As w velocity is dominant in this case, only w velocity is considered. We observe a significant difference in Fig. (**8.4**). The separation zone starts at $Z = 13.2h$ on the side and at $Z = 15.4h$ on the center. Apparently, ring-like vortices mitigate the flow separation in the central area.

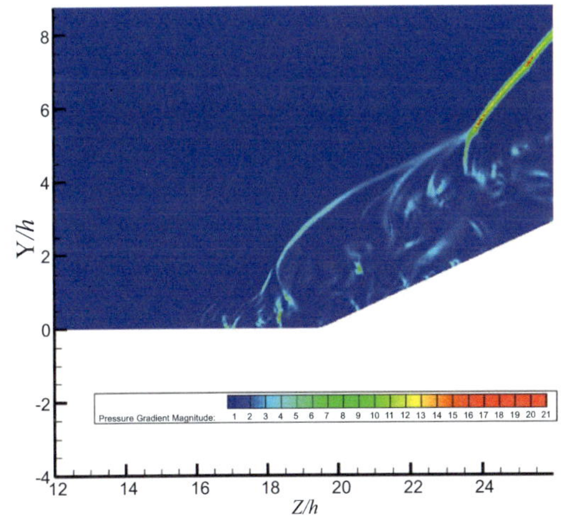

(a) $X = -3.45h$

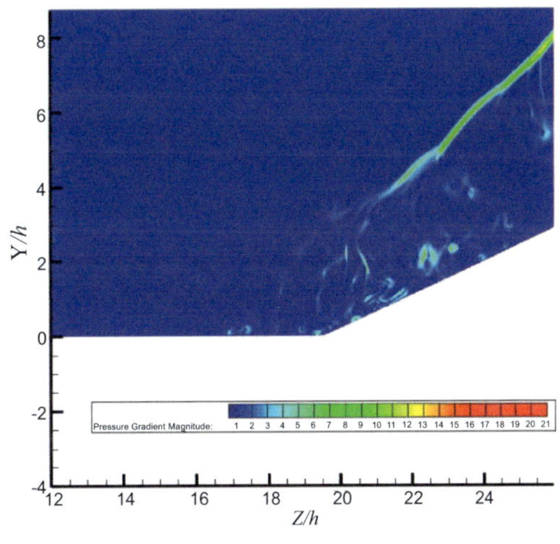

(b) $X = 0$

Fig. (8.3). $\|\nabla p\|$ distribution on two planes at $t = 1800T$.

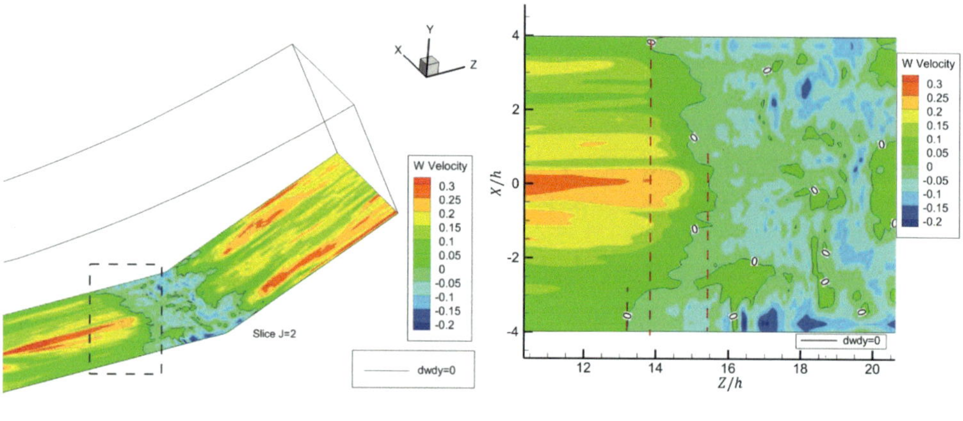

(a) (b)

Fig. (8.4). Distribution of streamwise velocity w and contour lines of $\frac{\partial w}{\partial y} = 0$ near the wall.

Eight sub-sequential time steps (1838T - 1852T) are investigated in details. Fig. (**8.5**) shows the process of vortex rings weakening and breaking up the shock wave on the central plane $X = 0$. The colored contour demonstrates $\|\vec{\omega}\|$, which is vorticity magnitude. The white curves show the pressure gradient ∇p, which give the position of shock waves. The typical high $\|\vec{\omega}\|$ value at point C in Fig. (**8.5**) represents the ring-like vortex head. When interaction of shock and vortex ring happens, the shock wave is altered, weakened and then undermined. However, shock wave keeps the shape well at the top, even though it suffers a severe destroy at the bottom by the vortex ring and shock interaction. When a ring-like vortex breaks into the shock, the shock is weakened, but the shock still has complete structure. However, the shock is smashed immediately after the ring passed through shock wave.

The point P(X = 0, Y = 2.9258h, Z = 20.9795h) is selected for quantitative analysis. It is marked by a pentagram in Fig. (**8.5**). We can see it is on the way of vortex ring head traveling. The changes of pressure gradient magnitude $\|\nabla p\|$, vorticity magnitude $\|\vec{\omega}\|$ and density gradient magnitude $\|\nabla p\|$ at point P are shown in Fig. (**8.6**). A severe reduction of $\|\nabla p\|$ and $\|\nabla p\|$ can be observed, which shows that the strength of shock wave at point P is weakened and even

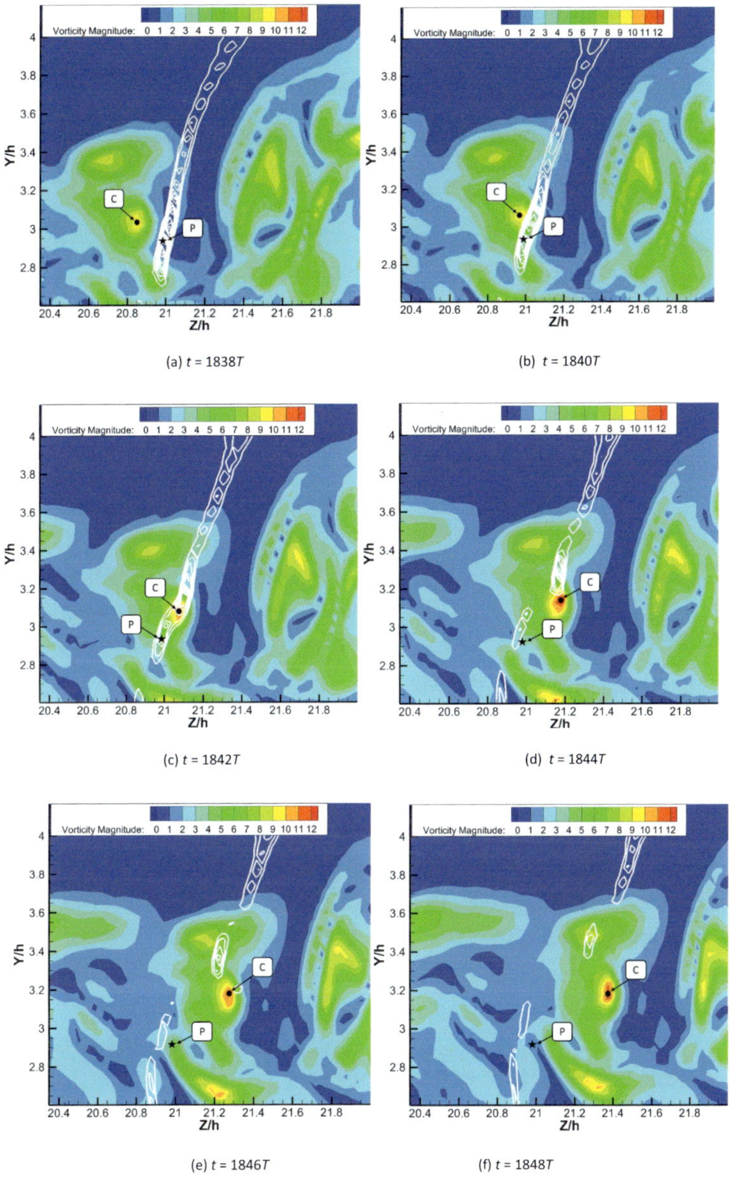

(a) t = 1838T

(b) t = 1840T

(c) t = 1842T

(d) t = 1844T

(e) t = 1846T

(f) t = 1848T

Fig. 8.5 contd.....

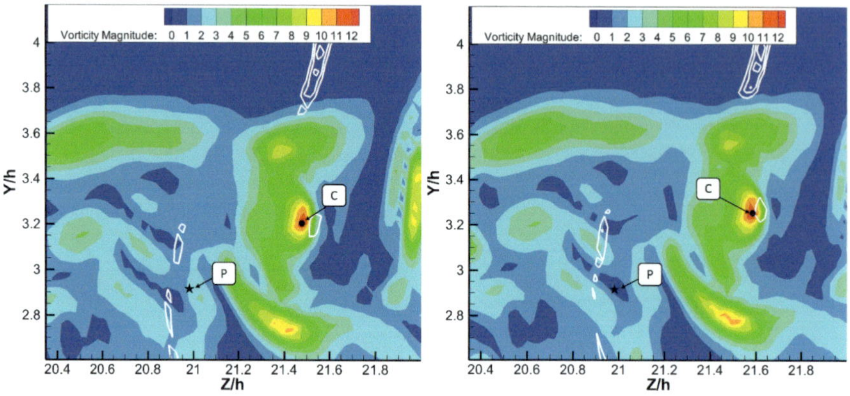

(g) t = 1850T (h) t = 1852T

Fig. (8.5). The process of vortex rings weakening and breaking up the shock wave on the central plane $X = 0$.

vanished. The peak of $\|\vec{\omega}\|$ represents the vortex ring head is passing through point P. The changes of density ρ, pressure p and w-velocity (streamwise) at different time steps at point P are given in Fig. (**8.7**). Density ρ and pressure p decend at first, resulting in reduction of $\|\nabla p\|$ and $\|\nabla p\|$, and $\|\nabla p\|$ rises back some while $\|\nabla p\|$ keeps reducing. The reason is, as the vortex ring head passed through this selected point, density ρ and pressure p rise back, but the change rate of density ρ and pressure p turns out to be smoother after vortex ring head passed through. This makes density gradient ∇p and pressure gradient ∇p keep decreasing. By Fig. (**8.7**), w velocity has the opposite changing pattern. And even though w velocity deceases, it still remains supersonic flow. Therefore, low density and low pressure play a dominant role in the vortex ring and shock interaction by breaking shock waves up.

Fig. (**8.8**) shows a three dimensional structure of ring-like vortex by iso-surfaces $\Omega = 0.52$. In Fig. (**8.8a**), the central plane ($X = 0$) is contoured by vorticity magnitude $\|\vec{\omega}\|$. White lines are contour lines of pressure gradient magnitude $\|\nabla p\|$, which indicate the position of shock waves. In Fig. (**8.8b**), a clear ring vortex structure is given alone with coordinates. Fig. (**8.9a**) shows the three dimensional vortex ring with low pressure zone, where the blue iso-surfaces are $p = 0.95$. Fig. (**8.9b**) shows the three dimensional vortex ring with low density zone, where the blue iso-surfaces are $\rho = 0.8$. By comparison, the locations of both low pressure and low density concentrate on the head part of vortex ring.

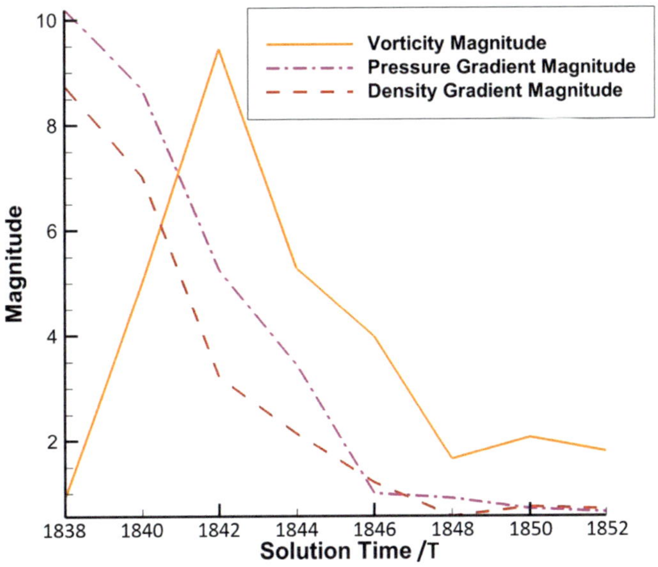

Fig. (8.6). The change of $\|\vec{\omega}\|$, $\|\nabla p\|$ and $\|\nabla p\|$ with solution time at fixed point P in Fig. **(8.5)**.

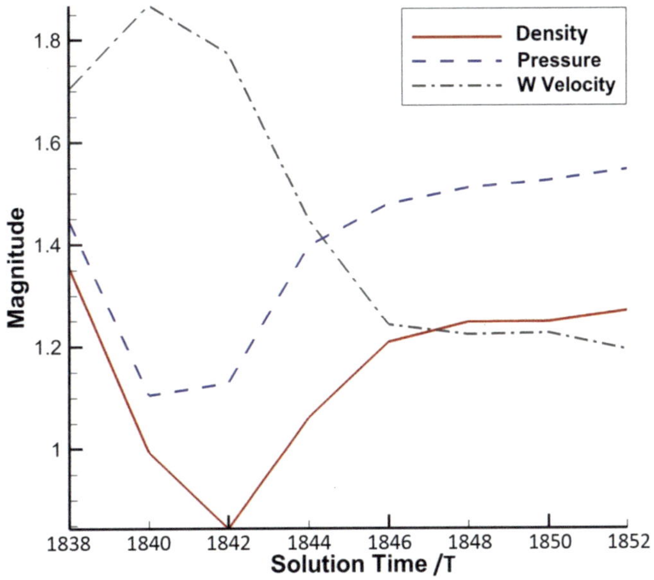

Fig. (8.7). The change of ρ, p and w velocity with solution time at fixed point P in Fig. **(8.5)**.

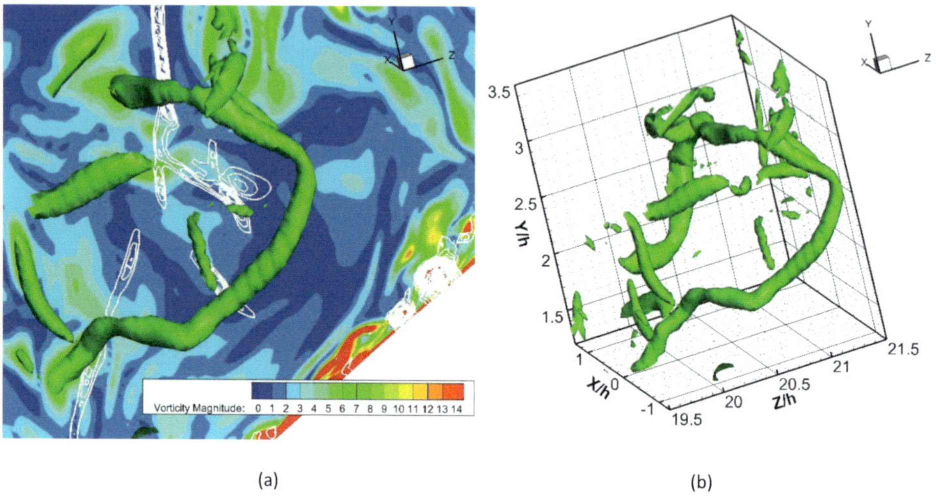

Fig. (8.8). Three dimensional structure of vortex ring (green iso-surfaces, $\Omega = 0.52$) (a) with the central plane $X = 0$ and (b) alone with coordinates.

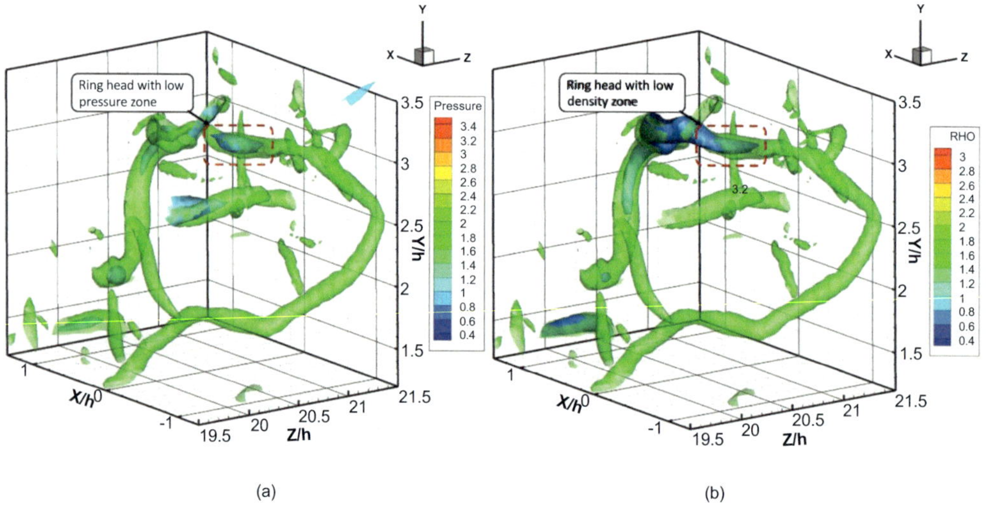

Fig. (8.9). Vortex ring (green iso-surfaces, $\Omega = 0.52$) with (a) low pressure zone (blue iso-surfaces, $p = 0.95$) and (b) low density zone (blue iso-surfaces, $\rho = 0.8$).

8.1.2. Vortex Rings Before and After the Shock

Some quantity changes are observed before and after the shock by tracing this vortex ring in several sequential steps. Fig. (**8.10**) shows the positions of the vortex ring at first and last time steps. Green iso-surfaces are $\Omega = 0.52$ iso-surfaces; Blue translucent contour indicates the vorticity magnitude $\|\vec{\omega}\|$ distribution on the central plane $X = 0$; Black contour lines present the pressure gradient ∇p . The vortex ring is located before the shock in Fig. (**8.10a**) and after the shock in Fig. (**8.10b**). Namely, the ring head locates at $Y = 3.03h$ and $Z = 20.86h$ at $1838T$ while $Y = 3.25h$ and $Z = 21.69h$ at $1852T$. By comparing Figs. (**8.10a** and **8.10b**), we could find that the structure of vortex is influenced little by the shock wave.

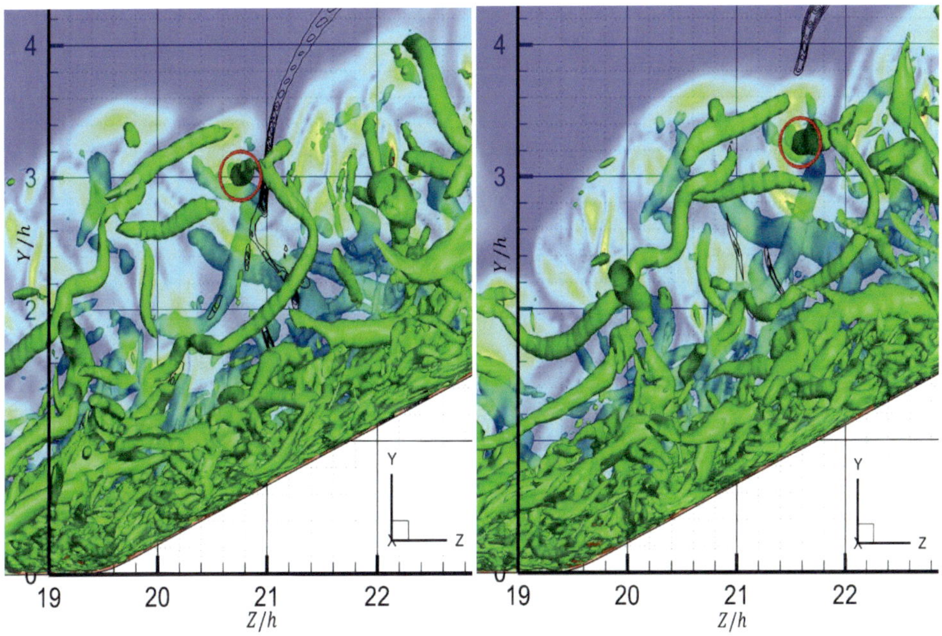

(a) $t = 1838T$ (b) $t = 1852T$

Fig. (8.10). Position of the same ring-like vortex at two different time steps. Red circle indicates the position of ring head.

The location of the center of ring head is identified as the location with maximum vorticity magnitude $\|\vec{\omega}\|$ on the central plane $X = 0$, which is marked as point C in Fig. (**8.5**) for every time step. The change of $\|\nabla p\|$ and $\|\nabla p\|$ at point C at 8 time steps is shown in Fig. (**8.11**). The peaks could be easily identified at time step $1842T$, which means the vortex ring head is passing through the shock wave. This can be proved from Fig. (**8.5c**). At time step $1842T$ in Fig. (**8.5c**), the ring head is penetrating the shock. Fig. (**8.12**) gives the change of vorticity magnitude $\|\vec{\omega}\|$ and *X*-vorticity ω_x at the center of ring head at different time steps. From Fig. (**8.12**), we can find $\|\vec{\omega}\|$ and ω_x have the same pattern and these two sets of data are very close. It can be seen that the vortex here is spanwise but not streamwise. And there is a significant increase in $\|\vec{\omega}\|$ when vortex ring is passing through the shock. The variation of density ρ, pressure p and w-velocity at point C in different time steps is given in Fig. (**8.13**). The pressure and density increase after vortex ring passed through shock wave but w-velocity decrease in this progress. This is caused by different distribution of pressure, density and velocity before and after the shock wave.

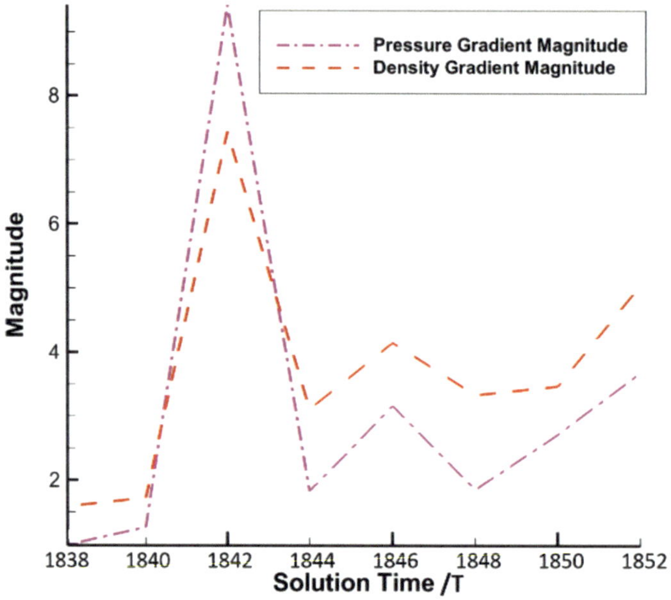

Fig. (8.11). $\|\nabla p\|$ and $\|\nabla p\|$ of change at the center of ring head with solution time.

8.2. HIGH-SPEED ROTATION AND HIGH-FREQUENCY VORTEX RINGS

Liu [189] addressed, as Ω is a quotient of vortical part over the total vorticity, the vorticity magnitude $\|\vec{\omega}\|$ can be further decomposed as a vortical part and non-vortical part by Ω, as follows:

$$\omega_{\mathrm{vor}} = \Omega\|\vec{\omega}\| \tag{8.4}$$

$$\omega_{\mathrm{nonvor}} = (1-\Omega)\|\vec{\omega}\| \tag{8.5}$$

where ω_{vor} and ω_{nonvor} present vortical part and non-vortical part respectively.

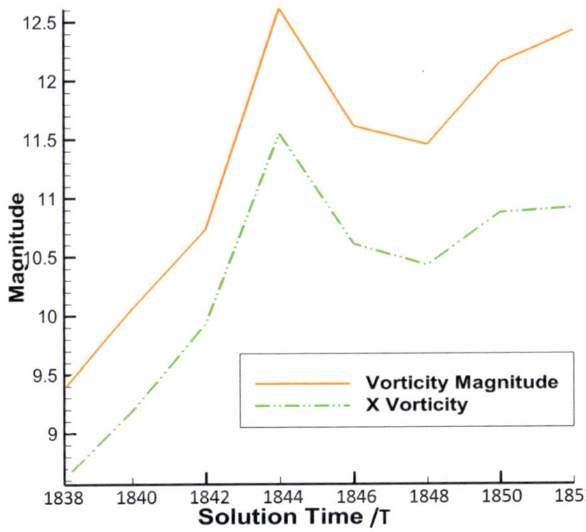

Fig. (8.12). Vorticity magnitude and X vorticity of change in the center of ring head with solution time.

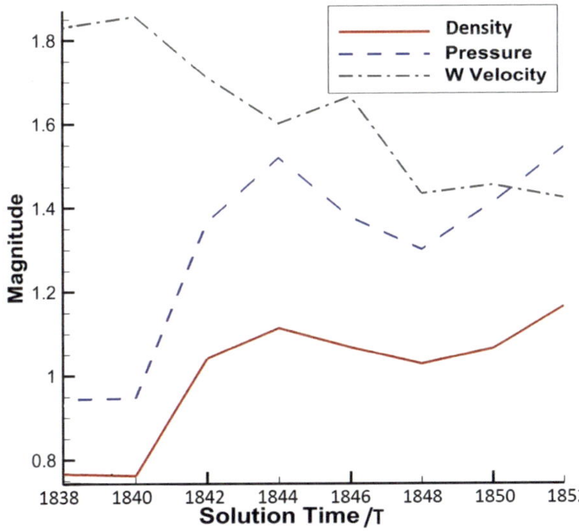

Fig. (8.13). Density, pressure and *w* velocity in the center of ring head.

A quantitative analysis is given on the vortex ring which is tracked in **section 8.1** at 1838T. Fig. (**8.14**) gives the $\|\vec{\omega}\|$ distribution at the vortex ring head. Point C denotes the location with maximum $\|\vec{\omega}\|$, which is 9.9654.

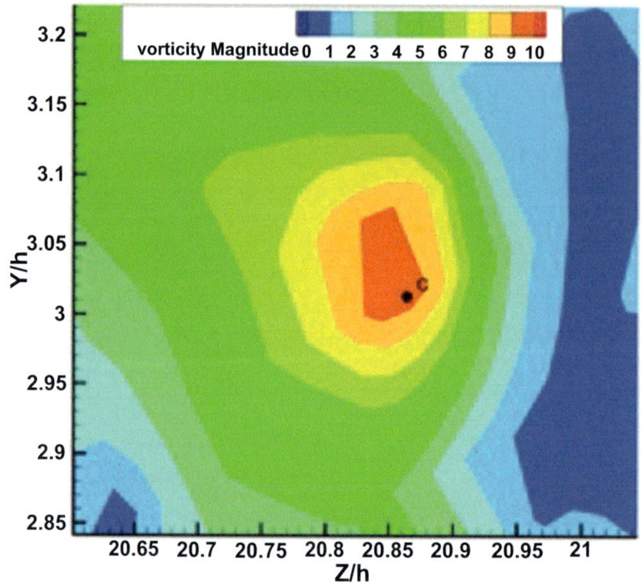

Fig. (8.14). $\|\vec{\omega}\|$ at vortex ring head at 1838T.

The velocity gradient tensor $\nabla\vec{V}$ at point C is

$$\nabla\vec{V} = \begin{pmatrix} \frac{\partial u}{\partial x} & \frac{\partial u}{\partial y} & \frac{\partial u}{\partial z} \\ \frac{\partial v}{\partial x} & \frac{\partial v}{\partial y} & \frac{\partial v}{\partial z} \\ \frac{\partial w}{\partial x} & \frac{\partial w}{\partial y} & \frac{\partial w}{\partial z} \end{pmatrix} = \begin{pmatrix} -0.6121 & 2.6205 & 0.8943 \\ -1.3580 & 0.3605 & -5.1198 \\ 0.1789 & 3.9889 & 0.0877 \end{pmatrix}$$

By Eq. (8.1), $\Omega = 0.95$. It shows that 95% vorticities are vortical vorticities at point C. According to Eq. (8.4), the vortical vorticity is $\omega_{vor} = 9.48$. Hence, the angular speed is $\psi = 1/2\omega_{vor} = 4.74$. The angular speed ψ is dimensionless and the dimensional rotation speed is $\psi \times \psi_{CH}$, where ψ_{CH} is characteristic angle speed. ψ_{CH} is calculated as

$$\psi_{CH} = \frac{1}{2\pi T} = 13528\ s^{-1}$$

where T is characteristic time.

Therefore, the dimensional rotation speed at point C is $\psi \times \psi_{CH} \approx 64138\ s^{-1}$. Namely, the vortex will rotate 64138 circles per second. This is an extremely high speed.

Three adjacent vortex rings are shown in Fig. (**8.15**). It also shows the distance Δs between two adjacent rings is $1.1h$. Fig. (**8.16**) demonstrates the distributions of $\|\vec{\omega}\|$ and w-velocity around the head of vortex ring 1. We can find the point P with maximal $\|\vec{\omega}\| = 14$ the center of vortex ring 1. Thus, the w-velocity at point P is taken as the speed of vortex ring 1, which is $w_{ring} = 1.54C$, where $C = 340m/s$ is sound speed. Therefore the dimensional frequency of vortex rings is given as following,

$$f = \frac{w_{ring}}{\Delta s} = 119000\ s^{-1}$$

Considering the streamwise velocity of the free stream $w_\infty = 2.5C$, the dimensionless frequency Strouhal number is

$$St = f\frac{h}{w_\infty} = 119000 \times \frac{0.004}{2.5 \times 340} = 0.56,$$

which is consistent with the experiments of Sun *et al* [191] and Wang *et al* [192].

In an intuitive understanding, the dimensional frequency $f = 119000s^{-1}$ means there are 119000 vortex rings generated per second. It is shown that the MVG generates vortex rings with a significantly high frequency.

8.3. MECHANISM OF REDUCTION OF SHOCK INDUCED FLOW SEPARATION BY RAMP-TYPE MVG

The vortex in this case rotates in a speed of 64,000 circles per second. It is an extremely high speed. The high-speed rotation generates low pressure region inside the vortex, as well as low density region, see Fig. (**8.9**). The low pressure and density region diminish the difference of pressure and density before and after shock waves, when vortex rings are passing through shock waves, see Fig. (**8.7**). Then shocks are distorted and even broking because shock waves are pressure and density discontinuity. The ring-like vortices are generated in a high frequency and keep shock waves damped and smashed successively. When shock waves are broken up, shock induced flow separation is reduced.

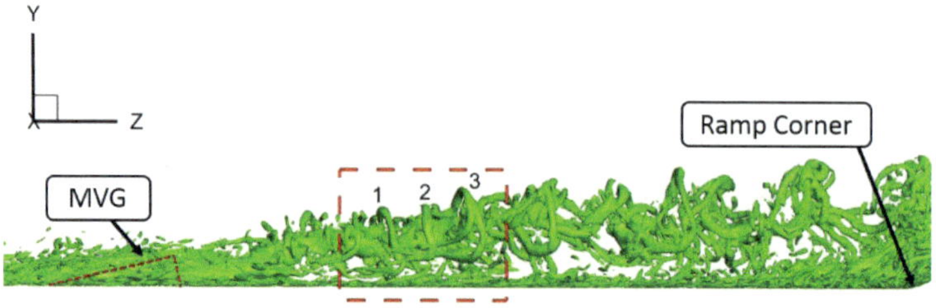

(a) global view

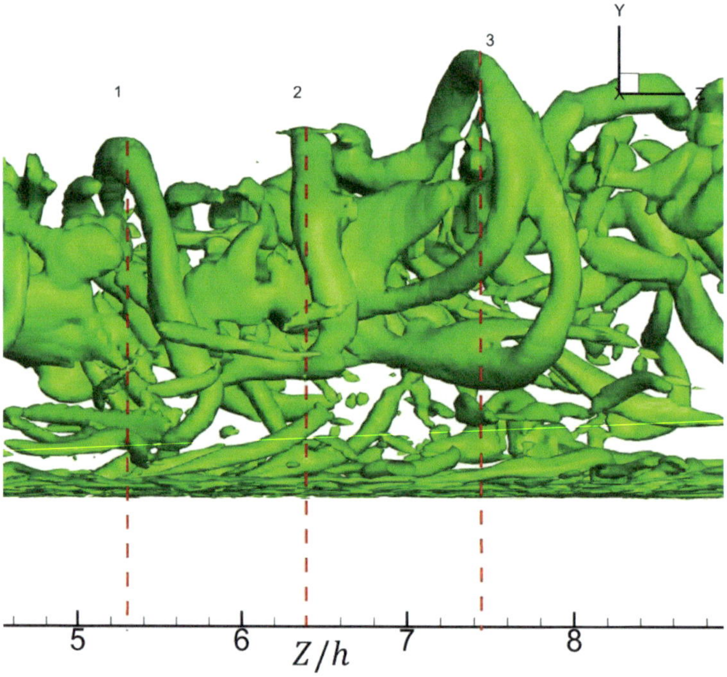

(b) local view

Fig. (8.15). Positions of three vortex rings: (a) global view; (b) local view.

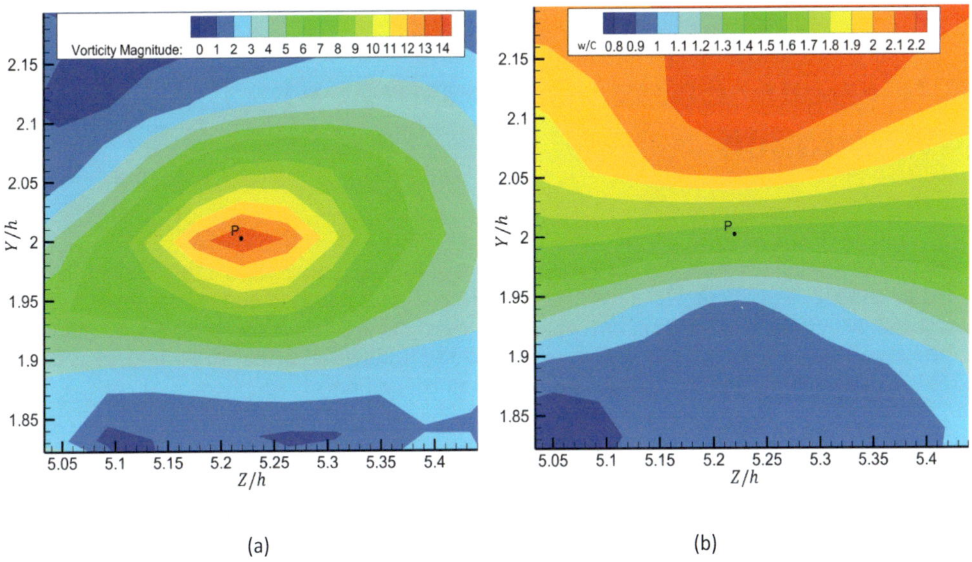

Fig. (8.16). Vorticity and *w*-velocity distribution around first ring head.

CONCLUSION

In a time-averaged view, the mechanism of micro vortex generators (MVGs) reducing of shock induced flow separation was considered as a result of streamwise vortex mixing. With mixing, the velocity profile has stronger resistance to separation. In instantaneous view, it is a different story according to our LES investigation. A chain of vortex rings is generated behind MVG due to deficit of momentum. When vortex rings travel through shock waves, the shock waves are distorted and undermined by high-speed rotating vortex rings. However, the shock waves have little effect on the structure of vortex. After passing through shock waves, the vortex keeps the same structure typology as beyond the shock. According to investigation, high-speed rotation generates low pressure and low density regions inside vortex rings and they diminish the pressure and density difference before and after shock waves. As a result, the structure of shock waves is distorted and smashed. We also investigate that the vortex rings are generated in a high frequency. One after another vortex ring makes the breaking and deforming successively. In a nutshell, the ramp-type MVG generates consecutive ring-like vortices, weaken, alter and smash shock waves and then shock-induced flow separations are reduced.

CHAPTER 9

Correlations Between Pressure Fluctuation and Vortex Motion

Abstract: In this chapter, the correlation between density and pressure fluctuation and vortex motion is studied. The correlation between flow separation and vortex motion is studied as well. The correlation clearly shows the vortex ring generation, size, frequencies, and strength are closely related to the fluctuation of density and pressure. This would provide a powerful tool for SBLI control and noise reduction.

Keywords: Correlation, Flow separation, Fluctuation, Mean value, Noise reduction, Shock, Shock and vortex interaction, Vortex ring, Vorticity.

9.1. INVESTIGATION ON SHOCK-VORTEX RING INTERACTION

A train of vortex rings is generated by the trailing edge of the MVG, see Fig. (**8.1**). These rings develop further downstream and generate the interaction with the impinging shock wave, eventually distorting the structure of the shock wave. Although the shock meets serious distortion, the vortex rings are stiff so that the shock cannot break up it during the interaction. In Fig. (**9.1**), the ring structures (green in figure) are recognized through the Omega methodproposed by Liu *et al.* [164, 189] at the iso-surface of $\Omega = 0.52$, and the shock is identified by the red iso-surface measured by the pressure gradient magnitude.

Fig. (9.1). Shock ring interaction behind the MVG.

Chaoqun Liu, Qin Li, Yonghua Yan, Yong Yang, Guang Yang, Xiangrui Dong

In order to investigate the flow properties and find the separated flow unsteadiness mechanism, only one typical vortex ring before, when and after it hits the shock front is tracked, and the starting time step is t = 1800T. The central plane on which the cross section is chosen since it gives the largest scale of the vortices. The ring head is tracked (see red circle) and the pressure gradient contour lines in black represent the location of the shock wave (Fig. **9.2**). The initial position of the ring head is at approximately (x, y) = (18, 2.5) in the upstream boundary layer.

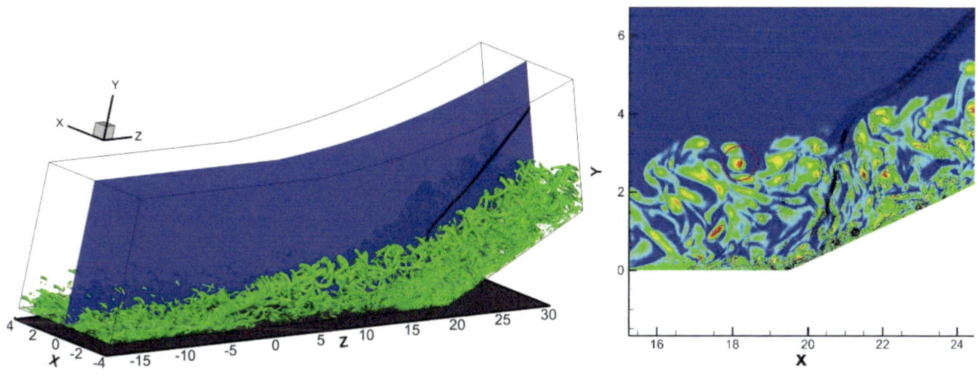

Fig. (9.2). The central plane (Z = 0) at t = 1800T.

9.2. INFLUENCE OF THE INTERACTION ON THE OBLIQUE SHOCK

Four typical time steps during the shock and vortex ring interaction are shown in Fig. (**9.3**). The vortex ring meets the shock leading edge at t = 1844T. When the ring passes through the shock, the shape and structure of the vortex is slightly distorted because of the deformation. The vortex ring doesn't break down after passing the strong shock and maintains its previous topology very well, which means, in the three-dimensional shock ring interaction, the vortex structure under the shear layer is basically not affected by the shock.

However, the shock starts its distortion at t = 1860T and soon break at t = 1884T. The reason is that at the very thin shock front the pressure is substantially high. When the vortex rings travel downstream, they carry relative low pressure and interact with the shock. The instantaneous pressure will drop dramatically at the interaction region, which results in the movement of the shock with the rings (Fig. **9.3c**). A qualitative analysis of the shock intensity changes as the ring passes by will be given below.

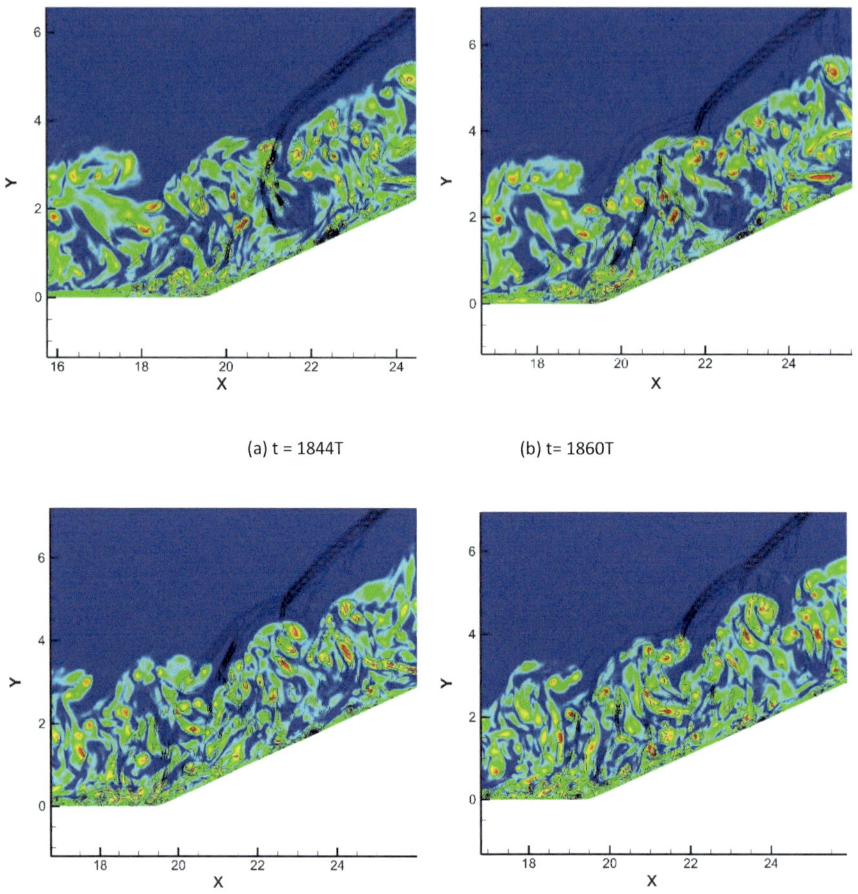

(a) t = 1844T (b) t= 1860T

(c) t = 1884T (d) t = 1904T

Fig. (9.3). Shock and vortex ring interaction.

When t = 1904T, the upper reflection shock quickly recovers after the previous ring leaves, however, the lower separation shock still gets distorted by a train of vortex rings coming from the upstream continuously. The propagation of the rings causes the shock foot to sweep back and forth. The outer region of the separation zone looks disordered as shown in Fig. (**9.3d**). As we can see, there are many weak shocks formed around the small scale rings near the wall because of the fast rotation of the rings (around 120000 cycles per second, see Yang *et al.*, 2016). The shock ring interaction in the downstream may in turn affect the separation bubble dynamics and low-frequency oscillations.

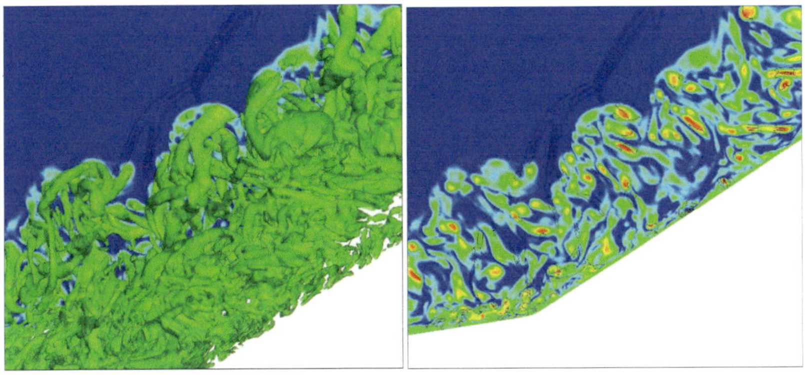

Fig. (9.4). The 3D vortex structures at t = 471000.

Fig. (**9.4**) shows the three-dimensional vortex structure with its two-dimensional counterpart at t = 1884. The large scale rings are dominant in the outer region of the separation zone and the passage of the upstream rings is directly responsible for the motion of the shock. It is possible that the low-frequency unsteadiness is due to the shock ring interaction. Some statistical analysis is needed to find the correlation between the strength of the rings and the separation shock motion. More information will be given in section 9.4.

9.3. INFLUENCE OF THE INTERACTION ON THE VORTEX RING

The ring head flow properties are discussed (circled in Fig. **9.2**) for 20 time steps starting from t = 1800T. The largest Omega value is recorded for each point at which the rotation is the most dominant. Fig. (**9.5**) shows the ring head location around the interaction. We see that the ring is moving downstream and lifting up.

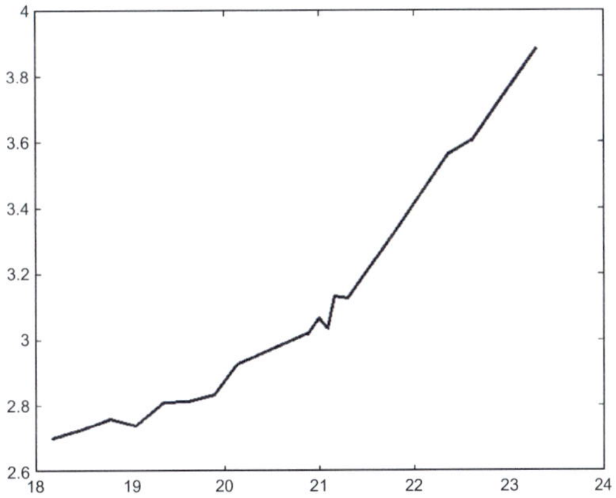

Fig. (9.5). The location of the vortex ring.

After the vortex rings crossing the shock wave, the properties of the flow like pressure p, density ρ, energy E, velocity u, v, w and vorticity w_x, w_y, w_z will change discontinuously. When the vortex rings develop downstream, the gradient of pressure in this area suddenly jumps from 1.5 to 5 at around X = 21 where interacting the shock (Fig. **9.6**). After it penetrates the shock, its pressure gradient drops quickly. Also a range of weak shocks are discovered around the vortex ring behind the stronger shock due to the fast rotation of the vortex ring with low pressure which can be treated as a stiff solid; when the flows surrounding the ring move to it, the pressure drops dramatically therefore weak shocks are formed.

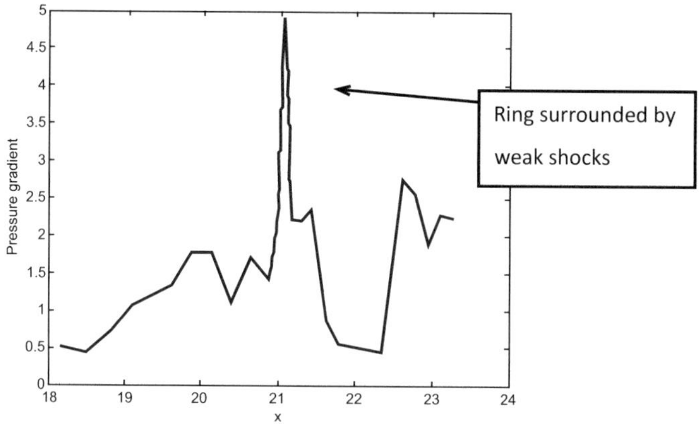

Fig. (9.6). The pressure gradient of the vortex ring.

Fig. (**9.7** and **9.8**) show the density and energy distributions of the ring *versus* streamwise location as time evolves. Because of the heat and energy transport during the shock ring interaction, both two quantities keep increasing. A jump can be seen at the position of the shock (X = 21).

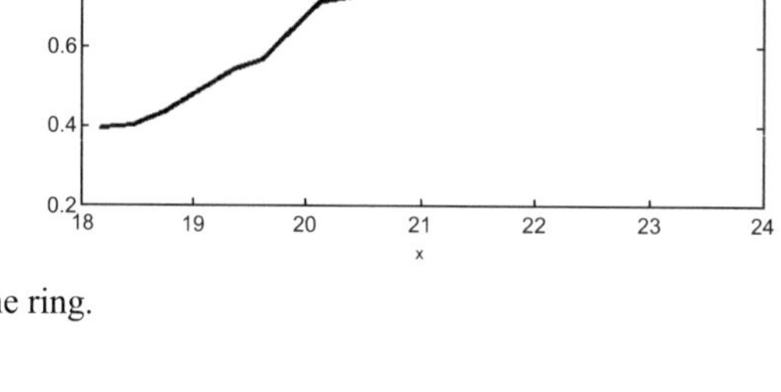

Fig. (9.7). Density of the ring.

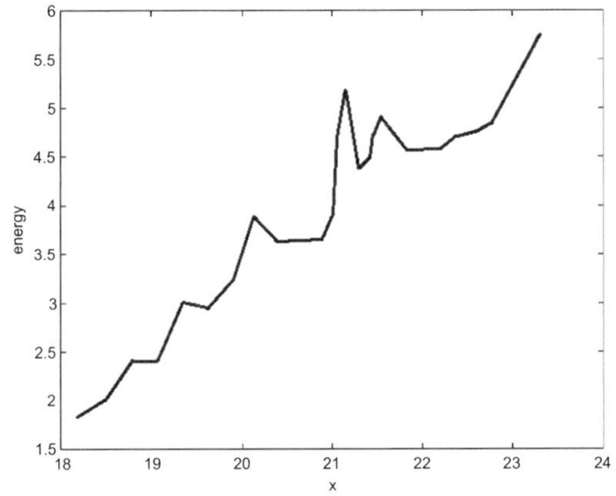

Fig. (9.8). Energy of the ring.

Fig. (**9.9**) displays the velocity profile around ring during the shock ring interaction. The streamwise velocity of the ring head is decreasing when interacting with the shock; it means that the ring is hold back by the shock. It can be seen that a discontinuity at the position of the shock presents, but the overall ring motion is not influenced by the action interacting with the shock. The wall normal velocity increase indicates that the ring is rolling up gradually.

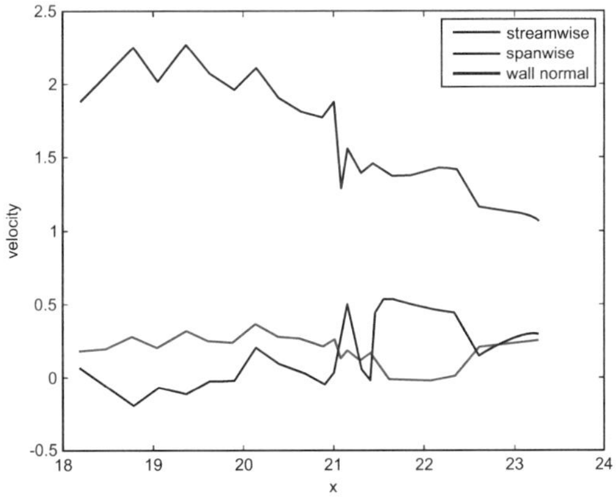

Fig. (9.9). Velocity profile.

Fig. (**9.10**) gives the vorticity distribution of the interaction. It is found that the spanwise vorticity of vortex ring is playing the leading role when moving downstream; due to the vorticity magnitude basically has the same track of the spanwise vorticity. It shows that the vorticity is related to the average angular velocity of a rotating body. Although the wall normal vorticity rises with time, the spanwise vorticity is dominant as a dropping trend, indicating that the rotation of the vortex ring is slower than before it hits the shock. This phenomenon is further confirmed by the Omega value distribution in Fig. (**9.11**). Omega measures the stiffness of the flow. From the Omega definition, larger value of Omega means the vorticity is more dominant. Therefore the drop of Omega indicates that the vortex ring rotation is not as strong as before.

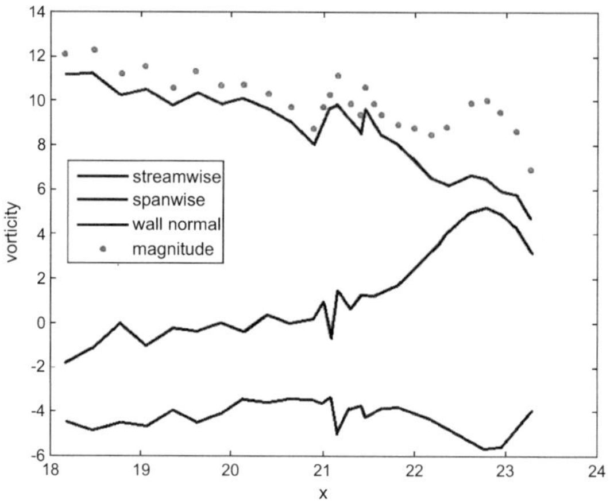

Fig. (9.10). Vorticity distribution.

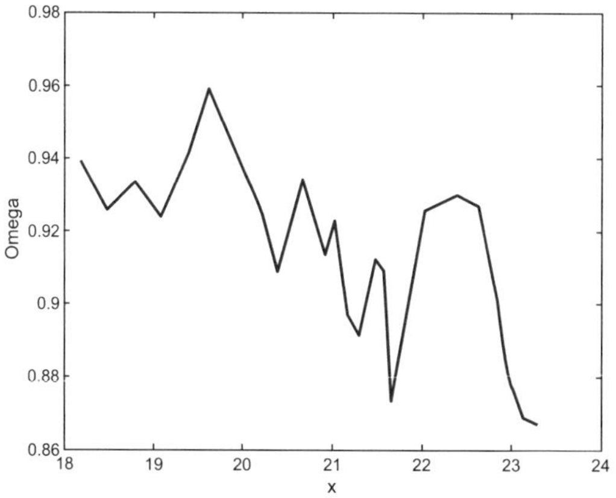

Fig. (9.11). Distribution of the Omega value.

9.4. INFLUENCE OF THE INTERACTION ON THE SEPARATION BUBBLE

As discussed in Section 9.2, the reflection shock gets distorted when a vortex ring passes through and recovers after it leaves. As more rings travelling from the upstream, the separation shock keeps being distorted and sweeps back and forth. The statistical relation between the movement of vortex rings and the shock motion is investigated.

To study the correlation between vortex rings and the shock motion, we fix a position in the interaction region at (x, y) = (21, 3) on the path of the travelling of vortex rings (Fig. **9.12** left). We record 20000 data from t = 1840T. During this time, there are thousands of rings coming from the upstream, the strength of which can be captured by the Omega values. The shock motion is characterized by the pressure fluctuations. In Fig. (**9.12**), the right gives the signals of Omega fluctuations and pressure fluctuations. As we can see, at most of time the peaks of the Omega correspond to the valleys of pressure and vice versa. The statistical correlation is found to be -0.7986, which indicates a strong negative correlation between the propagation of vortex rings and the shock motion.

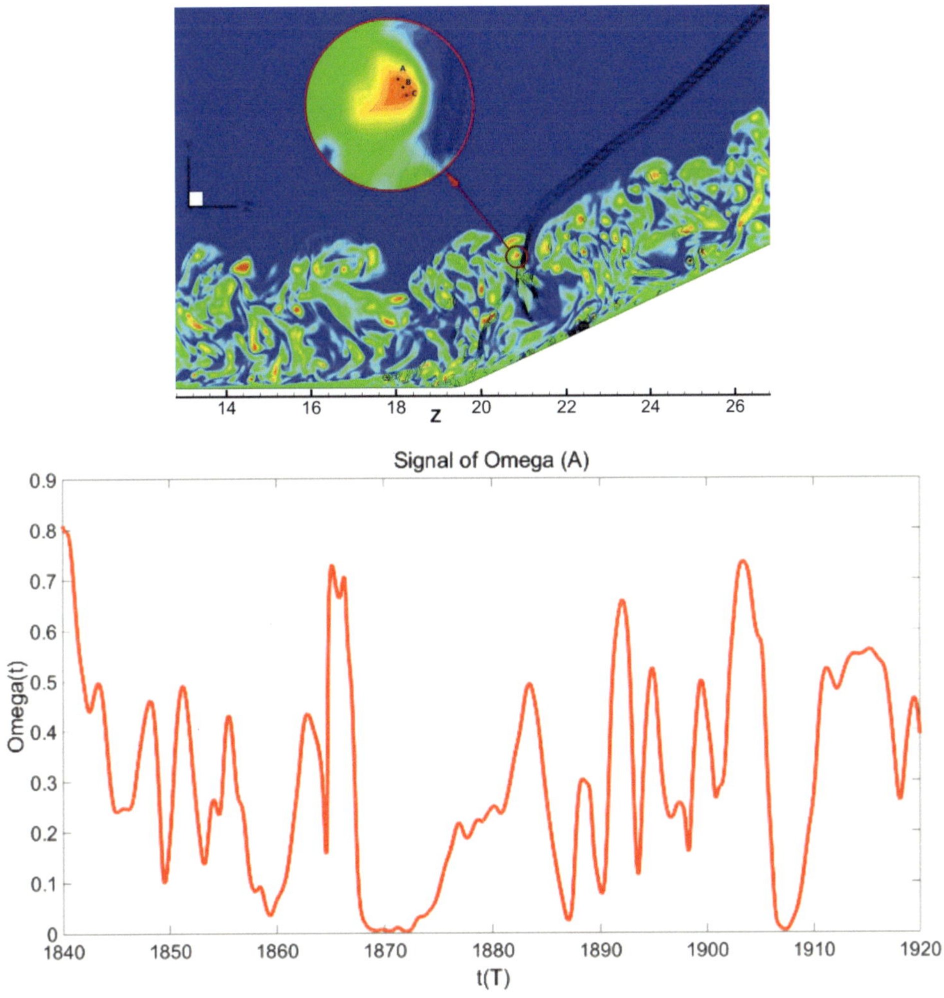

Fig. 9.12 contd.....

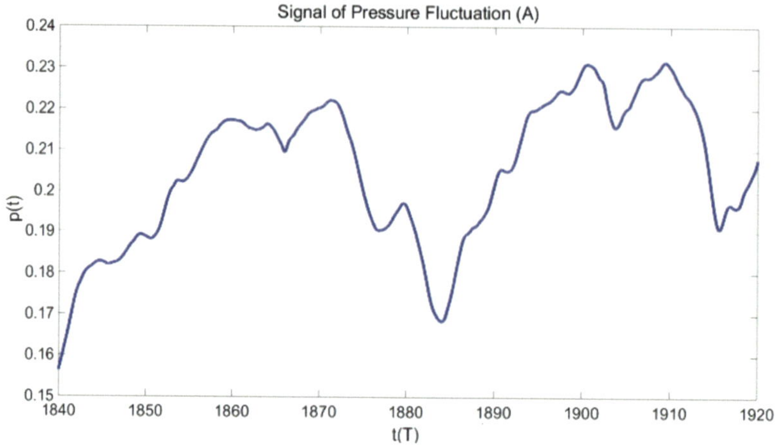

Fig. (9.12). The correlation between vortex rings and shock motion at the position (21, 3).

Fig. (**9.13**) shows the three-dimensional separation zone near the wall and the streamwise position of the separation point when $\partial u/\partial y = 0$ on the central plane. It is clear that the vortex rings reduce the flow separation in the central region. The propagation of the rings causes the shock foot to sweep back and forth, which will induce the separation bubble to shrink and grow over time. As shown in Fig. (**9.14**), at t = 1804T the previous rings have passed the shock and the new rings have not arrived, the separation bubble is shrinking; at t = 1844T the shock is pushed forward by the rings, the bubble is growing. To further reveal the relation between the shock motion and separation location, the power spectra are presented in Fig. (**9.15**). The pressure fluctuation has most of its energy at low frequencies around Strouhal number $St = fh/u_\infty = 0.018$ where h is the characteristic length, and these frequencies have a good agreement with the range of frequencies discovered in the separation point.

During the shock vortex-ring interaction, the statistical relation between the rings and the shock motion thus the separation bubble can also be investigated by means of the coherence quantities. The coherence between two time signals $x(t)$ and $y(t)$ is defined as

$$r_{xy}^2(f) = \frac{|P_{xy}(f)|^2}{P_{xx}(f)P_{yy}(f)} \qquad \textbf{(9.1)}$$

where $P_{xx}(f)$ is the power spectral density of signal $x(t)$, $P_{xy}(f)$ is the cross power spectral density between ese two signals $x(t)$ and $y(t)$. For all the frequency f, we have $0 \leq r^2_{xy}(f) \leq 1$. If $r_{xy} = 1$, then the signals $x(t)$ and $y(t)$ are linearly related at frequency f; if $r_{xy} = 0$, they are unrelated.

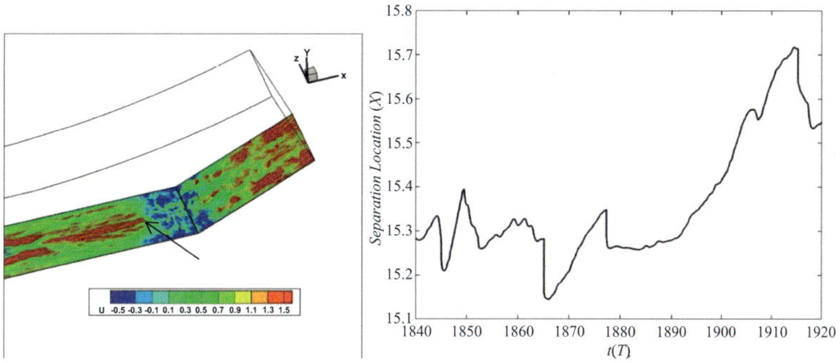

Fig. (9.13). The separation zone contoured by the streamwise velocity (left); Streamwise positon of the separation point on the central plane (right).

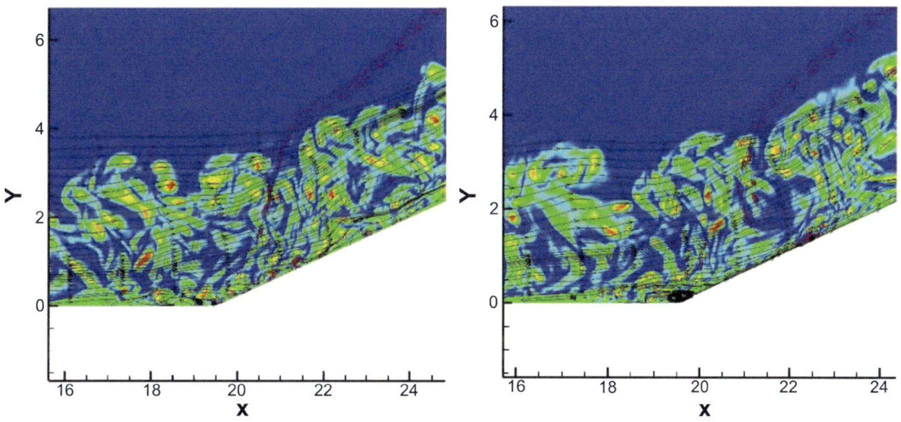

Fig. (9.14). Separation bubble.

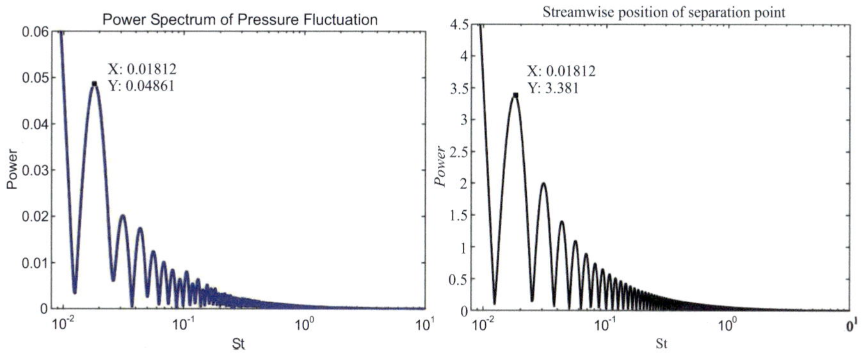

Fig. (9.15). Power spectra of pressure fluctuation and separation location.

Using the 20000 dataset starting from t = 1840T, the coherences are calculated and the results are shown in Fig. (**9.16**). The left figure gives the coherence distribution between the vortex rings and shock motion at different Strouhal numbers, while the right figure gives the coherence between the rings and separation point. Very high values of coherence are observed in both figures at low frequencies, indicating that the rings and separation signals are almost linearly related. The shapes of the two coherence distributions are almost the same except that the coherences between rings and shock motion are higher at $0.1 \leq St \leq 0.5$, because the vortex rings have a direct interaction with the shock but indirectly influence the separation bubble.

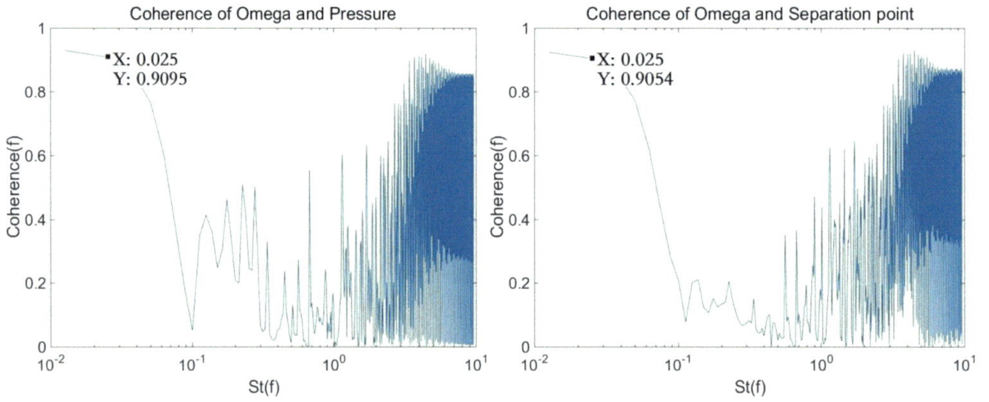

Fig. (9.16). Coherence between vortex rings and shock motion (left); Coherence between vortex rings and separation location (right).

CONCLUSION

As the flow going through the shock, its properties will change intermittently. The interaction between the vortex rings and the shock wave for the MVG controlled ramp flow is investigated using our LES data. By tracking a typical vortex ring when it passes through the shock, it can be found that the spanwise vorticity takes a leading role in flow propagation, the topology of vortex rings is kept well but not rotating as fast as before, the separation shock gets distorted after hit by rings and sweeps back and forth, and a number of weak shocks exist around ring heads in downstream.

CHAPTER 10

Other Respects on SBLI Controls

Abstract: An implicit Large Eddy Simulation (ILES) is carried out for a Micro Vortex Generator (MVG) controlled supersonic ramp flow at Ma = 2.5 and Re = 5760 based on inflow momentum boundary thickness. The results are validated against the TU Delft experiment on the same MVG geometry and Mach number. Several techniques are used to analyze the coherent structures in the MVG wake. The vortex system in the MVG wake is visualized using vortex identification method and it is found that two primary counter rotating streamwise vortex pair are induced by the MVG, which would further lead to a trail of vortex rings through K-H instability. The average distance between adjacent vortex rings is determined by a spatial auto-correlation of vorticity, which is estimated as 1.5 MVG height. Two dimensionality reduction algorithms, namely Proper Orthogonal Decomposition (POD), and Dynamic Mode Decomposition (DMD) are applied to a set of flow field sequences on the spanwise symmetry plane. POD modes identify structures containing most of the turbulent kinetic energy and DMD modes, capturing single frequency structures which are most essential to the unsteady dynamics.

Keywords: Dynamic Mode Decomposition (DMD), Modes Analysis, MVG, Proper Orthogonal Decomposition (POD).

10.1. INVESTIGATION OF COHERENT STRUCTURES IN MVG CONTROLLED SUPERSONIC RAMP FLOW

There is a strong statistical correlation found between the propagation of the vortex rings in the MVG's wake and the reflected shock motion exists near the interaction zone. The SBLI research community has shown that there is a high coherence found between the motion of rings and the pulsation of separation bubbles at low frequencies. It is possible that the vortex rings/shock wave interaction creates substantial low-frequency unsteadiness in the upstream region of the separation zone.

Since the boundary layer separation is induced by the shock wave, it is suggested that when the shock strength is reduced, the shock-induced flow separation would be alleviated accordingly. The study of shock ring interaction reveals a great deal about instantaneous flow behavior, and thus will provide an insight into flow control.

Chaoqun Liu, Qin Li, Yonghua Yan, Yong Yang, Guang Yang, Xiangrui Dong

Fig. (**10.1**) shows the time averaged streamwise velocity profile in wall unit at x = -8h (right before MVG), and it complies well with the log-law of wall and DNS result of Guarini *et al.* [173], which testifies the boundary layer before MVG is indeed fully developed turbulence. Fig. (**10.2**) gives the averaged streamwise velocity profile and Reynolds Shear Stress profile at x/h = 12.0 on the middle spanwise plane (z = 0), our numerical results have excellent agreement with similar MVG experiment conducted by Sun *et al.* [180] at TU Delft. Fig. (**10.3**) gives the averaged streamwise velocity profile at 5 different streamwise locations in the MVG wake region. All the profiles show large momentum deficit, which gradually recovers in the downstream. The defected velocity profile resulted in two inflection points. According to the Rayleigh instability theorem, the velocity inflection points are unstable and lead to K-H instability. This shear layer instability would produce a trail of spanwise vortex rings, which contribute a lot to the Reynolds shear stress generation. Fig. (**10.4a**) shows the averaged turbulent kinetic energy on the z = 0 plane, it can be seen that there are 3 regions with high turbulence level, the separation region behind the MVG, the high shear layer region and the vortex path in the MVG wake and the separation zone near the corner. Fig. (**10.4b**) shows the contour of Reynolds Shear Stress (RSS), in the MVG wake, there are two parallel streaks of RSS peak, opposite in sign, whose normal wall positions are in accordance with that of the inflection points in the defected velocity profiles. Fig. (**10.4c**) gives the RSS contour in the MVG wake by PIV measurement of Sun *et al.* [180], where similar structures, like the upper edge negative peak and down edge positive peak are also observed. All the above results testify the reliability of present simulation.

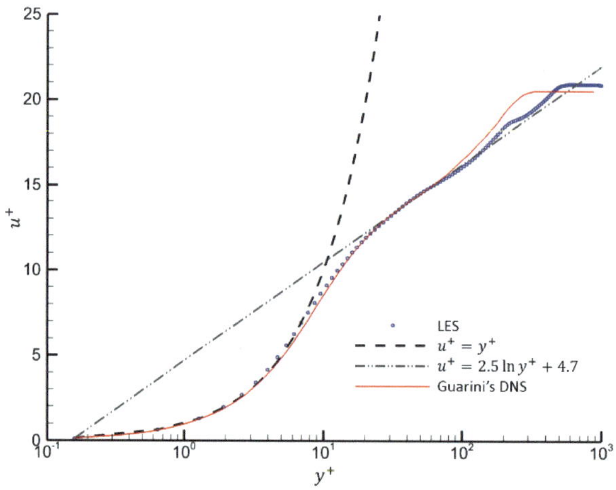

Fig. (10.1). The turbulent boundary layer profile in wall unit at x = -8h (right before MVG).

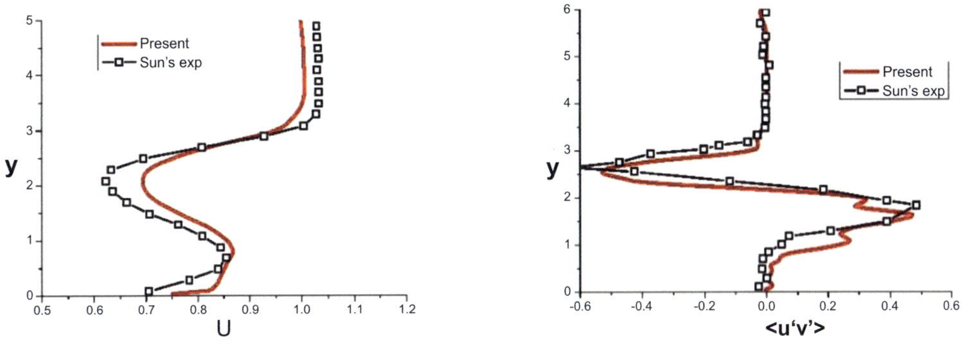

Fig. (10.2). (a)averaged streamwise velocity profile at x/h = 12.0 (b) Reynolds shear stress profile at x/h = 12.0.

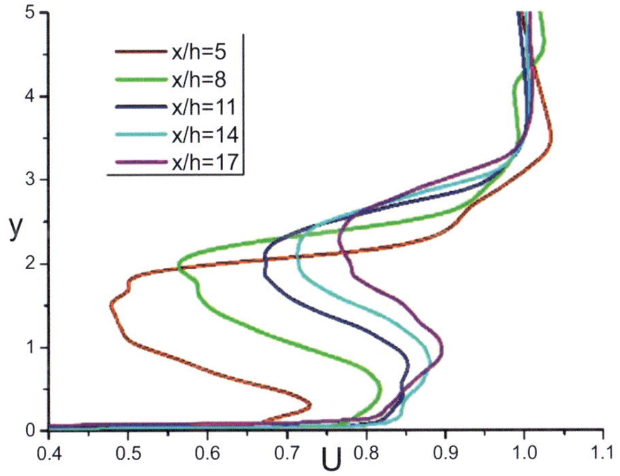

Fig. (10.3). Streamwise velocity profile at different stations.

10.1.1. Instantaneous Flow Field

The flow structures in the MVG wake flow are complicated, yet important to understand its control mechanism. Different models have been proposed to elaborate the coherent structures and vortex system in the MVG wake. Babinsky *et al.* [51] sketched the vortex structure behind micro ramp as shown in Fig. (**10.5a**) (half spanwise domain). By surface oil flow observation, they identified the footprint of two primary counter rotating vortices, and secondary vortices originating from the side-wall/floor junction of the ramp. Li and Liu *et al.* [98] proposed a dynamic vortex model composed of five vortices as shown in Fig. (**10.5b**).

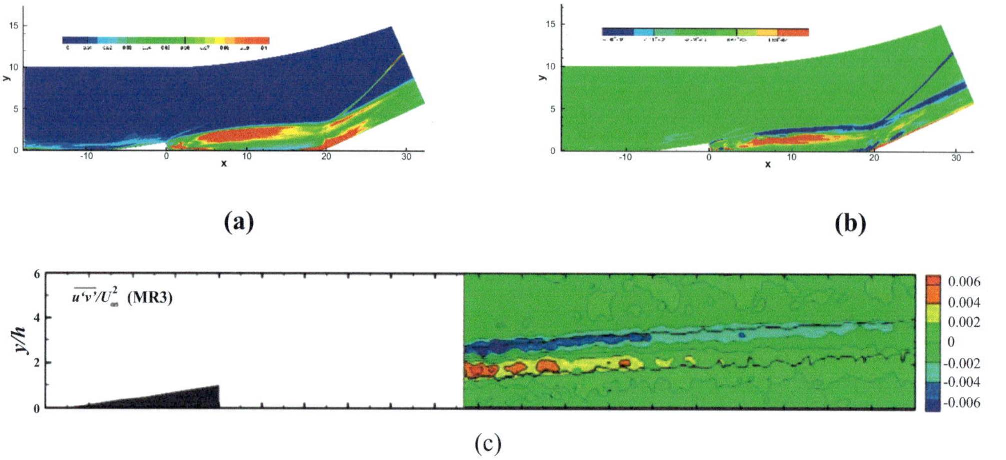

Fig. (10.4). (a)Turbulent Kinetic Energy (b)Reynolds Shear Stress <u'v'>(c) RSS <u'v'> measurement by Sun [180].

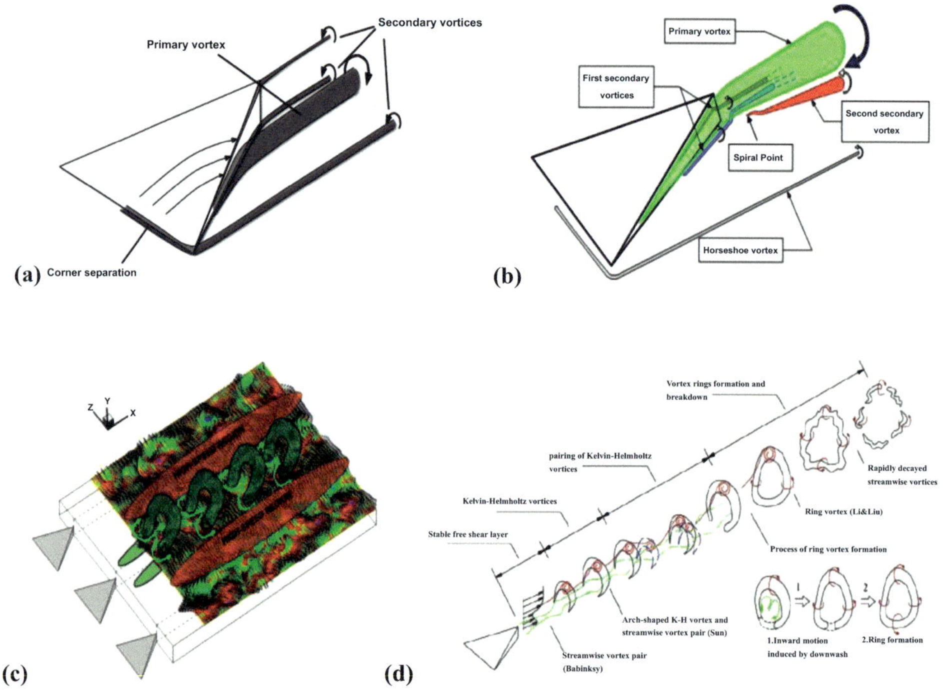

Fig. (10.5). (a)Sketch of vortex system behind MVG [51] .(b) The dynamic vortex model [98] (c) Conceptual hairpin model after MVG array [52] (d) complete vortex evolution scenario [191].

In their model, the secondary vortices are underneath the primary vortices. Blinde *et al.* [52] gave a conceptual sketch of the MVG wake flow as shown in Fig. (**10.5c**), based on their PIV measurement. In their description, there are alternated low speed region and high speed region behind the MVG arrays, and such a configuration would lead to the hairpin vortex generation just like the mechanism of turbulence boundary layer, where hairpin vortex resulted from streaks [193]. Sun [191] proposed a complete evolution process scenario of the MVG wake vortex, incorporating some of the previous models as partial stages or elements, as shown in Fig. (**10.5d**).

Fig. (**10.6**) shows the instantaneous flow field structures, the vortex are identified by λ_2 criterion and shown as the green iso-surface, while the shock surface is identified by pressure gradient and shown as the grey iso-surface. Turbulent coherent structures can be seen in the incoming boundary layer before MVG. Behind the MVG, there is a trail of vortex rings. As the vortex rings moved downstream, they are lifted up and with an increasing size. Then the vortex rings interact with the shock surface and cause a three- dimensional deformation of the shock surface. Both the fine scale turbulence, vortical structures and sharp in continuous shock are very well captured in the present numerical simulation.

Fig. (10.6). vortex identified by 2 iso-surface and shock surface.

Fig. (**10.7a**) shows the numerical schlieren in the middle spanwise plane, two weak shock waves are generated in the presence of MVG. In the wake region, the schlieren graph shows two parallel rows of staggered arrayed K-H type structures, as marked by the red oval. Above the compression corner, the shock wave is shapely detected, at the shock foot where upstream structures penetrate into the shock surface, the shock is broken to a set of much weaker shocklets. Fig. (**10.7b**) is the schlieren image from the MVG experiment of Rogier [194], around the MVG, flow structures similar to the numerical results are detected. Fig. (**10.7c**) is the acetone vapor screen visualization of the flow by Lu *et al.* [76], a trail of vortex rings after MVG is also clearly identified.

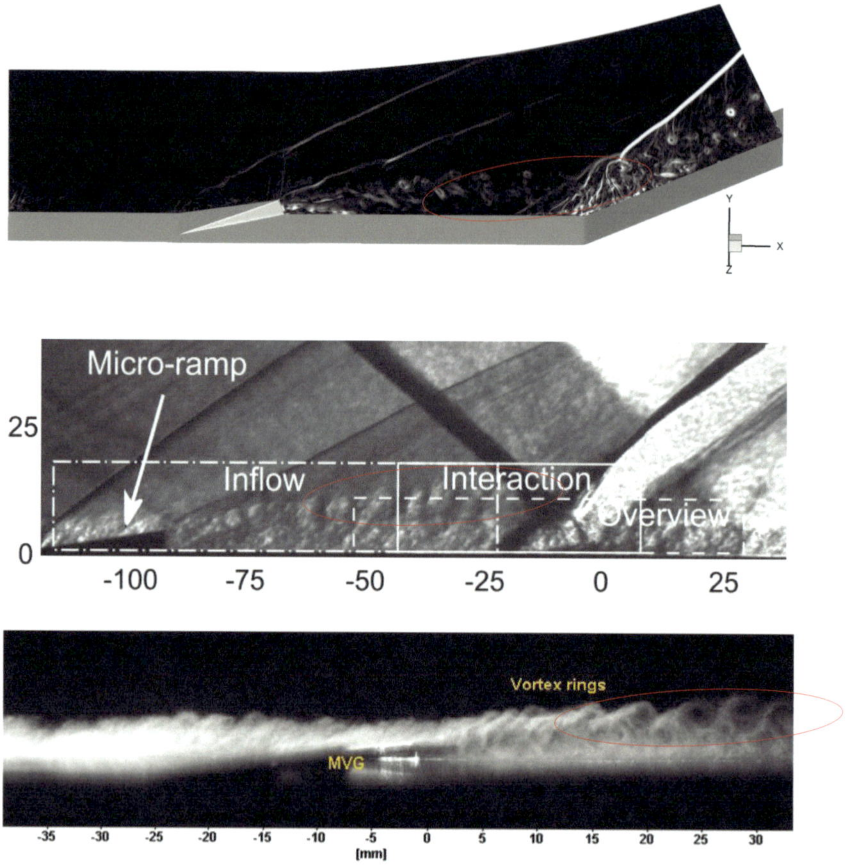

(a)

(b)

(c)

Fig. (10.7). Schlieren in the symmetrical plane of MVG (a) Numerical Schlieren in current simulation (b) Schlieren in the experiment of Roiger [194] (c) acetone vapor screen visualization [76].

Fig. (**10.8a**) shows the streamwise vorticity contour and streamlines on four slices across the MVG wake region, the existence of two counter rotating primary vortex is obviously identified. Beneath the primary vortex, a pair of secondary vortex can be also observed. Fig. (**10.8b**) gives the density contour and streamlines on two slices behind MVG by Ye *et al.*'s measurement in TU Delft [195], where flow structures are very similar to the numerical results. Both results testify that the secondary vortices should be underneath the primary vortices. From the streamlines, we can see that there is an upwash motion on the symmetry plane induced by the primary vortex, which is the origin of the pronounced velocity deficit, and this upwash motion also lifts up the generated vortex rings in the downstream. Fig. (**10.9**) shows the vortical structures in 3-D iso-surfaces, the red identifies large positive streamwise vorticity ω_x while the blue is the large negative ω_x, and the grey identifies the vortex rings by the contour of large circumferential vorticity $\sqrt{\omega_y^2 + \omega_z^2}$. The counter rotating streamwise vortex are wrapped by a trail of vortex rings. Similar structures can also be observed from the PIV measurements of Sun *et al.* in TU Delft, Fig. (**10.10a**) identifies the streamwise vortices and Fig. (**10.10b**) shows the resulted vortex ring from K-H instability using their PIV measured flow field data.

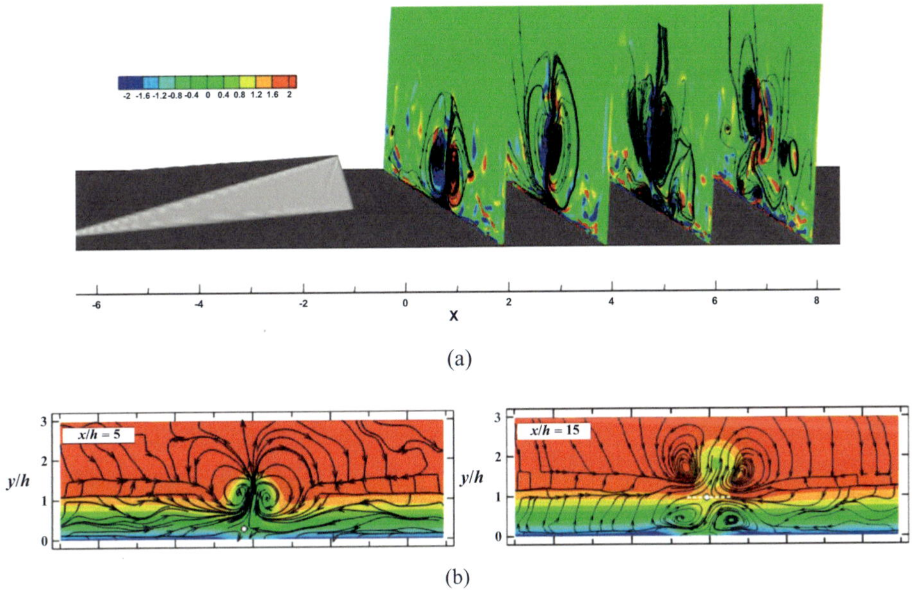

Fig. (10.8). (a)streamwise vorticity contour and streamlines.(b)streamlines on the planes x/h=5 and x/h=15.

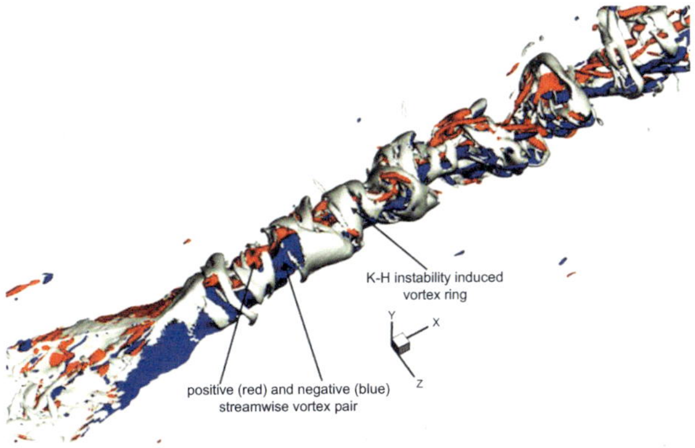

Fig. (10.9). counter rotating streamwise vortex pair (red and blue) and vortex ring (grey).

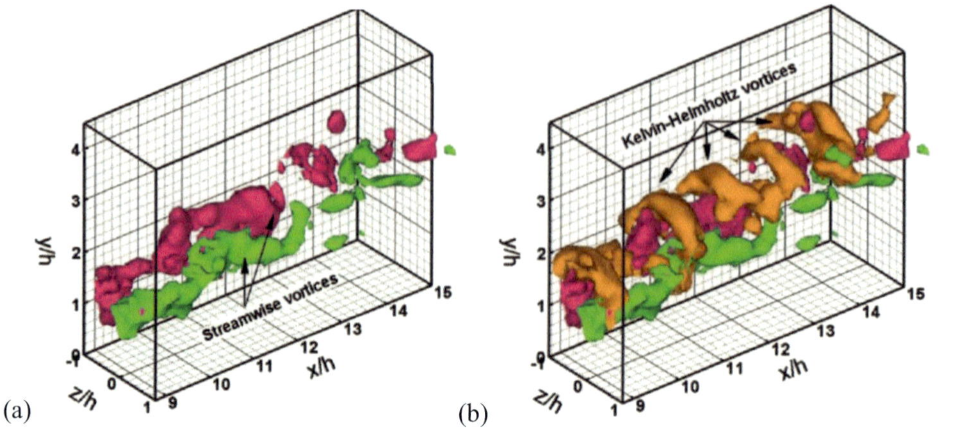

Fig. (10.10). Stereo PIV of Sun *et al.* (a) streamwise vortices (b)vortex ring by K-H instability.

As discussed above, a trail of vortex rings is produced through Kelvin-Helmholtz instability, thus the streamwise wavelength of such instability, *i.e.* the average distance between the generated vortex rings is of interest. Fig. (**10.11a**) shows the spanwise vorticity contour on the symmetry plane. The white line in the figure is the approximated path of vortex ring convection, and the spanwise vorticity on the line is drawn in Fig. (**10.14b**), where the periodically appeared local spikes identify the vortex ring position. To evaluate the averaged distance between two adjacent rings, a spatial auto-correlation analysis is carried out. The spatial auto-correlation function of spanwise vorticity is defined as:

$$R(\Delta x, \Delta y) = \frac{\iint \omega_z(x,y) * \omega_z(x+\Delta x, y+\Delta y)dxdy}{\iint \omega_z(x,y)^2 dxdy}$$

The spatial auto-correlation function is calculated for the MVG wake region. The shape of the function is shown in Fig. (**10.12a**), where a second peak can be observed. The position of this second peak gives average distance between adjacent rings. The 2D contour map of the auto-correlation function is shown in Fig. (**10.12b**), and it is found that the average distance between vortex ring is about 1.5 MVG height, same as the result given by Sun *et al.* [180]. As the average convection velocity at vortex rings center is about $0.8u_\infty$, the corresponding vortex passage frequency can be estimated as $f = u/\lambda = 0.8u_\infty/1.5h = 0.53u_\infty/h \approx 2683Hz$, close to the result given by Ye *et al.* [195].

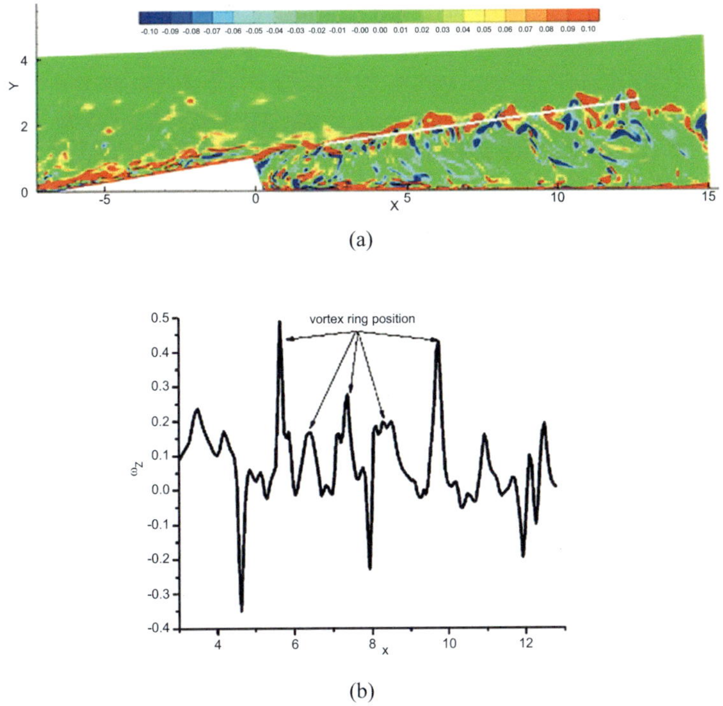

Fig. (10.11). (a)spanwise vorticity on the symmetric plane (b)spanwise vorticity on the while line.

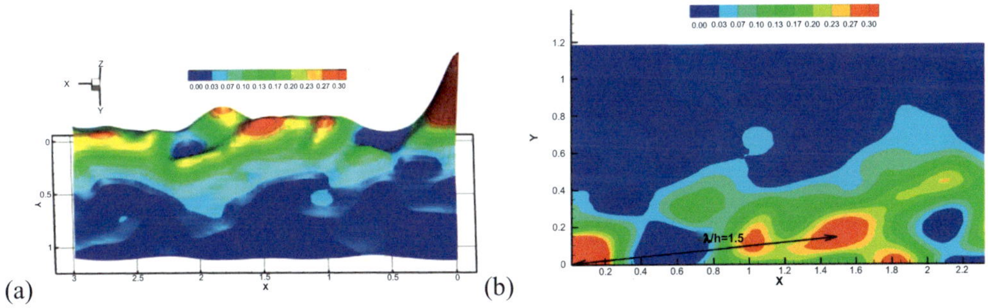

Fig. (10.12). correlation contour (a) 3-D shape (b) 2-D contour.

10.2. DIMENSIONALITY REDUCTION ANALYSIS

With the rapid development of CFD and PIV techniques, numerical computations and experimental measurements are conducted with much higher spatial and temporal resolution than ever. How to effectively analyze the huge database, extract flow structure and interpret the physical meaning behind the data becomes a burning question. According to Jimenez's idea, Coherent Structure can be viewed as a lower dimensional manifold in whose neighbor the dynamic system spends a large fraction of time [196]. Based on this idea, some dimensionality reduction techniques have been proposed to extract coherent structures from the high dimensional datasets, and proved to be successful, such as Proper Orthogonal Decomposition (POD) and Dynamic Mode Decomposition (DMD).

10.2.1. POD Analysis

Proper Orthogonal Decomposition (POD) is a widely used dimensionality reduction technique to find an optimal low-dimensional representation of a high-dimensional dataset. It has vast applications in many different scientific research areas, including image processing, data mining, and control theory, sometimes it bears other names such as Karhunen-Loève decomposition, or Principal Component Analysis (PCA) when appearing in different background. The mathematical essence of POD is to find the bases of a lower dimensional space that maximizes the projected data covariance, *i.e.* to preserve most of the data characteristics. In fluid dynamics, POD method identifies modes that store the largest possible amount of kinetic energy. Lumley *et al.* used POD method to extract coherent structures in turbulence [197].

In the case of dealing DNS or LES datasets, where the spatial samples are much larger than the temporal samples, the snapshot POD method proposed by Sirovich [198] is adopted to save memory. The mathematical procedures are briefly presented as follows:

Firstly, the fluctuating velocity is calculated in each time step by subtracting the mean flow and then arranged in a matrix U as:

$$\boldsymbol{U} = [\boldsymbol{u}_1 \ \boldsymbol{u}_2 \ \cdots \ \boldsymbol{u}_N] = \begin{bmatrix} u_1^1 & u_2^1 & \cdots & u_N^1 \\ \vdots & \vdots & \ddots & \vdots \\ u_1^M & u_2^M & \cdots & u_N^M \\ v_1^1 & v_2^1 & \cdots & v_N^1 \\ \vdots & \vdots & \ddots & \vdots \\ v_1^M & v_2^M & \cdots & v_N^M \\ w_1^1 & w_2^1 & \cdots & w_N^1 \\ \vdots & \vdots & \ddots & \vdots \\ w_1^M & w_2^M & \cdots & w_N^M \end{bmatrix}$$

where M is the number of spatial discrete points and N is the number of snapshots. The covariance matrix of fluctuating velocity is calculated as:

$$\boldsymbol{R} = \boldsymbol{U}^T\boldsymbol{U}$$

As the covariance R is a real symmetric matrix, we can get its eigenvalues and corresponding eigenvectors as:

$$\boldsymbol{R}\boldsymbol{A}_i = \lambda_i \boldsymbol{A}_i$$

and rank the eigenvalues in a descending order as:

$$\lambda_1 > \lambda_2 > \cdots > \lambda_N$$

then each POD mode is obtained by projecting matrix U onto each eigenvector and normalized by its 2-norm as:

$$\phi_i = \frac{\sum_{n=1}^{N}(\boldsymbol{A}_i^n \cdot \boldsymbol{u}_n)}{\left\|\sum_{n=1}^{N}(\boldsymbol{A}_i^n \cdot \boldsymbol{u}_n)\right\|}, i = 1,2,\cdots,N$$

after the POD modes are found, the original flowfield can be reconstructed using the POD modes as:

$$u(t,x) = \sum_{n=1}^{N} a_n(t)\,\phi_n(x)$$

$\boldsymbol{a}_n = \boldsymbol{\Phi}^T\boldsymbol{u}_n$ where $\boldsymbol{\Phi} = [\phi_1\ \phi_2\ \cdots\ \phi_N\,]$

a_n is the time coefficient of each POD modes ϕ_n.

Because most of the vortex ring dynamics can be reflected on the symmetry plane, the analysis is carried out for the MVG wake region on z = 0 plane, M = 72120 spatial points, and N = 200 time snapshots are used in the computation.

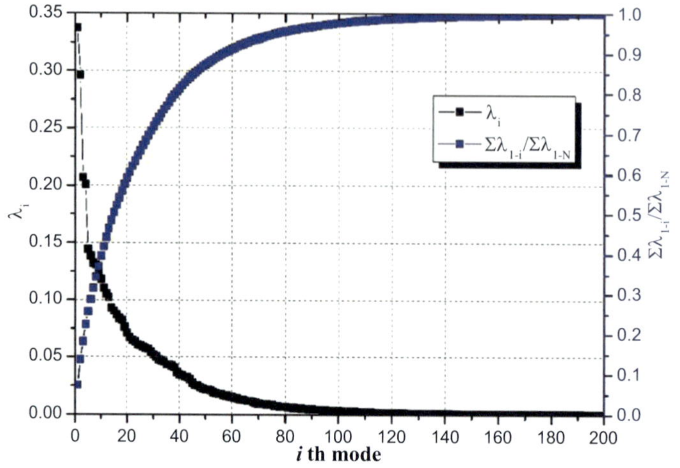

Fig. (10.13). Eigenvalues of the Covariance Matrix and their cumulative sums.

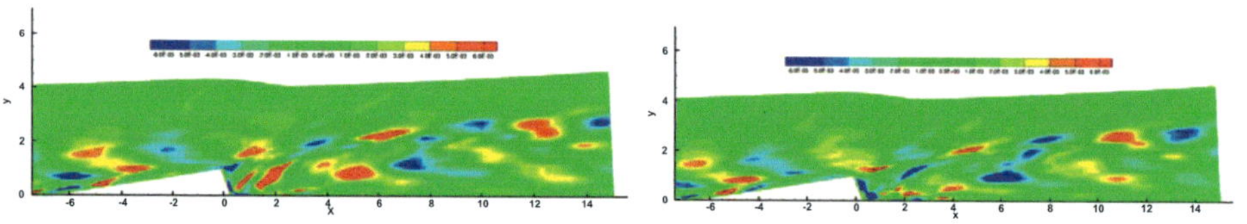

(a) (b)

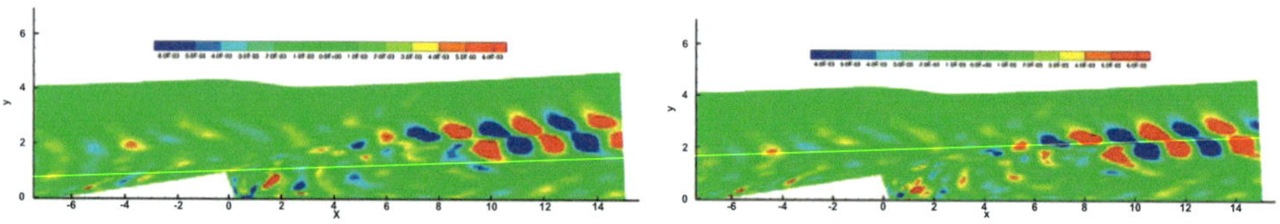

(c) (d)

Fig. (10.14). the first 4 primary POD modes.

Fig. (**10.13**) gives the eigenvalues of the covariance matrix (black) and the cumulative sum of the eigenvalues (blue). It can be seen that the first few eigenvalues are much larger than the rest. The cumulative sum of eigenvalues curve shows that the first 20% modes contain 80% of the energy. Fig. (**10.14**) gives the first 4 primary POD modes, contoured by streamwise fluctuating velocity and Fig. (**10.15**) gives the time coefficients of the first 4 POD modes. The first pair of POD modes in Fig. (**10.14(a-b)**) has large structures in the near wake of MVG, which might emerge from the fluctuation motion induced by the primary streamwise vortex. The second pair of POD modes in Fig. (**10.14(c, d)**) shows staggered arrayed structure similar to vortex street rolled from K-H instability, which is caused by the fluctuation motion induced by the vortex rings. Fig. (**10.15**) gives POD time coefficients of the first 4 POD modes. The coefficients exhibit an almost periodic characteristic. Also, as the first pair of POD modes is linked with larger structures than the second pair, the frequency is also lower accordingly. The first pair of POD modes has similar structures with a phase shift, which is reflected on the shape of their time coefficients, and the same phenomenon also holds true for the second pair of POD modes.

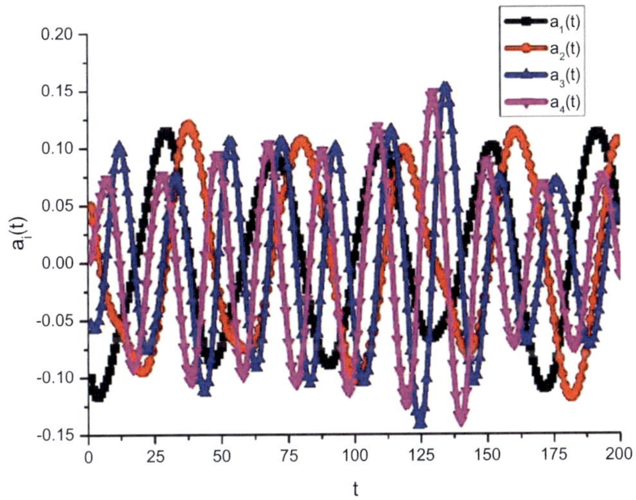

Fig. (10.15). time coefficients of the first 4 POD modes.

10.2.2. DMD Analysis

The Dynamic Modes Decomposition (DMD) is another powerful tool to analyze the coherent structures in fluid flow, recently proposed by Schmid *et al.* [199]. The DMD algorithm is based on a variant Arnoldi algorithm suggested by Ruhe [200]. Another closely linked mathematical method, the Koopman operator analysis of fluid flow was proposed earlier by Mezic [201]. More details on the spectral property of Koopman operator can be found in this review [202]. Rowley *et al.* revealed the relationship between Koopman mode decomposition and DMD, *i.e.* DMD modes constitute a subset of Koopman modes [203]. The combination of theoretical analysis and efficient numerical implementation [199, 201, 203] led to the growing popularity of this approach, especially for the unsteady wake flow [204, 205]. The procedure to carry out a DMD analysis is briefly presented as the following.

For a sequence of the flow field snapshot, given as,

$$V_1^N = \{v_1, v_2, \cdots, v_N\}$$

suppose there exists a linear operator $\boldsymbol{A}$, which maps the flow field v_j to the subsequent flow field v_{j+1},

$$v_{j+1} = \boldsymbol{A}v_j$$

This linear mapping assumption is essentially a linear tangent approximation of the nonlinear system and it allows us to formulate the sequence of flow field as a Krylov sequence,

$$V_1^N = \{v_1, \boldsymbol{A}v_1, \boldsymbol{A}^2 v_1, \cdots, \boldsymbol{A}^{N-1} v_N\}$$

For a sufficiently long sequence, the vectors in V_1^N become linearly dependent, and vector v_N can be expressed by the previous flow field vectors as,

$$v_N = a_1 v_1 + a_2 v_2 + \cdots + a_{N-1} v_{N-1} + r$$

or in the matrix form,

$$\boldsymbol{v}_N = \boldsymbol{V}_1^N \boldsymbol{a} + \boldsymbol{r}$$

with $\boldsymbol{a}^T = \{a_1, a_2, \ldots, a_{N-1}\}$ and $\boldsymbol{r}$ as the residual vector. It can be further derived as,

$$\boldsymbol{A}\{\boldsymbol{v}_1, \boldsymbol{v}_2, \cdots, \boldsymbol{v}_{N-1}\} = \{\boldsymbol{v}_2, \boldsymbol{v}_3, \cdots, \boldsymbol{v}_N\} = \{\boldsymbol{v}_1, \boldsymbol{v}_2, \cdots, \boldsymbol{V}_1^N \boldsymbol{a}\} + \boldsymbol{r}\boldsymbol{e}_{N-1}^T$$

or in the matrix form,

$$\boldsymbol{A}\boldsymbol{V}_1^{N-1} = \boldsymbol{V}_2^N = \boldsymbol{V}_1^{N-1}\boldsymbol{S} + \boldsymbol{r}\boldsymbol{e}_{N-1}^T$$

with $e_{N-1} \in R^{N-1}$ as the $(N - 1)$th unit vector,

$$\boldsymbol{S} = \begin{pmatrix} 0 & & & & a_1 \\ 1 & 0 & & & a_2 \\ & \ddots & \ddots & & \vdots \\ & & 1 & 0 & a_{N-2} \\ & & & 1 & a_{N-1} \end{pmatrix}$$

the eigenvalues of the companion matrix $\boldsymbol{S}$ is an approximation of the high-dimension matrix A, and captures the dominant dynamics described by A. The eigenvalues of S, referred to as the Ritz values are the approximated eigenvalues of $\boldsymbol{A}$. Given the Ritz values set $\{\lambda_i\}$, the DMD spectrum is obtained as $\mu = \log(\lambda)/\Delta t = \sigma + i\omega$, which determines the stability information about dynamic modes. The relationship between DMD modes and mean flow based global modes can be seen in Mezic [202]. The real part σ represents growth rate, and the imaginary part ω contains the frequency information.

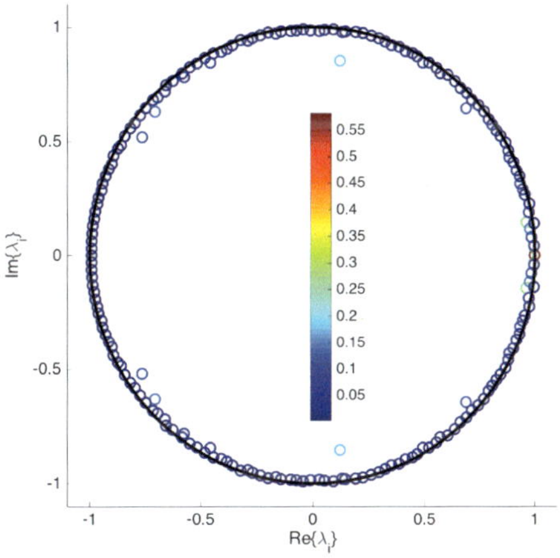

Fig. (10.16). Ritz values.

The temporal DMD analysis is carried out for 200 snapshots of the vorticity field in the MVG wake region on the symmetry plane. Fig. (**10.16**) gives the Ritz values, colored by corresponding DMD modes amplitudes. The Ritz values tend to be distributed on the unit circle, which is expected to be for this statistically stationary system. The DMD spectrum is shown in Fig. (**10.17**), whose real and imaginary parts represent the decay rate and the frequency of the modes, the scatters are sized by corresponding amplitude and colored by phase. The energy spectra are given in Fig. (**10.18**). The mode with largest energy has a zero imaginary part, *i.e.* frequency of 0, representing the averaged flow characteristics.

The contour map of the primary mode, *i.e.* average mode, is shown in Fig. (**10.19**). The near wall region and the shear layer in the MVG wake which have strong vorticity are identified. Contour map of the real parts of the first 4 unsteady DMD modes are shown Fig. (**10.20**), the imaginary parts are not shown as they have almost the same structure as the real parts. The decay rate and the frequency of each mode are also given. We can see that the (a) and (c) show similar structures with close frequency, which seems to identify the vortex and shear layer shedding dynamics, while (b) and (d) are associated with higher frequency structures and have higher decay rates, which might reflect the turbulent fluctuation.

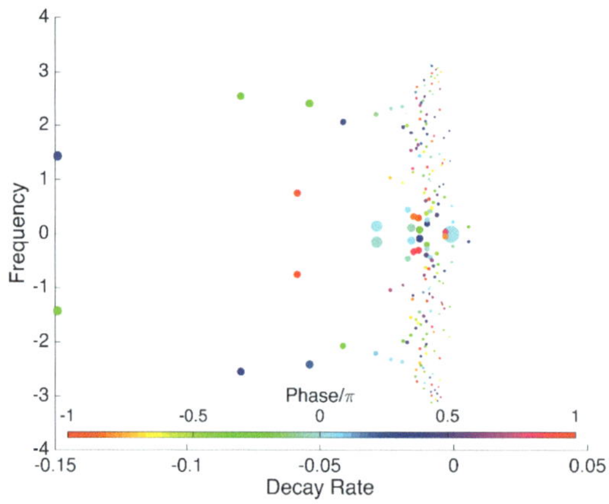

Fig. (10.17). DMD Spectra.

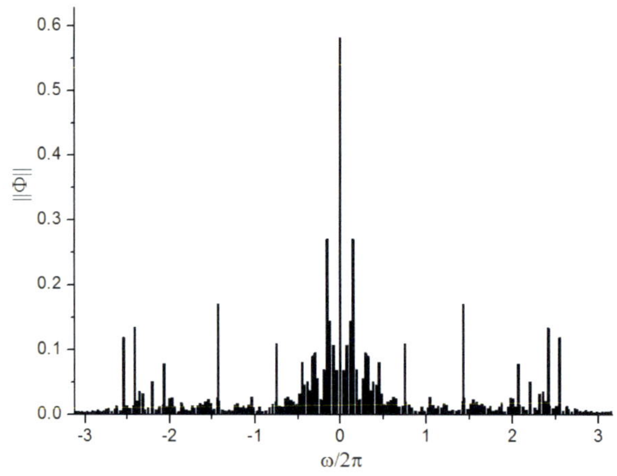

Fig. (10.18). Energy Spectra.

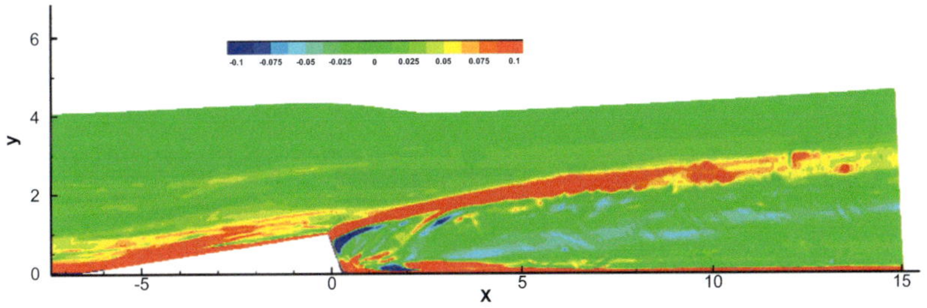

Fig. (10.19). the 0th modes with 0 frequency.

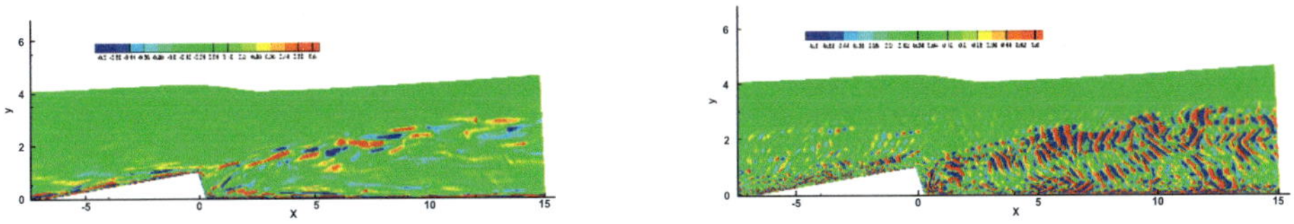

(a) $\mu_r = -0.0286, \mu_i = 0.1488$ (b) $\mu_r = -0.1490, \mu_i = 1.4301$

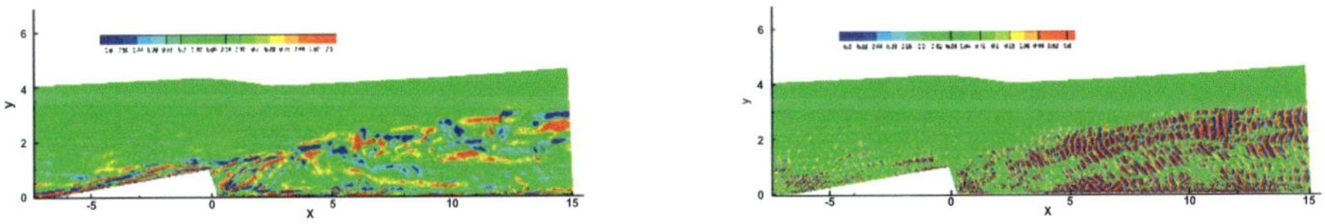

(c) $\mu_r = -0.0157, \mu_i = 0.1185$ (d) $\mu_r = -0.0538, \mu_i = 2.4126$

Fig. (10.20). contour of first 4 DMD modes.

10.3. UNSTEADINESS NEAR THE CORNER

The above discussions are mainly focused on the dynamics in the MVG wake region. In this section, the flow near the corner is analyzed. Fig. (**10.21**) shows the instantaneous spanwise vorticity contour and the shock wave position on the symmetry plane. It can be seen that the shock ends right above the vortex rings. As it has been widely testified, in shock wave boundary layer interaction, the separation shock would have an unsteady oscillation linked with the so-called "breath" motion of separation zone [8]. Thus how the MVG control would affect the shock oscillation is of interest. We set two monitor points on the symmetry plane, point A is near the shock foot, and point B is on the route of vortex rings passage. The time series of pressure signal at point A and spanwise vorticity signal at point B is recorded. Thus, a Fourier transformation is carried out for these signals and Fig. (**10.22**) gives the Amplitude frequency graph in frequency domain. The peak frequency is in accordance with that of the vortex rings passage frequency, namely $0.53u_{\infty}/h$, given in the above section. Also, it can be seen that the variance trend and peak frequency positions of these two signals are very consistent, which means that the frequency of the unsteady shock motion is tuned by the MVG generated vortex rings.

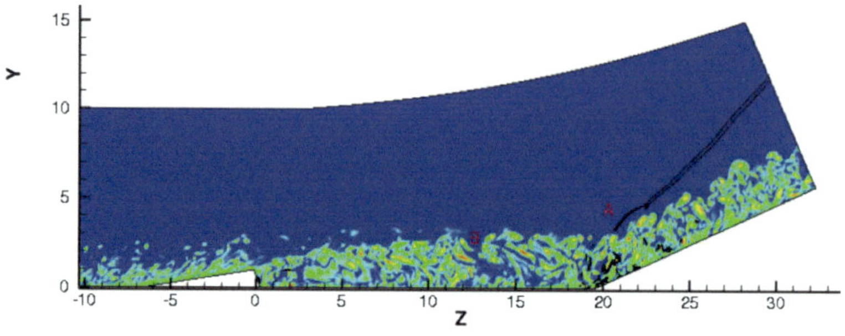

Fig. (10.21). instantaneous spanwise vorticity and shock wave.

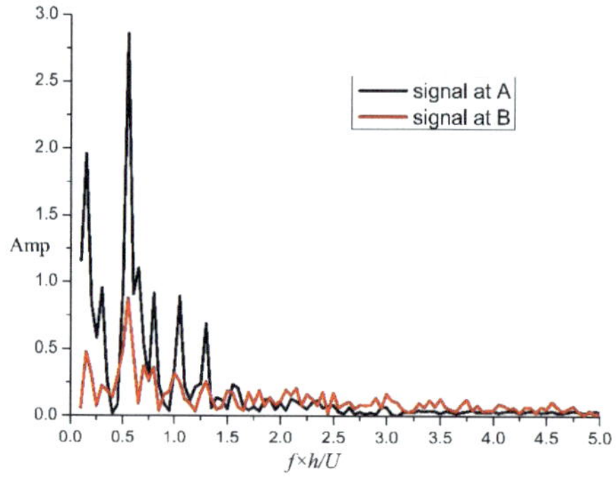

Fig. (10.22). Fourier transformation of the pressure signal at A and spanwise vorticity signal at B.

Fig. (**10.23**) shows the zoomed view of vortex rings/shock surface interaction near the corner. The right figure gives the shock surface separately and it can be seen that there is a hole on the shock surface where vortex rings penetrate. Fig. (**10.24**) gives a sketch of the mechanism behind this scenario, a vortex ring train would be generated from the K-H instability of shear layer induced by the primary streamwise vortices. While convected downstream, the vortex ring would be tilted by the streamwise velocity gradient and tend to be parallel with the shock surface. In the middle of the ring, there is an induced backward streamwise velocity, which would further cause a low speed defect region in the velocity profile. This low speed region exists where the shock strength is weakened and thus an empty hole is left on the shock surface.

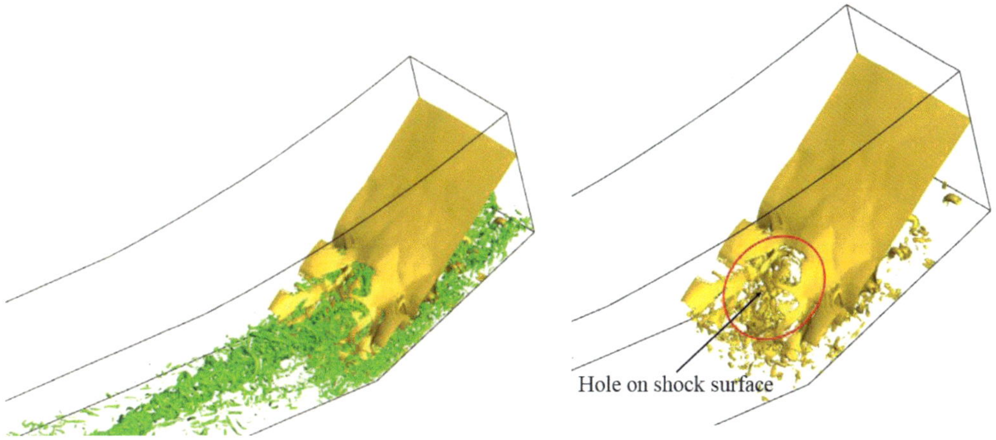

Fig. (10.23). vortex ring and shock interaction and the broken shock surface.

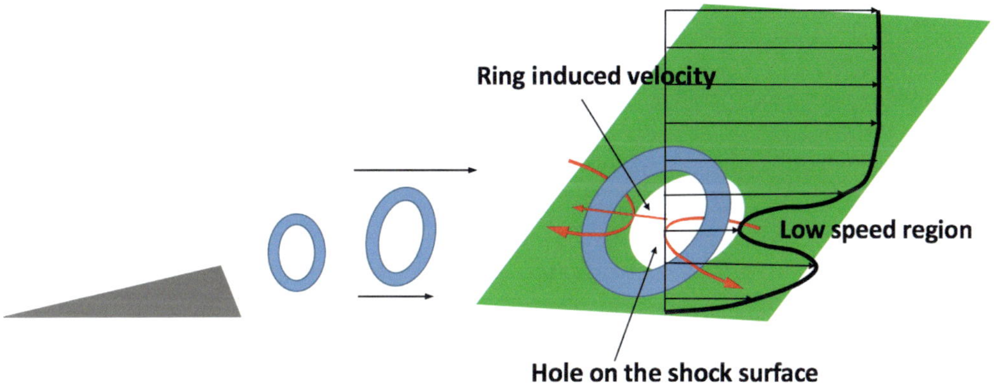

Fig. (10.24). sketch of ring/shock interaction mechanism.

10.4. STAGGERED DOUBLE-ROW MVGS

The above result indicates that using single-row MVG array, the separation region straightly downstream the MVG can be diminished, while the separation region downstream the gaps between MVGs remain. This illuminates the idea that using a second row of MVG, which is located at the gaps between first row of MVG, might improve the control effect of a single row MVG. In this section, a staggered double-row MVG configuration is applied in the control of oblique shock wave/boundary layer interaction by using an immersed boundary method.

The SBLI base flow is an oblique shock-wave generated by an 8° wedge impinging onto a Mach 2.3 turbulent boundary layer, the flow condition of which is consistent with the experiment performed by Dupont *et al.* [23]. Fig. (**10.25**) gives a schematic map of the computation setting. Dashed box is the computational domain. Periodic boundary conditions are used in the spanwise direction. At the wall, the no-slip condition is enforced. Furthermore, the wall is isothermal with a temperature close to the upstream adiabatic value (assuming a recovery factor of 1). The top (free-stream) and outflow boundaries make use of the characteristic non-reflecting boundary condition [206] in order to minimize unwanted reflections from the computational-box boundaries. The oblique shock-wave is introduced at the top boundary using the Rankine–Hugoniot relationships. The turbulent inflow turbulence is generated using a digital filter method [207].

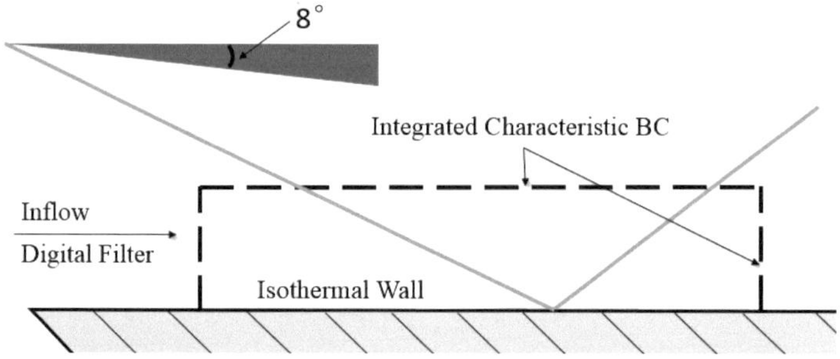

Fig. (10.25). schematic of SBLI flow.

The flow parameters are listed in Table 1. The Reynolds number is based on the boundary layer displacement thickness at interaction point without shock-wave impinging. The same flow has been previously studied by Garnier *et al.* [208] and Touber and Sandham [209], also using LES.

Table 1. Flow parameters

Mach Number	Wedge Angle (°)	Stagnation Pressure (bar)	Stagnation Temperature (K)	Reynolds Number	Sutherland's Law C Value
2.3	8.0	0.5	300.0	20000.0	0.76

The size of the computation domain is (Lx, Ly, Lz) = (256mm, 51mm, 59mm), along the streamwise, the wall-normal and the spanwise directions, respectively. (The streamwise domain range is 148mm, 404mm], to be consistent with the physical domain in the experiment.) The inviscid shock-wave impinging point is at x = 336mm. The grid number in each direction is (Nx, Ny, Nz) = (451, 151, 281), uniform in the streamwise and the spanwise directions while stretched in the wall normal direction. The corresponding grid resolution in each direction is estimated as (Δx^+, Δy^+, Δz^+) = (33, 1.3, 12), which satisfies the general LES requirement [210].

Fig. (**10.26**) gives the averaged streamwise velocity and Reynolds Shear Stress profiles at several streamwise stations, both agrees well with Dupont *et al.*'s PIV measurement. Fig. (**10.27**) shows the time averaged density field, and the impinging shock-wave, reflected shock-wave and the expansion fan can be clearly recognized. The white line is the average sonic line and the red line is the contour line of $u = 0$ that marks the average separation bubble boundary. Fig. (**10.28a**) gives the skin friction coefficient, and Fig. (**10.28b**) is the wall pressure distribution (normalized by incoming free-stream static pressure). The results are validated against the LES results of Touber and Sandham [209]. We can see from the figure that the flow separates at x = 295.32 mm and reattaches at x = 334.67 mm, resulting a mean separation bubble length of about 39.4 mm, showing the instantaneous numerical Schlieren (by using density gradient magnitude) and the streamwise velocity fluctuation at the plane of $y_+ = 15$. The shock-wave system is clearly captured and the typical low-speed streaks of the equilibrium turbulent boundary layer can be clearly observed before entering the interaction region. The streaks are broken in the interaction region and gradually recovered after the reattachment. By all the above validation against high fidelity experimental measurement or numerical simulation, the ability of the present LES solver to reproduce the complex turbulent flowfield of SBLI is verified.

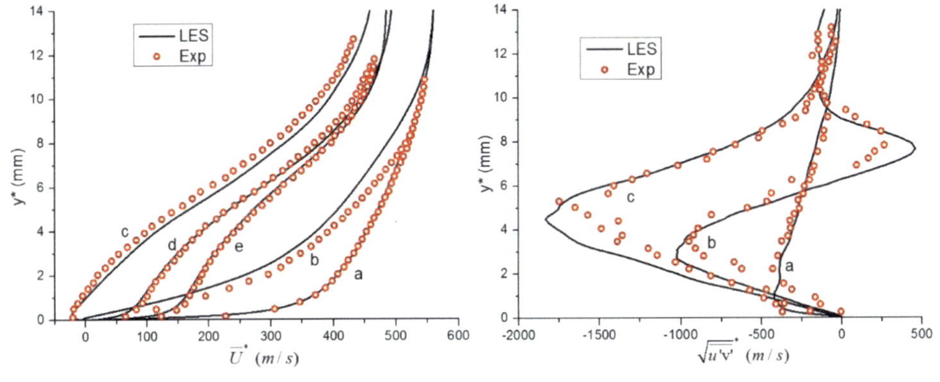

Fig. (10.26). Averaged streamwise velocity and Reynolds Shear Stress at several different streamwise stations.

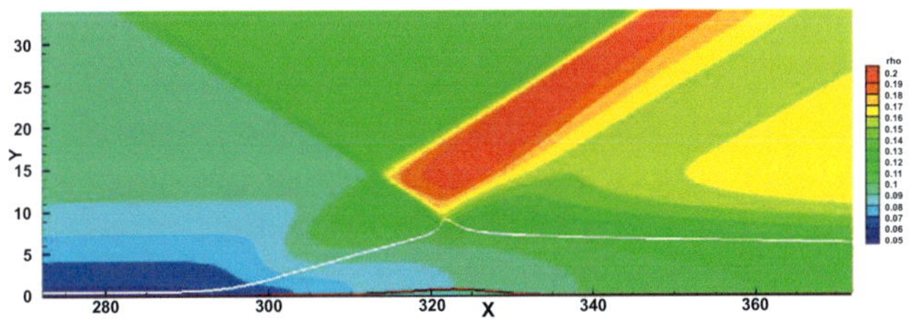

Fig. (10.27). Averaged density flowfield.

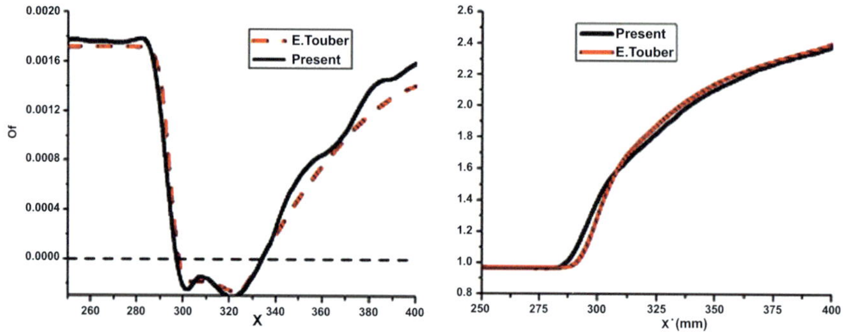

Fig. (10.28). Averaged skin friction and wall pressure.

Fig. (**10.29**) shows the schematic of such a staggered double-row arrangement. Such a configuration has been investigated by several researchers and found to have a better control effect than the single-row MVG array. Blinde *et al.* [52] carried out a Stereo-PIV measurements to investigate the effects of both single row and a staggered double-row array on a SBLI flow, and found that probability of reverse flow occurrence decreased by 20% and 30% for each configuration. Galbraith *et al.* [54] studied the staggered double-row MVG's control efficiency numerically by high-fidelity RANS and confirmed that the second row can further reduce the flow separation and improve the boundary health.

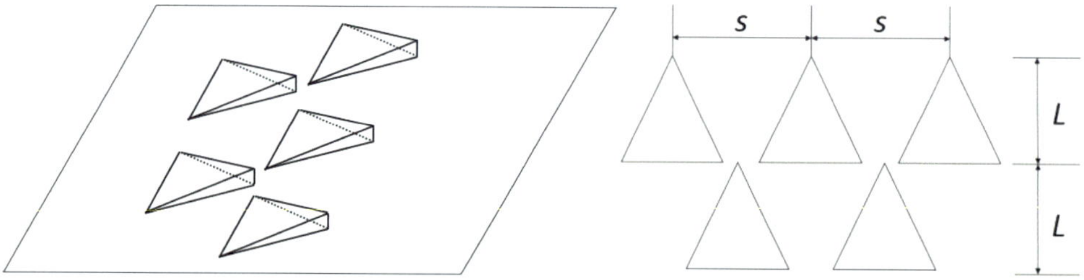

Fig. (10.29). schematic of double-row staggered MVG.

The streamwise length of the computational domain is extended to Lx = 400mm (range from 0 to 400) to accommodate the double-row MVG, and the streamwise grid number is added to Nx = 1001. The other two dimensions are kept unchanged. The double row arrangement here is the same as that in Blinde's experiment [52], where the second row is in immediate proximity to the first row. However, the optimal distance between two rows is still an open issue, and the arrangement where there is a gap between two rows is also reported by other researchers [54, 211].

Fig. (**10.30**) gives the vortical structures of an instantaneous flowfield. The wake flow structures downstream each MVG developed in parallel. Fig. (**10.31**) shows the streamwise vorticity contour and streamlines on the slice of X = 270mm, and the counter-rotating primary vortex after the first row and second row are clearly visualized. As can be observed from the streamlines, the primary vortex pair generated by a MVG tend to weaken the primary vortex generated by adjacent MVG due to alternate rotating direction. And the introduction of the second row would aggravate this issue as the vortex pairs become closer to each other. Further study is required to understand the mutual influence of neighboring vortex pairs in the wake flow.

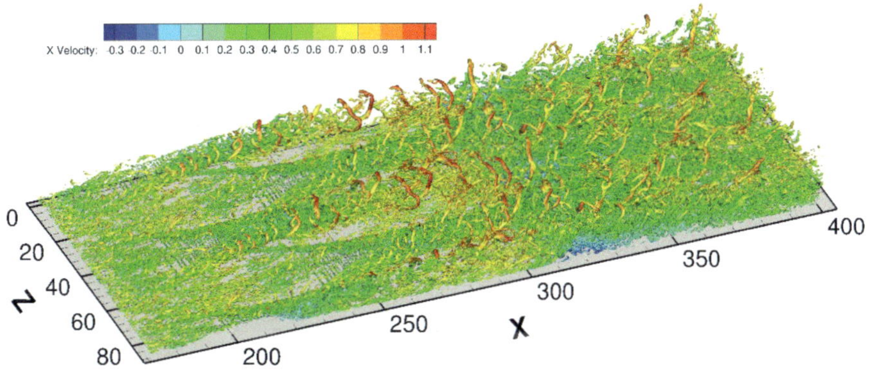

Fig. (10.30). Instantaneous flow structures.

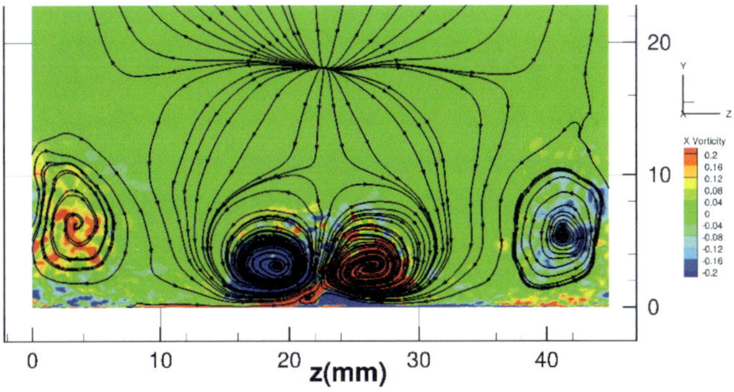

Fig. (10.31). streamwise vorticity contour and streamlines projection at X = 270mm.

Fig. (**10.32**) gives the time-averaged separation zone (colored by dark) on the wall surface with separation and reattachment lines of the uncontrolled case (red dashed lines). Although the flow separation is still not fully eradicated, the separation area is further decreased compared with the single-row case.

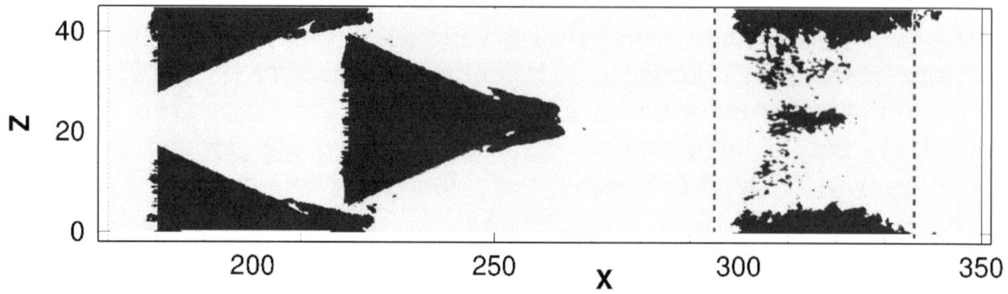

Fig. (10.32). Separation zone.

CONCLUSION

In the present study, an implicit LES is conducted for a MVG controlled supersonic ramp flow. The mean flow profiles and instantaneous structures are validated against experimental measurements. The vortex visualization shows that there are two primary counter rotating streamwise vortices in the MVG wake flow, underneath which are two secondary vortices. The normal momentum transportation induced by the streamwise vortices leads to formation of strong shear layer, whose instability would produce a vortex ring train. The vortex ring would convect to the downstream and interact with the shock wave. A spatial auto-correlation of the spanwise vorticity on the symmetry plane is calculated to determine the distance between adjacent vortex rings and it is found that the average distance is about 1.5 MVG height. Two dimensionality reduction techniques, namely POD and DMD, are applied to 200 flow field snapshots on the spanwise symmetry plane to extract coherent structures in the MVG wake flow. POD identifies structures carrying most of the turbulent kinetic energy. The first pair of POD modes seem to be related to the fluctuation motion induced by the streamwise vortex on this plane, while the second pair of POD modes show K-H type vortex structure, related with the vortex rings. DMD analysis is carried out for the spanwise vorticity field on the symmetry plane. The Ritz values, DMD spectrum and energy spectra of each mode are obtained. Unlike POD, DMD analysis considers both temporal and spatial orthogonality and identifies structure essential to the unsteady evolution. Each DMD mode has a single frequency. The four unsteady modes capture the vortex rings train and high frequency oscillation. Vortex ring/shock interaction near the corner is discussed. It is found that the shock motion frequency can be tuned by the MVG wake frequency and the new control mechanism based on this interaction is illustrated.

By adding a second row of MVG in those gaps can further depress the flow separation. Regarding the MVG control mechanism, the momentum transportation by primary vortex leads to a fuller boundary layer profile which is more resistant to adverse pressure gradient. The control efficiency should be a function of configuration parameters, the optimization of the MVG height, first row/second row distance, distance to the interaction zone needs further study.

REFERENCES

[1] D.S. Dolling, "Fifty years of shock-wave/boundary-layer interaction research: what next?", *AIAA J.*, vol. 39, pp. 1517-1531.

[2] J. Delery, and J. Marvin, "Shock-Wave Boundary Layer Interactions", No. AGARD-AG-280. Advisory group for aerospace research and development neuilly-sur-seine (France), FRANCE.

[3] D.S. Dolling, "Fluctuating loads in shock-wave/boundary layer interaction: tutorial and update", *Presented at AIAA Aerosp. Sci. Meet. Exhib., 31st, Reno, NV, 1993, AIAA Paper,* p. 1993-0284.

[4] A.J. Smits, and J.P. Dussauge, *Turbulent Shear Flow in Supersonic Flow.,* Springer: New York.

[5] A. Zheltovodov, "Shock wave/turbulent boundary layer interactions: fundamental studies and applications", *Presented at AIAA Fluid Dyn. Conf., 27th, New Orleans, 1996, AIAA Paper,* p. 1996-1977.

[6] Y. Andreopoulos, J.H. Agui, and G. Briassulis, "Shock wave—turbulence interactions", *Annu. Rev. Fluid Mech.,* vol. 32, no. 1, pp. 309-345.

[7] B.H. Lee, "Self-sustained shock oscillations on airfoils at transonic speeds", *Prog. Aerosp. Sci.,* vol. 37, no. 2, pp. 147-196.

[8] N.T. Clemens, and V. Narayanaswamy, "Low-frequency unsteadiness of shock wave/turbulent boundary layer interactions", *Annu. Rev. Fluid Mech.,* vol. 46, pp. 469-492.

[9] K.J. Plotkin, "Shock wave oscillation driven by turbulent boundary layer fluctuations", *AIAA J.,* vol. 13, pp. 1036-1040.

[10] J. Andreopoulos, and K.C. Muck, "Some new aspects of the shock-wave boundary layer interaction", *J. Fluid Mech.,* vol. 32, pp. 309-345.

[11] M.E. Erengil, and D.S. Dolling, "Effects of sweepback on unsteady separation in Mach 5 compression ramp interactions", *AIAA J.,* vol. 31, no. 2, pp. 302-311.

[12] M.E. Erengil, "Physical causes of separation shock unsteadiness in shock wave/turbulent boundary layer interactions", *Presented at AIAA Fluid Dyn. Conf., 24th, Orlando, FL, 1993. AIAA Paper,* p. 1993-3134.

[13] B. Ganapathisubramani, N.T. Clemens, and D.S. Dolling, "Effects of upstream boundary layer on the unsteadiness of shock-induced separation", *J. Fluid Mech.,* vol. 585, pp. 369-394.

[14] B. Ganapathisubramani, N.T. Clemens, and D.S. Dolling, "Low-frequency dynamics of shock-induced separation in a compression ramp interaction", *J. Fluid Mech.,* vol. 636, pp. 397-436.

[15] R.A. Humble, G.E. Elsinga, F. Scarano, and B.W. van Oudheusden, "Experimental investigation of the three-dimensional structure of a shock wave/turbulent boundary layer interaction", *16th Australasian Fluid Mechanics Conference (AFMC),* pp. 729-736.

[16] R. A. Humble, G. E. Elsinga, F. Scarano, and B. W. Van Oudheusden, "Investigation of the instantaneous 3D flow organization of a shock wave/turbulent boundary layer interaction using tomographic PIV", *AIAA Paper,* p. 2007-4112.

[17] R.A. Humble, F. Scarano, and B.W. Van Oudheusden, "Particle image velocimetry measurements of a shock wave/turbulent boundary layer interaction", *Exp. Fluids,* vol. 43, no. 2-3, pp. 173-183.

[18] R.A. Humble, G.E. Elsinga, F. Scarano, and B.W. Van Oudheusden, "Three-dimensional instantaneous structure of a shock wave/turbulent boundary layer interaction", *J. Fluid Mech.,* vol. 622, pp. 33-62.

[19] R.J. Adrian, C.D. Meinhart, and C.D. Tomkins, "Vortex organization in the outer region of the turbulent boundary layer", *J. Fluid Mech.,* vol. 422, pp. 1-54.

[20] F.O. Thomas, C.M. Putnam, and H.C. Chu, "On the mechanism of unsteady shock oscillation in shock wave/turbulent boundary layer interactions", *Exp. Fluids,* vol. 18, no. 1, pp. 69-81.

[21] E. Touber, and N.D. Sandham, "Large-eddy simulation of low-frequency unsteadiness in a turbulent shock-induced separation bubble", *Theor. Comput. Fluid Dyn.,* vol. 23, no. 2, pp. 79-107.

[22] P. Dupont, C. Haddad, J.P. Ardissone, and J.F. Debieve, "Space and time organisation of a shock wave/turbulent boundary layer interaction", *Aerosp. Sci. Technol.,* vol. 9, no. 7, pp. 561-572.

[23] P. Dupont, C. Haddad, and J.F. Debieve, "Space and time organization in a shock-induced separated boundary layer", *J. Fluid Mech.,* vol. 559, pp. 255-277.

[24] S. Piponniau, J.P. Dussauge, J.F. Debieve, and P. Dupont, "A simple model for low-frequency unsteadiness in shock-induced separation", *J. Fluid Mech.,* vol. 629, pp. 87-108.

[25] M. Wu, and M.P. Martin, "Analysis of shock motion in shockwave and turbulent boundary layer interaction using direct numerical simulation data", *J. Fluid Mech.,* vol. 594, pp. 71-83.

[26] B. Anderson, "The statistic nature of unsteady flows", 9th Annual Shock Wave/Boundary Layer Interaction Technical Interchange Meeting, Ohio Aerospace Institute (OAI) Cleveland, Ohio.

[27] S.S. Collis, R.D. Joslin, A. Seifert, and V. Theofilis, "Issues in active flow control: theory, control, simulation, and experiment", *Prog. Aerosp. Sci.,* vol. 40, no. 4, pp. 237-289.

[28] J. Délery, "Shock wave/turbulent boundary layer interaction and its control", *Prog. Aerosp. Sci.,* vol. 22, pp. 209-280.

[29] R. D. Joslin, and D. Miller, "Fundamentals and Applications of Modern Flow Control", Prog Astron Aeron 231.

[30] S. Raghunathan, "Passive control of shock-boundary layer interaction", *Prog. Aerosp. Sci.,* vol. 25, no. 3, pp. 271-296.

[31] K.A. Ahmad, W. McEwan, J.K. Watterson, and J.S. Cole, "Numerical Study of a vibrating sub-boundary layer vortex generator", *AIAA Pap.,* pp. 2005-4648.

[32] B.G. Allan, C.S. Yao, and J.C. Lin, "Numerical simulations of vortex generator vanes and jets on a flat plate", *AIAA Pap.,* pp. 2002-3160.

[33] P. Ashill, J. Fulker, and K. Hackett, "Research at DERA on sub boundary layer vortex generators (SBVGs", *39th Aerospace Sciences Meeting and Exhibit,* pp. 2001-0887.

[34] P. R. Ashill, J. L. Fulker, and K. C. Hackett, "Studies of flows induced by sub boundary layer vortex generators (SBVGs)", *AIAA paper, AIAA,* p. 2002-0968.

[35] J. Casper, J. C. Lin, and C. S. Yao, "Effect of sub-boundary layer vortex generators on incident turbulence", *AIAA paper, AIAA,* p. 2003-4162.

[36] A. Jirásek, "Vortex-generator model and its application to flow control", *J. Aircr.,* vol. 42, no. 6, pp. 1486-1491.

[37] J.C. Lin, S.K. Robinson, R.J. McGhee, and W.O. Valarezo, "Separation control on high-lift airfoils via micro-vortex generators", *J. Aircr.,* vol. 31, no. 6, pp. 1317-1323.

[38] M. P. Patel, R. Carver, F. J. Lisy, T. S. Prince, and T. Ng, "Detection and control of flow separation using pressure sensors and micro-vortex generators", *AIAA paper, AIAA ,* p. 2002-0268.

[39] E. Wik, and S. T. Shaw, "Numerical simulation of micro vortex generators", *AIAA paper, AIAA ,* p. 2004-2697.

[40] L. Jenkins, S. A. Gorton, and S. Anders, "Flow control device evaluation for an internal flow with an adverse pressure gradient", *AIAA paper, AIAA ,* p. 2002-0266.

[41] M. Meunier, and V. Brunet, "High-lift devices performance enhancement using mechanical and air-jet vortex generators", *J. Aircr.,* vol. 45, no. 6, pp. 2049-2061.

[42] R.F. Osborn, S. Kota, J.A. Hetrick, D.E. Geister, C.P. Tilmann, and J. Joo, "Active flow control using high-frequency compliant structures", *J. Aircr.*, vol. 41, no. 3, pp. 603-609.

[43] M.P. Patel, T.S. Prince, R. Carver, J.M. DiCocco, F.J. Lisy, and T.T. Ng, "Deployable flow effectors for phantom yaw control of missiles at high alpha", *1st Flow Control Conference,* pp. 2002-2827.

[44] M. Patel, T. Prince, R. Carver, J. DiCocco, F. Lisy, and T. Ng, "Control of aircraft stall via embedded pressure sensors and deployable flow effectors", *1st Flow Control Conference,* pp. 2002-3170.

[45] A. J. Rae, S. A. Galpin, and J. Fulker, "Investigation into scale effects on the performance of sub boundary-layer vortex generators on civil aircraft high-lift devices", *AIAA paper, AIAA* , p. 2002-3274.

[46] T.C. Tai, "Effect of midwing vortex generators on V-22 aircraft forward-flight aerodynamics", *J. Aircr.,* vol. 40, no. 4, pp. 623-630.

[47] B. H. Anderson, J. Tinapple, and L. Surber, "Optimal control of shock wave turbulent boundary layer interactions using micro-array actuation", *AIAA paper, AIAA* , p. 2006-3197.

[48] B. H. Anderson, J. L. Mace, and M. Mani, "Active 'fail safe'micro-array flow control for advanced embedded propulsion systems", *AIAA paper, AIAA* , p. 2009-00741.

[49] H. Babinsky, and H. Ogawa, "SBLI control for wings and inlets", *Shock Waves,* vol. 18, no. 2, pp. 89-96.

[50] H. Babinsky, N. J. Makinson, and C. E. Morgan, "Micro-vortex generator flow control for supersonic engine inlets", *AIAA paper, AIAA* , p. 2007-521.

[51] H. Babinsky, Y. Li, and C.W. Pitt Ford, "Microramp control of supersonic oblique shock-wave/boundary-layer interactions", *AIAA J.,* vol. 47, no. 3, pp. 668-675.

[52] P.L. Blinde, R.A. Humble, B.W. Van Oudheusden, and F. Scarano, "Effects of micro-ramps on a shock wave/turbulent boundary layer interaction", *Shock Waves,* vol. 19, no. 6, pp. 507-520.

[53] N. D. Domel, D. Baruzzini, and D. N. Miller, "CFD results for shock-boundary layer flow control with micro-ramps at various grid densities", *AIAA paper, AIAA* , p. 2009-4016.

[54] M.C. Galbraith, P.D. Orkwis, and J.A. Benek, "Multi-row micro-ramp actuators for shock wave boundary-layer interaction control", *47th AIAA Aerospace Sciences Meeting,* pp. 2009-0321.

[55] S. Ghosh, J.I. Choi, and J.R. Edwards, "Numerical simulations of effects of micro vortex generators using immersed-boundary methods", *AIAA J.,* vol. 48, no. 1, pp. 92-103.

[56] T. Herges, E. Kroeker, G. Elliott, and C. Dutton, "Microramp flow control of normal shock/boundary-layer interactions", *AIAA J.,* vol. 48, no. 11, pp. 25-29.

[57] H.A. Holden, and H. Babinsky, "Effect of microvortex generators on seperated normal shock/ boundary layer interactions", *J. Aircr.,* vol. 44, no. 1, pp. 170-174.

[58] A. Jirasek, "Development and application of design strategy for design of vortex generator flow control in inlets", *AIAA Paper, AIAA* , p. 2006-1050.

[59] S. Lee, and E. Loth, "Supersonic boundary-layer interactions with various micro-vortex generator geometries", *Aeronaut. J.,* vol. 113, no. 1149, pp. 683-697.

[60] S. Lee, E. Loth, C. Wang, and S. Kim, "LES of Supersonic Turbulent Boundary Layers with μVG's", *25th AIAA Applied Aerodynamics Conference,* pp. 2007-3916.

[61] S. Lee, E. Loth, N.J. Georgiadis, and J.R. DeBonis, "Effect of Mach number on flow past microramps", *AIAA J.,* vol. 49, no. 1, pp. 97-110.

[62] S. Lee, M.K. Goettke, E. Loth, J. Tinapple, and J. Benek, "Microramps upstream of an oblique-shock/boundary-layer interaction", *AIAA J.,* vol. 48, no. 1, pp. 104-118.

[63] H. Ogawa, H. Babinsky, M. Patzold, and T. Lutz, "Shock-wave/boundary-layer interaction control using three-dimensional bumps for transonic wings", *AIAA J.,* vol. 46, no. 6, pp. 1442-1452.

[64] M. Rybalko, E. Loth, R. V. Chima, S. M. Hirt, and J. R. DeBonis, "Micro-ramps for external compression low-boom inlets", *AIAA paper, AIAA* , p. 2009-4206.

[65] A. F. Shinn, S. P. Vanka, M. Mani, A. Dorgan, and T. Michal, "Application of BCFD unstructured grid solver to simulation of micro-ramp control of shock/boundary layer interactions", *AIAA paper, AIAA* , p. 2007-3914.

[66] J.W. Barter, and D.S. Dolling, "Reduction of fluctuating pressure loads in shock/boundary-layer interactions using vortex generators", *AIAA J.,* vol. 33, no. 10, pp. 1843-1849.

[67] C.M. Velte, M.O. Hansen, and V.L. Okulov, "Helical structure of longitudinal vortices embedded in turbulent wall-bounded flow", *J. Fluid Mech.,* vol. 619, pp. 167-177.

[68] O. Lögdberg, J.H. Fransson, and P.H. Alfredsson, "Streamwise evolution of longitudinal vortices in a turbulent boundary layer", *J. Fluid Mech.,* vol. 623, pp. 27-58.

[69] K.P. Angele, and F. Grewe, "Instantaneous behavior of streamwise vortices for turbulent boundary layer separation control", *J. Fluids Eng.,* vol. 129, no. 2, pp. 226-235.

[70] T. Duriez, J.L. Aider, and J.E. Wesfreid, "Self-sustaining process through streak generation in a flat-plate boundary layer", *Phys. Rev. Lett.,* vol. 103, no. 14, p. 144502.

[71] F. Laporte, and A. Corjon, "Direct numerical simulations of the elliptic instability of a vortex pair", *Phys. Fluids,* vol. 12, no. 5, pp. 1016-1031.

[72] F. Laporte, and T. Leweke, "Elliptic instability of counter-rotating vortices: experiment and direct numerical simulation", *AIAA J.,* vol. 40, no. 12, pp. 2483-2539.

[73] Q. Li, and C. Liu, "LES for Supersonic Ramp Flow Control Using MVG at M_infty= 3 and Re_theta= 2400", *AIAA Paper, AIAA* , p. 2010-0592.

[74] Q. Li, and C. Liu, *"Numerical investigations on the effects of the declining angle of the trailing-edge of MVG". AIAA paper, Vol.714.,* AIAA, pp. 2010-0714.

[75] F. K. Lu, A. J. Pierce, Y. Shih, C. Liu, and Q. Li, "Experimental and numerical study of flow topology past micro vortex generators", *AIAA Paper, AIAA* , p. 2010-4463.

[76] F. K. Lu, A. J. Pierce, and Y. Shih, "Experimental study of near wake of micro vortex generators in supersonic flow", *AIAA Paper, AIAA* , p. 2010-4623.

[77] P.G. Drazin, and W.H. Reid, *Hydrodynamic stability.,* Cambridge university press, p. 14.

[78] A.J. Pierce, F.K. Lu, D.S. Bryant, and Y. Shih, "New developments in surface oil flow visualization", *27th AIAA Aerodynamic Measurement Technology and Ground Testing Conference,* pp. 2010-4353.

[79] Q. Li, and C. Liu, "Numerical Investigations on the Effects of the Declining Angle of the Trailing-Edge of MVG", *J. Aircr.,* vol. 47, no. 6, pp. 2086-2095.

[80] M. Tobak, and D.J. Peake, "Topology of 3-dimensional separated flows", *Annu. Rev. Fluid Mech.,* vol. 14, pp. 61-85.

[81] V. A. Andrushchenko, and M. V. Meshcheryakov, "Motion of trains of vortex rings", *Journal of Engineering Physics and Thermophysics,* vol. 74, no. 1, pp. 20-25.

[82] F. Grasso, and S. Pirozzoli, "Simulations and analysis of the coupling process of compressible vortex pairs: free evolution and shock induced coupling", *Phys. Fluids,* vol. 13, no. 5, pp. 1343-1366.

[83] O. Inoue, "Sound generation by the leapfrogging between two coaxial vortex rings", *Phys. Fluids,* vol. 14, no. 9, pp. 3361-3364.

[84] H. Zare-Behtash, K. Kontis, N. Gongora-Orozco, and K. Takayama, "Compressible vortex loops: effect of nozzle geometry", *Int. J. Heat Fluid Flow,* vol. 30, no. 3, pp. 561-576.

[85] R.R. Kerswell, "Elliptical instability", *Annu. Rev. Fluid Mech.,* vol. 34, no. 1, pp. 83-113.

[86] J. Jeong, and F. Hussain, "On the identification of a vortex", *J. Fluid Mech.,* vol. 285, pp. 69-94.

[87] G.S. Settles, and H.Y. Teng, "Flow visualization methods for separated three-dimensional shock wave/turbulent boundary-layer interactions", *AIAA J.,* vol. 21, no. 3, pp. 390-397.

[88] G.S. Settles, and F.K. Lu, "Conical similarity of shock/boundary-layer interactions generated by swept and unswept fins", *AIAA J.,* vol. 23, no. 7, pp. 1021-1027.

[89] G.S. Settles, S.M. Bogdonoff, and I.E. Vas, "Incipient separation of a supersonic turbulent boundary layer at high Reynolds numbers", *AIAA J.,* vol. 14, no. 1, pp. 50-56.

[90] D. R. Chapman, D. M. Kuehn, and H. K. Larson, "Investigation of separated flows in supersonic and subsonic streams with emphasis on the effect of transition", NACA Report 1356.

[91] R. Sedney, "A survey of the effects of small protuberances on boundary-layer flows", *AIAA J.,* vol. 11, no. 6, pp. 782-792.

[92] S. Lee, E. Loth, and H. Babinsky, "Normal shock boundary layer control with various vortex generator geometries", *Comput. Fluids,* vol. 49, no. 1, pp. 233-246.

[93] F.K. Lu, "Surface oil flow visualization", *Eur. Phys. J. Spec. Top.,* vol. 182, no. 1, pp. 51-63.

[94] P.J. Mason, and B.R. Morton, "Trailing vortices in the wakes of surface-mounted obstacles", *J. Fluid Mech.,* vol. 175, pp. 247-293.

[95] A. Sau, R.R. Hwang, T.W. Sheu, and W.C. Yang, "Interaction of trailing vortices in the wake of a wall-mounted rectangular cylinder", *Phys. Rev. E,* vol. 68, no. 5, p. 056303.

[96] J.D. Crawford, and E. Knobloch, "Symmetry and symmetry-breaking bifurcations in fluid dynamics", *Annu. Rev. Fluid Mech.,* vol. 23, no. 1, pp. 341-387.

[97] D.H. Bridges, "Toward a theoretical description of vortex wake asymmetry", *Prog. Aerosp. Sci.,* vol. 46, no. 2, pp. 62-80.

[98] Q. Li, Y. Yan, P. Lu, A. Pierce, C. Liu, and F. Lu, "Numerical and experimental studies on the separation topology of the MVG controlled flow at $M = 2.5$", *AIAA Paper, AIAA ,* p. 2011-0072.

[99] F. Lu, Q. Li, and C. Liu, "Micro Vortex Generators in High-Speed Flow", *Journal of Progress in Aerospace Science,* vol. 53, pp. 30-45.

[100] P. Lu, Y. Yan, and C. Liu, "Numerical investigation on mechanism of multiple vortex rings formation in late boundary-layer transition", *Comput. Fluids,* vol. 71, pp. 156-168.

[101] R. Martinuzzi, and C. Tropea, "The flow around surface-mounted, prismatic obstacles placed in a fully developed channel flow", *J. Fluids Eng.,* vol. 115, no. 1, pp. 85-92.

[102] R. Bur, D. Coponet, and Y. Carpels, "Separation control by vortex generator devices in a transonic channel flow", *Shock Waves,* vol. 19, no. 6, pp. 521-530.

[103] S. Julien, J.M. Chomaz, and J.C. Lasheras, "Three-dimensional stability of periodic arrays of counter-rotating vortices", *Phys. Fluids,* vol. 14, no. 2, pp. 732-743.

[104] Q. Li, and C. Liu, Implicit LES for supersonic microramp vortex generator: new discoveries and new mechanisms.*Modelling and Simulation in Engineering.,* Hindawi Publishing Corporation, p. 2011-9349.

[105] Y. Yan, Q. Li, C. Liu, and F. K. Lu, "Numerical, experimental and theoretical studies on mechanism of K-H instability and ring generation behind supersonic MVG", *AIAA paper, AIAA,* p. 2011-676.

[106] T.T. Lim, "On the role of Kelvin–Helmholtz-like instability in the formation of turbulent vortex rings", *Fluid Dyn. Res.,* vol. 21, no. 1, pp. 47-56.

[107] M. Jiang, R. Machiraju, and D.S. Thompson, *"Detection and Visualization of Vortices", Visualization Handbook.,* Elsevier, pp. 295-309.

[108] A.G. Panaras, *Aerodynamic Principles of Flight Vehicles.,* AIAA: Reston, Virginia.

[109] T.J. Garrison, G.S. Settles, N. Narayanswami, and D.D. Knight, "Structure of crossing-shock

wave/turbulent-boundary-layer interactions", *AIAA J.,* vol. 31, no. 12, pp. 2204-2211.

[110] N. Titchener, and H. Babinsky, "Microvortex generators applied to flowfield containing a normal shock wave and diffuser", *AIAA J.,* vol. 49, no. 5, pp. 1046-1056.

[111] B. Konig, M. Patzold, T. Lutz, E. Kramer, H. Rosemann, K. Richter, and H. Uhlemann, "Numerical and experimental validation of three dimensional shock control bumps", *J. Aircr.,* vol. 46, no. 2, pp. 675-682.

[112] S. Pirozzoli, "Dynamics of ring vortices impinging on planar shock waves", *Phys. Fluids,* vol. 16, no. 5, pp. 1171-1185.

[113] Z. Sun, F. Scarano, B. Oudheusden, F. Schrijer, X. Wang, Y. Yan, and C. Liu, "Numerical and experimental investigation of the flow behind a supersonic Micro-ramp", *AIAA paper, AIAA* , p. 2013-0954.

[114] D.C. Reda, and J.D. Murphy, "Shock wave/turbulent boundary-layer interactions in rectangular channels", *AIAA J.,* vol. 11, no. 2, pp. 139-140.

[115] P.J. Bruce, D.M. Burton, N.A. Titchener, and H. Babinsky, "Corner effect and separation in transonic channel flows", *J. Fluid Mech.,* vol. 679, pp. 247-262.

[116] J.L. Souverein, and J.F. Debieve, "Effect of air jet vortex generators on a shock wave boundary layer interaction", *Exp. Fluids,* vol. 49, no. 5, pp. 1053-1064.

[117] D.A. Dickmann, and F.K. Lu, "Shock/boundary-layer interaction effects on transverse jets in crossflow over a flat plate", *J. Spacecr. Rockets,* vol. 46, no. 6, pp. 1132-1141.

[118] G.G. Stokes, "On the theories of the internal friction of fluids in motion, and of the equilibrium and moption of elastic fluids", *Transactions of the Cambridge Philosophical Society,* vol. 8, p. 287.

[119] W. Sutherland, "LII. The viscosity of gases and molecular force", *Lond. Edinb. Dublin Philos. Mag. J. Sci.,* vol. 36, no. 223, pp. 507-531.

[120] J. Steger, and R. Warming, "Flux vector splitting of the inviscid gasdynamic equations with applications to finite-difference methods", *J. Comput. Phys.,* vol. 40, pp. 263-293.

[121] S.P. Spekreijse, "Elliptic grid generation based on Laplace equations and algebraic transformations", *J. Comput. Phys.,* vol. 118, no. 1, pp. 38-61.

[122] J.F. Thompson, A survey of composite grid generation for general three-dimensional regions.*Numerical Methods for Engine-Airframe Integration, Ed. Murthy, SNB and Paynter, GC, AIAA*. pp. 1986

[123] J.F. Thompson, "A general three-dimensional elliptic grid generation system on a composite block structure", *Comput. Methods Appl. Mech. Eng.,* vol. 64, no. 1-3, pp. 377-411.

[124] S. K. Godunov, "A finite difference method for the computation of discontinuous solutions of the equations of fluid dynamics", *Sbornik: Mathematics,* vol. 74, no. 8-9, pp. 357-393.

[125] P.L. Roe, "Approximate Riemann solvers, parameter vectors, and difference schemes", *J. Comput. Phys.,* vol. 135, no. 2, pp. 250-258.

[126] L.B. Van, "Towards the ultimate conservative difference scheme. V. A second order sequel to Godunov's scheme", *J. Comput. Phys.,* vol. 32, pp. 101-136.

[127] A. Harten, "High resolution schemes for hyperbolic conservation laws", *J. Comput. Phys.,* vol. 49, pp. 357-393.

[128] A. Harten, B. Engquist, B. Osher, and S.R. Charkravarthy, "Uniformly high order accurate essentially non-oscillatory schemes III", *J. Comput. Phys.,* vol. 71, pp. 231-303.

[129] C.W. Shu, and S. Osher, "Efficient implementation of essentially non-oscillatory shock-capturing scheme", *J. Comput. Phys.,* vol. 77, pp. 439-471.

[130] C.W. Shu, and S. Osher, "Efficient implementation of essentially non-oscillatory shock-capturing

schemes II", *J. Comput. Phys.,* vol. 83, pp. 32-78.

[131] D. Liu, S. Osher, and T. Chan, "Weighted essentially nonoscillatory schemes", *J. Comput. Phys.,* vol. 115, pp. 200-212.

[132] G.S. Jiang, and C.W. Shu, "Efficient implementation of weighted ENO scheme", *J. Comput. Phys.,* vol. 126, no. 1, pp. 202-228.

[133] S.K. Lele, "Compact finite difference schemes with spectral-like resolution", *J. Comput. Phys.,* vol. 103, pp. 16-42.

[134] M. Visbal, and D. Gaitonde, "On the use of higher-order finite-difference schemes on curvilinear and deforming mashes", *J. Comput. Phys.,* vol. 181, pp. 155-158.

[135] D. Kim, and J. Kwon, "A high-order accurate hybrid scheme using a central flux scheme and a WENO scheme for compressible flowfield analysis", *J. Comput. Phys.,* vol. 210, pp. 554-583.

[136] B. Costa, and W.S. Don, "High order Hybrid central-WENO finite difference scheme for conservation laws", *J. Comput. Appl. Math.,* vol. 204, pp. 209-218.

[137] Y. Ren, M. Liu, and H. Zhang, "A characteristic-wise hybrid compact-WENO scheme for solving hyperbolic conservation laws", *J. Comput. Phys.,* vol. 192, pp. 365-386.

[138] Q. Zhou, Z. Yao, F. He, and M.Y. Shen, "A new family of high-order compact upwind difference schemes with good spectral resolution", *J. Comput. Phys.,* vol. 227, no. 2, pp. 1306-1339.

[139] M. Oliveira, P. Lu, X. Liu, and C. Liu, "New Shock/Discontinuity Detector", *Int. J. Comput. Math.,* no. July, p. 29.

[140] A. Harten, "The artificial compression method for computation of shocks and contact discontinuities: III. self-adjusting hybrid schemes", *Math. Comput.,* vol. 32, no. 142, pp. 363-389.

[141] A. Jameson, W. Schmidt, and E. Turkel, "Numerical solutions of the Euler equations by finite volume methods using Runge-Kutta time-stepping schemes", *AIAA pape,* p. 1981, 1981.

[142] L. Jiang, H. Shan, and C. Liu, "Weighted compact scheme for shock capturing", *Int. J. Comput. Fluid Dyn.,* vol. 15, pp. 147-155.

[143] R. Vichnevetsky, and J.B. Bowles, *Fourier analysis of numerical approximations of hyperbolic equations.,* . Society for Industrial and Applied Mathematics.

[144] H. Fu, Z. Wang, and C. Liu, "Modified Weighted Compact Scheme for Shock Capturing, J. of Computers and Fluids", *Comput. Fluids,* vol. 96, pp. 165-176.

[145] J.D. Anderson, and J. Wendt, *Computational fluid dynamics.,* McGraw Hill: New York, .

[146] P. Xie, M. Oliveira, and C. Liu, "Weighted Compact Schemes for Shock / Boundary Layer Interaction", *5th International Conference on Computational Fluid Dynamics*. Seoul, Korea.

[147] G. Degrez, C.H. Boccadoro, and J.F. Wendt, "The interaction of an oblique shock wave with a laminar boundary layer revisited. An experimental and numerical study", *J. Fluid Mech.,* vol. 177, pp. 247-263.

[148] M. Wu, and M.P. Martin, "Direct Numerical Simulation of Supersonic Turbulent Boundary Layer over a Compression Ramp", *AIAA J.,* vol. 45, no. 4, pp. 879-889.

[149] T.S. Lund, X. Wu, and K.D. Squires, "Generation of turbulent inflow data for spatially-developing boundary layer simulations", *J. Comput. Phys.,* vol. 140, no. 2, pp. 233-258.

[150] L. Chen, X. Liu, M. Oliveira, D. Tang, and C. Liu, "Vortical structure, sweep and ejection events in transitional boundary layer", *Science China, Series G, Physics, Mechanics, Astronomy,* vol. 39, no. 10, pp. 1520-1526.

[151] L. Chen, X. Liu, M. Oliveira, and C. Liu, "DNS for ring-like vortices formation and roles in positive spikes formation", *AIAA paper, AAIA ,* p. 2010-1471.

[152] L. Chen, D. Tang, P. Lu, and C. Liu, "Evolution of the vortex structures and turbulent spots at the late-stage of transitional boundary layers", *Sci. China Phys. Mech. Astron.,* vol. 54, no. 5, pp. 986-990.

[153] L. Chen, and C. Liu, "Numerical study on mechanisms of second sweep and positive spikes in transitional flow on a flat plate", *Comput. Fluids,* vol. 40, no. 1, pp. 28-41.

[154] L. Chen, D. Tang, and C. Liu, "Numerical studies on evolution of secondary streamwise vortices in transitional boundary layers", *Appl. Math. Mech.,* vol. 32, no. 4, pp. 449-458.

[155] C. Liu, and L. Chen, "Study of mechanism of ring-like vortex formation in late flow transition", *AIAA Paper, AIAA,* p. 2010-1456.

[156] X. Liu, Z. Chen, and C. Liu, "Late-stage vortical structures and eddy motions in a transitional boundary layer", *Chin. Phys. Lett.,* vol. 27, no. 2, p. 024706.

[157] X. Liu, L. Chen, M. Oliveira, D. Tang, and C. Liu, "DNS for late stage structure of flow transition on a flat-plate boundary layer", *AIAA Paper, AIAA,* pp. 2010-1470.

[158] C. Liu, L. Chen, and P. Lu, "New Findings by High Order DNS for Late Flow Transition in a Boundary Layer", *Journal of Modeling and Simulation in Engineering.*

[159] C. Liu, and L. Chen, "Parallel DNS for vortex structure of late stages of flow transition", *Journal of Computers and Fluids,* vol. 45, pp. 129-137.

[160] C. Liu, "Numerical and theoretical study on vortex breakdown", *Int. J. Comput. Math.,* vol. 88, no. 17, pp. 3702-3708.

[161] C. Liu, L. Chen, P. Lu, and X. Liu, "Study on multiple ring-like vortex formation and small vortex generation in late flow transition on a flat plate", *Theor. Comput. Fluid Dyn.,* vol. 27, no. 1, pp. 41-70.

[162] C. Liu, Y. Yan, and P. Lu, "Physics of turbulence generation and sustenance in a boundary layer", *Comput. Fluids,* vol. 102, pp. 353-384.

[163] C. Liu, New Theory of Turbulence Generation and Sustenance, TGS2015 Short Course, Tsinghua University, Beijng, China, June 4-6

[164] C. Liu, Y. Wang, Y. Yang, and Z. Duan, "New omega vortex identification method", *Sci. China Phys. Mech. Astron.,* vol. 59, no. 8, pp. 1-9.

[165] P. Lu, Z. Wang, L. Chen, and C. Liu, "Numerical study on U-shaped vortex formation in late boundary layer transition", *Comput. Fluids,* vol. 55, pp. 36-47.

[166] P. Lu, and C. Liu, "Numerical study on mechanism of small vortex generation in boundary layer transition", *AIAA Paper, AIAA ,* p. 2011-0287.

[167] P. Lu, and C. Liu, "DNS study on mechanism of small length scale generation in late boundary layer transition", *Physica D,* vol. 241, no. 1, pp. 11-24.

[168] Y. Yan, C. Chen, H. Fu, and C. Liu, "DNS study on Λ-vortex and vortex ring formation in flow transition at Mach number 0.5", *J. Turbul.,* vol. 15, no. 1, pp. 1-21.

[169] L. Jiang, and C. Liu, "Direct numerical simulation for flow separation control with pulsed vortex generator jets", *AIAA Paper, AIAA ,* p. 2003-0611.

[170] C. Lee, and R. Li, "Dominant structure for turbulent production in a transitional boundary layer", *J. Turbul.,* vol. 8, no. 55, pp. 1-34.

[171] H. Guo, V.I. Borodulin, Y.S. Kachanov, C. Pan, J.J. Wang, Q.X. Lian, and S.F. Wang, "Nature of sweep and ejection events in transitional and turbulent boundary layers", *J. Turbul.,* vol. 11, pp. 1468-5248.

[172] H.F. Fasel, U. Rist, and U. Konzelmann, "Numerical investigation of the three-dimensional development in boundary- layer transition", *AIAA J.,* vol. 28, no. 1, pp. 29-37.

[173] S.E. Guarini, R.D. Moser, K. Shariff, and A. Wray, "Direct numerical simulation of a supersonic

turbulent boundary layer at Mach 2.5", *J. Fluid Mech.,* vol. 414, no. 1, pp. 1-33.

[174] Z. Sun, "Micro Ramps in Supersonic Turbulent Boundary Layers: An experimental and numerical study", *Ph.D. Thesis, TU Delft, Delft University of Technology.*

[175] M.J. Ringuette, M. Wu, and M.P. Martin, "Low Reynolds Number Effects in a Mach 3 Shock/Turbulent-Boundary-Layer Interaction", *AIAA J.,* vol. 46, no. 7, pp. 1884-1887.

[176] M.J. Ringuette, M. Wu, and M.P. Martin, "Coherent Structures in Direct Numerical Simulation of Turbulent Boundary Layers at Mach 3", *J. Fluid Mech.,* vol. 594, pp. 59-69.

[177] M.P. Martin, "Direct Numerical Simulation of Hypersonic Turbulent Boundary Layers. Part 1. Initialization and Comparison with Experiments", *J. Fluid Mech.,* vol. 570, pp. 347-364.

[178] M.R. Saad, H. Zare-Behtash, A. Che-Idris, and K. Kontis, "Micro-ramps for hypersonic flow control", *Micromachines (Basel),* vol. 3, no. 2, pp. 364-378.

[179] S. Ghosh, I. Calizo, D. Teweldebrhan, E.P. Pokatilov, D.L. Nika, and A.A. Balandin, "Extremely high thermal conductivity of graphene: Prospects for thermal management applications in nanoelectronic circuits", *Appl. Phys. Lett.,* vol. 92, no. 15, p. 151911.

[180] Z. Sun, F.F. Schrijer, F. Scarano, and B.W. Van Oudheusden, "The three-dimensional flow organization past a micro-ramp in a supersonic boundary layer", *Physics of Fluids,* vol. 24, no. 5, p. 055105.

[181] Y. Yan, C. Chen, P. Lu, and C. Liu, *"Study on shock wave-vortex ring interaction by the MVG controlled ramp flow with turbulence inlet from DNS". AIAA paper, Vol.47.,* AIAA, pp. 2012-0047.

[182] Q. Li, Y. Yan, X. Wang, and C. Liu, "The interaction between vortex rings and oblique shocks by the MVG controlled ramp flow at M = 2.5", *AIAA Paper, AIAA ,* p. 2011- 861.

[183] Y. Yan, C. Chen, X. Wang, and C. Liu, "LES Study on the Mechanism of Vortex Rings behind Supersonic MVG with Turbulent Inflow", *AIAA Paper, AIAA ,* p. 2012-1093.

[184] Y. Yan, Q. Li, C. Liu, A. Pierce, F. Lu, and P. Lu, "Numerical discovery and experimental confirmation of vortex ring generation by microramp vortex generator", *Appl. Math. Model.,* vol. 36, no. 11, pp. 5700-5708.

[185] Y. Yan, and C. Liu, "Shear Layer Stability Analysis in Later Boundary Layer Transition and MVG controlled flow", *AIAA Paper, AIAA ,* p. 2013- 0531.

[186] D.S. Dosanjh, and T.M. Weeks, "Interaction of a starting vortex as well as a vortex street with a traveling shock wave", *AIAA J.,* vol. 3, pp. 216-223.

[187] O. Inoue, and Y. Hattori, "Sound generation by shock-vortex interactions", *J. Fluid Mech.,* vol. 380, pp. 81-116.

[188] F. Grasso, and S. Pirozzoli, "Shock-wave-vortex interactions: shock and vortex deformations, and sound production", *Theor. Comput. Fluid Dyn.,* vol. 13, no. 6, pp. 421-456.

[189] C. Liu, Y. Wang, and J. Tang, "New Vortex Identification Method and Vortex Ring Development Analysis in Boundary Layer Transition", *54th AIAA Aerospace Sciences Meeting.*

[190] Y. Yan, and C. Liu, "Further Investigation on Shock Wave -Vortex Ring Interaction by the MVG Controlled Ramp Flow", *AIAA Paper, AIAA ,* p. 2013-0401.

[191] Z. Sun, F. Scarano, B.W. van Oudheusden, F.F. Schrijer, Y. Yan, and C. Liu, "Numerical and experimental investigations of the supersonic microramp wake", *AIAA J.,* vol. 52, no. 7, pp. 1518-1527.

[192] W. Bo, L. Weidong, Z. Yuxin, F. Xiaoqiang, and W. Chao, "Experimental investigation of the micro-ramp based shock wave and turbulent boundary layer interaction control", *Phys. Fluids,* vol. 24, no. 5, p. 55110.

[193] R.L. Panton, "Overview of the self-sustaining mechanisms of wall turbulence", *Prog. Aerosp. Sci.,* vol.

37, no. 4, pp. 341-383.

[194] R.H. Giepman, F.F. Schrijer, and B.W. Van Oudheusden, "Flow control of an oblique shock wave reflection with micro-ramp vortex generators: Effects of location and size", *Physics of Fluids,* vol. 26, no. 6, p. 066101.

[195] Q. Ye, F.F. Schrijer, and F. Scarano, "Boundary layer transition mechanisms behind a micro-ramp", *J. Fluid Mech.,* vol. 793, pp. 132-161.

[196] J. Jimenez, "Coherent structures and dynamical systems", *Studying Turbulence Using Numerical Simulation Databases.*

[197] G. Berkooz, P. Holmes, and J.L. Lumley, "The proper orthogonal decomposition in the analysis of turbulent flows", *Annu. Rev. Fluid Mech.,* vol. 25, no. 1, pp. 539-575.

[198] L. Sirovich, "Turbulence and the dynamics of coherent structures. I. Coherent structures", *Q. Appl. Math.,* vol. 45, no. 3, pp. 561-571.

[199] P.J. Schmid, "Dynamic mode decomposition of numerical and experimental data", *J. Fluid Mech.,* vol. 656, pp. 5-28.

[200] A. Ruhe, "Rational Krylov sequence methods for eigenvalue computation", *Linear Algebra Appl.,* vol. 58, pp. 391-405.

[201] I. Mezić, "Spectral properties of dynamical systems, model reduction and decompositions", *Nonlinear Dynamics,* vol. 41, no. 1, pp. 309-325.

[202] I. Mezić, "Analysis of Fluid Flows via Spectral Properties of the Koopman Operator", *Annu. Rev. Fluid Mech.,* vol. 45, no. 1, pp. 357-378.

[203] C.W. Rowley, I. Mezić, S. Bagheri, P. Schlatter, and D.S. Henningson, "Spectral analysis of nonlinear flows", *J. Fluid Mech.,* vol. 641, pp. 115-127.

[204] G. Gilka, D.M. Luchtenburg, F. Thiele, and M. Morzynski, "Dynamic characterization of an actuated bluff body wake", *V European Conference on Computational Fluid Dynamics, ECCOMAS CFD.*

[205] C. Pan, D. Yu, and J. Wang, "Dynamical mode decomposition of Gurney flap wake flow", *Theoretical and Applied Mechanics Letters,* vol. 1, no. 1, p. 12002.

[206] T.J. Poinsot, and S.K. Lelef, "Boundary conditions for direct simulations of compressible viscous flows", *J. Comput. Phys.,* vol. 101, no. 1, pp. 104-129.

[207] E. Touber, and N. D. Sandham, "Oblique Shock Impinging on a Turbulent Boundary Layer: Low-Frequency Mechanisms", *AIAA paper, AIAA* , p. 2008-4170.

[208] E. Garnier, P. Sagaut, and M. Deville, "Large Eddy Simulation of Shock/Boundary-Layer Interaction", *AIAA J.,* vol. 40, no. 10, pp. 1935-1944.

[209] E. Touber, and N.D. Sandham, "Large-eddy simulation of low-frequency unsteadiness in a turbulent shock-induced separation bubble", *Theor. Comput. Fluid Dyn.,* vol. 23, no. 2, pp. 79-107.

[210] E. Garnier, N. Adams, and P. Sagaut, *Large eddy simulation for compressible flows.,* Springer Science & Business Media.

[211] A.M. Schreyer, D. Sahoo, and A.J. Smits, "Experiments on the influence of a microramp array on a hypersonic shock turbulent boundary layer interaction", *41st AIAA Fluid Dynamics Conference and Exhibit,*. Honolulu, Hawaii.

SUBJECT INDEX

A

B

C

D

Chaoqun Liu, Qin Li, Yonghua Yan, Yong Yang, Guang Yang, Xiangrui Dong

E

F

G

H

I

J

K

L

M

N

O

P

R

S

T

U

V

W

Z

Made in the USA
Las Vegas, NV
04 November 2026

57425977R00193